S0-ANX-472

Topics in Applied Physics
Volume 93

Available online at
SpringerLink.com

Topics in Applied Physics is part of [SpringerLink] service. For all customers with standing orders for Topics in Applied Physics we offer the full text in electronic form via [SpringerLink] free of charge. Please contact your librarian who can receive a password for free access to the full articles by registration at:

springerlink.com → Orders

If you do not have a standing order you can nevertheless browse through the table of contents of the volumes and the abstracts of each article at:

springerlink.com → Browse Publications

There you will also find more information about the series.

Springer
Berlin
Heidelberg
New York
Hong Kong
London
Milan
Paris
Tokyo

Topics in Applied Physics

Topics in Applied Physics is a well-established series of review books, each of which presents a comprehensive survey of a selected topic within the broad area of applied physics. Edited and written by leading research scientists in the field concerned, each volume contains review contributions covering the various aspects of the topic. Together these provide an overview of the state of the art in the respective field, extending from an introduction to the subject right up to the frontiers of contemporary research.

Topics in Applied Physics is addressed to all scientists at universities and in industry who wish to obtain an overview and to keep abreast of advances in applied physics. The series also provides easy but comprehensive access to the fields for newcomers starting research.

Contributions are specially commissioned. The Managing Editors are open to any suggestions for topics coming from the community of applied physicists no matter what the field and encourage prospective editors to approach them with ideas.

See also: springeronline.com

Managing Editors

Dr. Claus E. Ascheron

Springer-Verlag Heidelberg
Topics in Applied Physics
Tiergartenstr. 17
69121 Heidelberg
Germany
Email: ascheron@springer.de

Dr. Hans J. Koelsch

Springer-Verlag New York, Inc.
Topics in Applied Physics
175 Fifth Avenue
New York, NY 10010-7858
USA
Email: hkoelsch@springer-ny.com

Assistant Editor

Dr. Werner Skolaut

Springer-Verlag Heidelberg
Topics in Applied Physics
Tiergartenstr. 17
69121 Heidelberg
Germany
Email: skolaut@springer.de

Hiroshi Ishiwara Masanori Okuyama
Yoshihiro Arimoto (Eds.)

Ferroelectric Random Access Memories

Fundamentals and Applications

With 125 Figures
and 12 Tables

Prof. Hiroshi Ishiwara

Tokyo Institute of Technology
Frontier Collaborative Research Center
4259 Nagatsuta, Midoriku
226-8503 Yokohama
Japan
ishiwara@pi.titech.ac.jp

Prof. Masanori Okuyama

Osaka University
Graduate School of Engineering Science
Department of Systems Innovation
1-3 Machikaneyama-cho, Toyonaka
560-8531 Osaka
Japan
okuyama@ee.es.osaka-u.ac.jp

Dr. Yoshihiro Arimoto

Fujitsu Laboratories Ltd.
Silicon Technology Laboratories
243-0197 Atsugi
Japan
arimoto@jp.fujitsu.com

Library of Congress Cataloging in Publication Data
Ferroelectric random access memories: fundamentals and applications/ Hiroshi Ishiwara, Masanori Okuyama, Yoshihiro Arimoto (eds.). p. cm. – (Topics in applied physics; v. 93) Includes bibliographical references and index. ISBN 3-540-40718-9 (alk. paper) 1. Ferroelectric storage cells. I. Ishiwara Hiroshi, 1945- II. Okuyama, Masanori, 1946- III. Arimoto yoshihiro, 1952- IV. Series. TK7895.M4F38 2004 621.39'73–dc22 2003059057

Physics and Astronomy Classification Scheme (PACS): 77.84.B, 68.55.J

ISSN print edition: 0303-4216
ISSN electronic edition: 1437-0859
ISBN 3-540-40718-9 Springer-Verlag Berlin Heidelberg New York

This work is subject to copyright. All rights are reserved, whether the whole or part of the material is concerned, specifically the rights of translation, reprinting, reuse of illustrations, recitation, broadcasting, reproduction on microfilm or in any other way, and storage in data banks. Duplication of this publication or parts thereof is permitted only under the provisions of the German Copyright Law of September 9, 1965, in its current version, and permission for use must always be obtained from Springer-Verlag. Violations are liable for prosecution under the German Copyright Law.

Springer-Verlag is a part of Springer Science+Business Media

springeronline.com

© Springer-Verlag Berlin Heidelberg 2004
Printed in Germany

The use of general descriptive names, registered names, trademarks, etc. in this publication does not imply, even in the absence of a specific statement, that such names are exempt from the relevant protective laws and regulations and therefore free for general use.

Typesetting: DA-TEX · Gerd Blumenstein · www.da-tex.de
Cover design: *design & production* GmbH, Heidelberg

Printed on acid-free paper 56/3141/mf 5 4 3 2 1 0

Preface

As modern portable electronic devices such as mobile phones and notebook computers become more and more popular, there is a confirmed increase in the demand for nonvolatile memories. The ferroelectric random access memory (FeRAM) is one of the most promising candidates for satisfying this demand, because its power consumption is the lowest among the various semiconductor memories, and it also possesses nonvolatile and random access characteristics. Thus, research and development for this memory are being conducted actively in many semiconductor companies, and FeRAMs up to 256 kb have already been mass-produced for RF tag and computer game applications. Furthermore, production of Mb-scale FeRAMs is also being prepared.

On the basis of this background, the 1st International Meeting on Ferroelectric Memories was held in Gotemba City, Japan, in November 2001. At this meeting, the publication of a new book on FeRAM was discussed and the outline of the book was determined. After the meeting, the individual chapters were written, mainly by the attendees of the meeting. Thus, I believe that each chapter in this book has been written by one of the best authors, who knows the specific field very well. The book consists of five parts – (I) ferroelectric thin films, (II) deposition and characterization methods, (III) the fabrication process and circuit design, (IV) advanced-type memories, and (V) applications and future prospects – and each part is further divided into several chapters. I hope that this book will contribute to the research and development of future FeRAMs.

Finally, I am grateful to Dr. C. E. Ascheron of Springer-Verlag for his encouragement and patience during the chapter collection and reviewing periods. I am also grateful to Dr. S. Yamamoto (Research Associate) and Ms. I. Sugita (Secretary) for their advice and efforts in converting almost all of the manuscripts into LaTeX style.

Tokyo, *Hiroshi Ishiwara*
Osaka, *Masanori Okuyama*
Atsugi, February 2004 *Yoshihiro Arimoto*

Contents

Part I Ferroelectric Thin Films

Part III The Fabrication Process and Circuit Design

Part I

Ferroelectric Thin Films

Overview

James F. Scott

Symetrix Centre for Ferroics, Earth Sciences Department, Cambridge University, Cambridge CB2 3EQ, UK
jsco99@esc.cam.ac.uk

Abstract. A review on ferroelectric thin films used for nonvolatile random access memories is given. Particular attention is paid to fundamental limitations on the materials. Optimization of ferroelectric films by impurity doping and grain-size control is first discussed; then size effects are considered (both thickness and lateral dimensions) from the point of view of both depolarization field instabilities and electrical breakdown mechanisms. Finally, dynamic characteristics such as polarization switching and retention are discussed, in which a theory on polarization reversal is presented and three characteristic fields, the breakdown field, the coercive field, and the activation field, are compared.

1 Introduction

The recent development of ferroelectric random access memories (FeRAMs) has shown that extremely high-density ferroelectric memory devices (ULSI) will become commercially feasible within the next few years [1, 2, 3]. Despite this progress in commercializing these integrated ferroelectric devices, there has been little emphasis in the published literature on the fundamental limitations of such micro-electric memories. In this chapter, after reviewing typical ferroelectric materials used for FeRAMs, I address some questions concerning mainly SBT thin films, which are important in the fabrication of high-density FeRAMs. Basic questions include: What is the ultimate switching speed in a ferroelectric thin film and what is the rate-limiting parameter? What is the ultimate minimum thickness for a ferroelectric film before depolarization fields creep in from the surfaces and destroy the switching properties? What is the ultimate breakdown field or voltage for a dielectric film? How small in lateral area can one construct a ferroelectric capacitor cell without fringing field limitations [4]?

2 Materials for FeRAMs

2.1 Conditions Desired for FeRAMs

A ferroelectric crystal exhibits a polarization (an electric dipole moment per unit volume) even in the absence of an external electric field, and the direction of the spontaneous polarization can be reversed by an external electric

H. Ishiwara, M. Okuyama, Y. Arimoto (Eds.): Ferroelectric Random Access Memories,
Topics Appl. Phys. **93**, 3–16 (2004)
© Springer-Verlag Berlin Heidelberg 2004

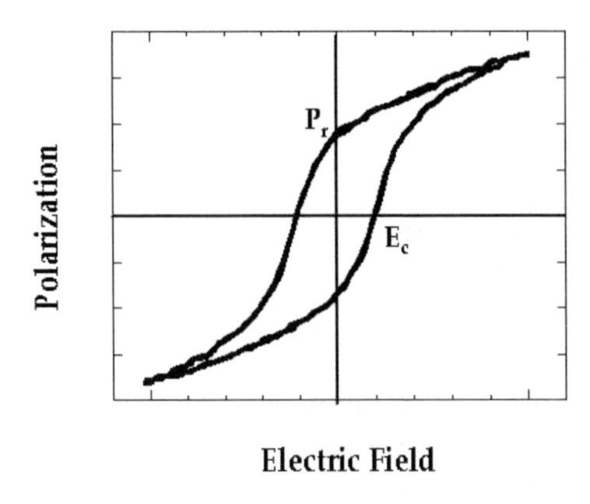

Fig. 1. A schematic drawing of a P–E hysteresis loop in a ferroelectric film

field. In the ferroelectric state the center of the positive charge of the crystal does not coincide with the center of negative charge. A typical plot of polarization versus electric field (P–E) for the ferroelectric state is shown in Fig. 1, in which the coercive field E_C is the reverse field necessary to bring the polarization to zero and the remanent polarization P_r is the value of P at $E = 0$.

In an FeRAM, information is stored by the polarization direction in a ferroelectric film and the stored datum is read out using the polarization reversal current. Thus, the following characteristics are desired for a ferroelectric thin film. The remanent polarization should be large, so that a large polarization reversal current can be derived from a small-area capacitor. The dielectric constant should be low, because a high dielectric constant material produces a large displacement current (linear response) and hinders detection of the polarization reversal current. The coercive field should be low for low-voltage operation of the FeRAM. Fatigue is a term describing the fact that the remanent polarization becomes small when a ferroelectric film experiences a large number of polarization reversals. Traditional US specifications, derived from magnetic memory parameters, are that the film is fatigue-free for switching over 10^{15} cycles when a 10-yr operation is assumed. However, in the present FeRAM technology the actual test cycles are about 10^{12}. This number seems to be determined by limited test time as well as an expectation that a single cell is not used continuously in the normal use of a memory, but various cells are used randomly. The retention time must be longer than 10 yr. Imprint is a measured phenomenon in which the polarization of a ferroelectric film and its response to opposite-polarity applied voltages are not symmetric; it depends upon the prior switching history and results in asymmetric switching times and incomplete polarity reversal. This effect should be as small as possible.

2.2 Typical Materials

So far, many ferroelectric materials have been investigated under the above conditions, and at present the following three materials are known to be most important for the fabrication of FeRAMs: $PbZr_XTi_{1-X}O_3$ (PZT), $SrBi_2Ta_2O_9$ (SBT), and $(Bi,La)_4Ti_3O_{12}$ (BLT). Their characteristics are summarized in Tab. 1. PZT is a typical ferroelectric crystal with a perovskite structure and its large P_r value is advantageous for the fabrication of future high-density FeRAMs. The crystallization temperature of PZT films is lower than 650 °C, which is suitable for implementing PZT capacitors on CMOS logic circuits. Some of the largest problems with PZT were fatigue and imprint, which were conspicuous with Pt electrodes. However, these problems have almost been solved at present by the use of oxide electrodes such as IrO_2, RuO_2, and $SrRuO_3$.

Table 1. The properties of typical ferroelectric thin films used for FeRAMs

Materials	P_r (μC/cm^2)	E_C (kV/cm)	Crystallization temperature (°C)
$Pb(Zr,Ti)O_3$ (PZT)	25	60	600
$SrBi_2Ta_2O_9$ (SBT)	10	40	750
$(Bi,La)_4Ti_3O_{12}$ (BLT)	15	80	700

SBT and BLT are typical Bi-layer structured ferroelectrics (BLSF). The crystal structure of SBT is shown in Fig. 2, in which the spontaneous polarization is directed along the a-axis of the crystal. One of the largest advantages of an SBT film is that it does not show the fatigue phenomenon up to 10^{13} switching cycles, even if Pt electrodes are used. The imprint and retention characteristics at high temperatures are also known to be superior to those of PZT. On the contrary, it is disadvantageous that the crystallization temperature of BLSF is generally higher than 700 °C. In some cases, Nb atoms are added to SBT up to 20–30 % (SBTN). The Nb increases the switched charge density $2P_r$ from about 18 μC/cm^2 to 24 μC/cm^2. There is also an increase in the dielectric constant. Unfortunately, the coercive field also increases with added Nb, typically from 40 to 63 kV/cm. For similar reasons, Sr-deficient (20–30 % less than nominal stoichiometry) and Bi-rich (10–15 %) compositions are sometimes used to increase the remanent polarization and the switched charge [5].

2.3 Doping Effects

The detailed studies on dopants in ferroelectric films thus far have been by *Dey* et al. [6] and *Wouters* et al. [7]. Dey found that for $(Pb,La)TiO_3$ on Pt electrodes, doping with the transition metals Ni, Cr, and Ti all produced ohmic contacts, whereas doping with the noble metals Ag, Au, and

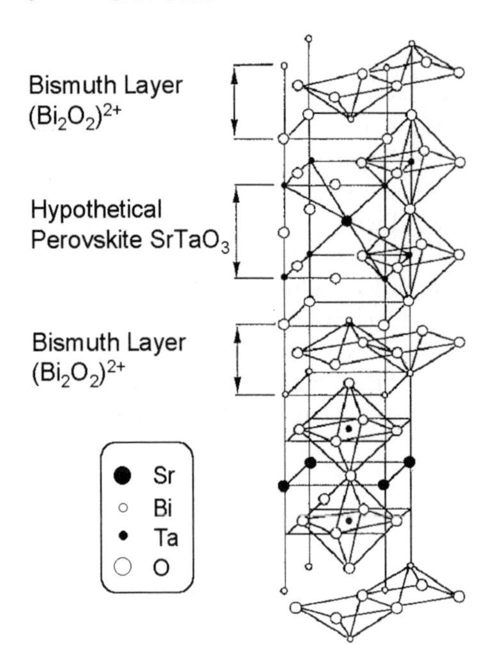

Fig. 2. The crystal structure of SBT

Pt produced Schottky contacts. In the case of Ni, Cr, and Ti, the dopants were all nonisoelectronic, with Ni^{3+}, Cr^{3+}, or Ti^{3+} replacing Pb^{2+} or Ti^{4+}. *Wouters* et al. examined the role of Nb and Na doping on PZT. They found that there was normally an inversion region near the electrode interface in the PZT (previously established by *Scott* et al. [8]), resulting in a p–n junction within the ferroelectric, as initially postulated by *Waser* et al. [9, 10]. For positive polarity this p–n junction blocks the electrode leakage current (a reversed-bias p–n junction), and the only remaining current flow is ionic, which can be further reduced through the influence of donor or acceptor doping on the oxygen vacancy concentration in the PZT.

2.4 Grain Sizes

Most techniques [sol-gel spin-on, liquid-source MOD (metalorganic deposition), and sputtering] produce SBT or SBNT with an average grain size of 40–50 nm. Larger grains can be produced intentionally with slow deposition on heated surfaces, but large grains are generally undesirable. It is important that each capacitor has many grains. This is useful to insure cell-to-cell reproducibility but is attractive in other ways too: fine-grained ceramic perovskite oxides have lower electrical conductivities than do single crystals of the same substances, due to the inter-grain depletion regions. Thus, even if a single crystal SBT thin-film ferroelectric capacitor could be fabricated, its leakage currents would be worse (higher) and its breakdown voltage worse (lower) than those in a 40 nm diameter grain size ceramic. PZT ceramics crystallize with columnar grains that usually extend all the way from anode to cathode. However, SBT grains are globular polyhedra. A 0.2 μm × 0.2 μm SBT cell will contain at least 25 grains of 40 nm in size, depending upon the thickness.

In some ferroelectrics, domain walls cannot move through grain boundaries, but *Gruverman* et al. have recently demonstrated [11] that domain walls in oxide perovskite ferroelectrics move through grain boundaries. The walls do this by extending finger-like tips into neighboring grains at oblique angles to the grain boundaries. Very recently, *Gruverman* et al. have shown that domain stability is a simple function of the distance to the nearest grain boundary.

3 Size Effects in Ferroelectric Capacitors

3.1 Lateral Area

The fundamental limitation in lateral area probably arises from the fringing field. One can calculate the fringing field correction for a thin parallel plate capacitor from conformal integral techniques. The result is that for an aspect ratio of 10 : 1 a capacitance correction of about 7 % is found. However, the correction increases nonlinearly as the aspect ratio approaches unity; a 20 % correction is found for the 5 : 1 aspect ratio now used commercially (0.7 μm width and 1.0 μm diagonal on a 0.2 μm thickness). For aspect ratios smaller than 5 : 1 the fringing field corrections are not small perturbation (the lines of force are not normal to the capacitor surfaces), and one should rather solve Laplace's equation for a box of the geometry representing the actual capacitor shape. Unfortunately, since the main interest in this problem is switching kinetics and the speed of the sensed displacement current, this method is useless, because Laplace's equation is not satisfied in the time-varying case (except for a few special examples, such as uniform coaxial cables and TEM waves). The general problem of switching behavior in a sub-micron ferroelectric capacitor is therefore not immediately accessible: numerical solutions starting with Maxwell's equations are required.

Simple approximations show, however, that the switched charge for a small-area capacitor of length b and width a will be proportional not to $a \cdot b$, but to $b \cdot \log a$. Here, a is the smaller dimension, and it is assumed that $b \gg a$. Hence, the switched charge might be expected to vary logarithmically, as observed in SBT by *Amanuma* and *Kunio* [12].

3.2 Thickness Dependence

As a ferroelectric film gets thinner, depolarization fields from the surface penetrate and destroy the ferroelectricity. For a very thin film, the polarization state is no longer bistable. Early estimates by *Batra* and *Silverman* [13] suggested that "very thin" meant 400 nm on semiconducting electrodes and 4 nm on metals; we know now that the threshold for this depolarization instability is about 6–9 nm and at most 20 nm [*Fridkin* (private communication) reports switching in copolymer Langmuir–Blodgett films down to 0.9 nm (two monolayers)]. Since ferroelectric nonvolatile memories will be about 80–200 nm thick, this intrinsic problem is primarily of academic interest.

Breakdown in ferroelectric thin films is generally avalanche-like, in that it is electronically initiated but characterized by thermal runaway when the leakage currents reach certain thresholds. As such it is best understood by careful studies of leakage currents. In SBT, like PZT and $(Ba,Sr)TiO_3$ (BST), the leakage is Schottky-like up to about 400 kV/cm at 295 K, above which Fowler–Nordheim tunneling becomes very important.

3.3 Electrodes

The primary consideration for electrode choice (assuming that only robust, chemically inert metals are considered) is the leakage current and consequent breakdown voltage. In this regard, it is useful to pick an electrode metal with the highest work function; Pt with a work function of about 5.3 eV is probably the best choice. It gives a maximum breakdown field of about 2.8 MV/cm in SBT and a leakage current of approximately 1 nA/cm^2 at 2.7 V operating voltage across 180 nm thickness. For other metal electrodes, the breakdown field can be calculated using a formula originally proposed by *von Hippel* in 1935 [14]:

$$q\lambda E_{\mathrm{B}} = \Phi_{\mathrm{H}} \,. \tag{1}$$

Here, q is the electron charge; λ is the electron mean free path in a ferroelectric (typically 0.1 nm); E_{B} is the breakdown field; and Φ_{H} is a quantized but unspecified characteristic energy which I take as the contact potential at the interface (metal work function minus ferroelectric electron affinity). Considering only this criterion for electrodes, Pt would be the best choice; Pd or W would be acceptable; and Ag or Au would be worst.

It is important to note that most ferroelectric perovskite oxides are p-type wide-bandgap semiconductors with bandgaps between 3.2 and 4.3 eV. E_{g} in SBT is 4.2 eV [15]. However, despite being nominally p-type, the thin films behave as fully depleted devices. All the conduction carriers are electrons injected from the cathodes, not holes. Hole concentration (double injection) occurs only under illumination or at very high voltages (possibly at 12 V across 180 nm in SBT). PZT and BST exhibit n-type inversion within 20 nm of the Pt electrodes [8], but such an inversion is smaller in SBT.

A simple abrupt Schottky model of SBT/Pt is shown [16] in Fig. 3, but the actual band model of the Pt interface is not compatible with a completely abrupt Schottky model. The Pt work function Φ_{M} is 5.3 eV. The measured Schottky barrier height is 0.9 eV (*Watanabe* et al. [17]). But the SBT electron affinity χ is 3.4 eV (XPS data, *Hartmann* et al. [15]). The Schottky barrier

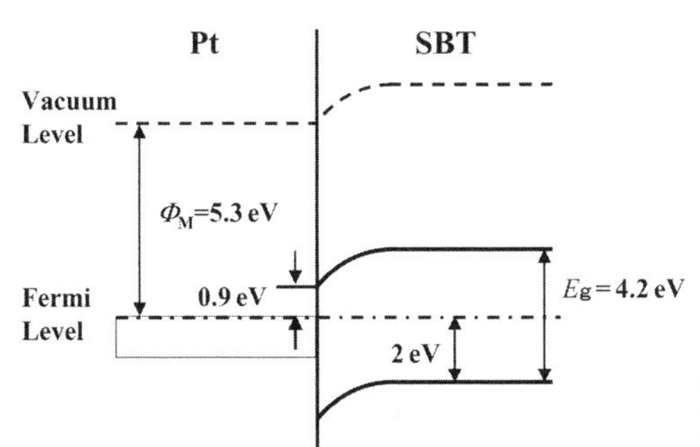

Fig. 3. An abrupt Schottky model of SBT/Pt

height Φ_{Bn} should be simply $\Phi_M - \chi$ for an abrupt Schottky barrier. Therefore exactly 1.0 eV needs to be accounted for, via either a graded junction or an insulating defect layer between the Pt and the SBT. For PZT the problem is worse, with a large 20 nm damage layer at the electrode interface.

More detailed measurements and calculations have been made recently for SBT/Pt, as shown in Fig. 3. Calculation of the Schottky barrier height in SBT on Pt is an interesting problem, because although the index $S = \mathrm{d}\Phi_{Bn}/\mathrm{d}X$ (where X is the electronegativity of the ferroelectric) of the interface behavior for such an ionic ABO_3 perovskite is nearly unity, the very large density of surface states compensates for the ionicity, resulting in an interface condition that is midway between the covalent, trap-limited case and the more ionic, trap-free situation often considered as a limiting case [18].

The Schottky barrier height is given approximately by

$$\Phi_{Bn} = c(\Phi_M - \chi) + (1 - c)(E_g - \Phi_0) - \Delta\Phi . \tag{2}$$

Here,

$$c = 2q\varepsilon N_A \delta/\varepsilon_i^2 , \tag{3}$$

where ε is the dielectric constant of SBT (about 500); N_A is the density of acceptors; δ is the width of the interfacial layer between Pt and the SBT; and ε_i is the dielectric constant of that layer (which may be BiPt or Bi_2Pt). One can see in (3) that a large trap density N_A can be compensated for linearly by a large interfacial dielectric constant ε_i, provided that ε_i is comparable to the dielectric constant of the ferroelectric semiconductor itself, ε.

With an estimate of 5×10^{14} surface traps per cm^2, from the measurements of *Mihara* et al. [19], we can approximate $c = 0.7$ for SBT/Pt. The numerical value of the barrier height is not very sensitive to this value; $c = 1.0$ for purely ionic models with no surface-state Fermi level pinning, and $c = 0$ for a purely covalent semiconductor with a very high surface trap density ($c = 0.27$ in Si). Hence c in SBT must be much more than 0.27 and less than 1.0. $\Phi_M = 5.3$ eV for Pt; we measure $\chi = 3.4$ eV for SBT via XPS; $E_g = 4.2$ eV from UV absorption; and Φ_0 is the energy level at the surface (the SBT Fermi level versus the SBT valence band level), which from our XPS data is 2.1 eV – approximately half the bandgap. By comparison, the values of Φ_0 in Si, GaP, or GaAs are always around $0.3\,E_g$, increasing to $0.6\,E_g$ in more ionic II–VI compounds such as CdS [20], so this result is reasonable.

This gives $\Phi_{Bn} = 0.7 \times 1.9\,\text{eV} + 0.3 \times 2.1\,\text{eV} - \Delta\Phi = 1.96 - \Delta\Phi$, where $\Delta\Phi$ is the image field barrier reduction energy. It is given by $[qE/(4\pi\varepsilon_{op})]^{1/2}$, where E is the electric field and ε_{op} is not the DC dielectric constant [because the electrons move through the interface in a time $t = (6\,\text{nm})/(10^5\,\text{m/s}) = 10^{-13}\,\text{s}$] much faster than the dielectric relaxation time, so that the ions are not perturbed and the response is purely electronic, but approximately $n^2 = \varepsilon_{op}$, where n is the index of refraction in the visible region, about 2.4. For this value of the dielectric constant, the Schottky barrier width is about 2 nm. At the measuring voltages of about 2 V, $\Delta\Phi = 0.8 \pm 0.3\,\text{eV}$ (for comparison, from *Mead* and *Spitzer* [20], it is $0.55 \pm 0.22\,\text{eV}$ in Si at the same field levels, the value being lower by the square root of the ratio of ε_{op}, which is 11.7 in Si and $2.4 \times 2.4 = 5.8$ in SBT); and hence the SBT/Pt barrier height is calculated as $1.1 \pm 0.3\,\text{eV}$.

Experimentally, *Watanabe* et al. [17] measure 0.83 eV Schottky barrier height, and *Lee* et al. report 0.9–1.1 eV, which is in complete agreement with this calculation. Extremely similar values are obtained by a careful fit of the Schottky emission in BST on Pt. In this case, *Joshi* et al. [21] showed that the zero-voltage Schottky barrier height is 1.5 eV (compared with our value of 1.96 eV in SBT above) and that the barrier reduction value is 0.8 eV at 3 V (across 150 nm), exactly as we calculate above for SBT, so that the reduced barrier height is $0.7 \pm 0.2\,\text{eV}$ for BST, in close agreement with our calculated value of $1.1 \pm 0.3\,\text{eV}$ for SBT and even closer to Watanabe's experimental value of 0.83 eV. Similarly, *Waser* gives 1.1 eV experimentally for $SrTiO_3$/Pt [22]. Thus, we see that band model calculations work well for these materials and that BST, PZT, and SBT films for DRAM and FeRAM capacitors all behave as wide-bandgap p-type semiconductors with n-type inversion layers at the Pt interface, thermally populated at an ambient temperatures. Under normal operating voltages of about 3 V, they act as fully depleted devices, so that all of the charge carriers are electrons injected from the cathode. This was first established by *Melnick* et al. [23] and later confirmed by *Wouters* et al. [7].

4 Dynamic Characteristics

4.1 Domain Structure

Because SBT is a highly uniaxial ferroelectric ($c/a \gg 1$), it does not generally exhibit 90-degree domains. The domain structure consists of ribbon-like rectangles of 180-degree domains. The domain size depends critically on the grain size and film thickness. As the grain diameter (or film thickness) decreases, the domain size also decreases. In general, for a bulk ferroelectric the stable ground state is a single large domain; but as the film gets thinner or the

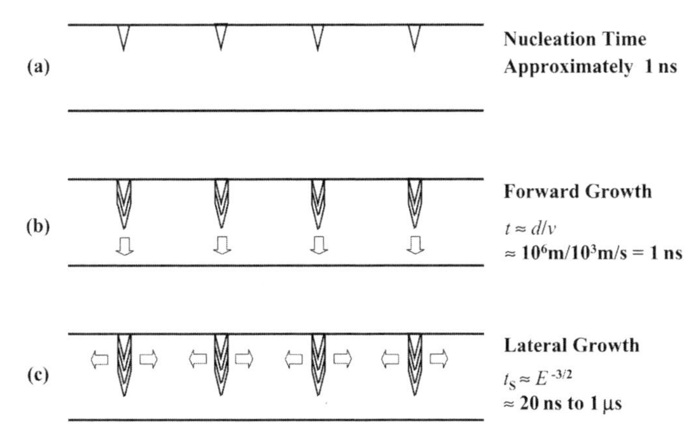

Fig. 4. The switching kinetics of a ferroelectric thin film. (**a**) Inhomogeneous nucleation, usually at anode or cathode. (**b**) The forward growth regime; needle-like domains move at the speed of sound parallel or antiparallel to the applied field. (**c**) Sideways growth in which the needle-like domains spread out laterally. The rate-limiting parameter can be (**a**), (**b**), or (**c**) in different materials

grain smaller, the energy cost (the savings in surface energy) in forming many small domains is compensated for by strain relief. This idea was developed in a series of papers by *Roytburd* [24] and extended to incorporate surface stress by *Kwak* et al. [25]. Usually, the domain size varies as the square root of the grain size (or the film thickness for very thin films). Typical values for PZT are 45 nm × 70 nm rectangular domains in medium-sized grains from 0.2 to 0.6 μm. SBT domains follow a similar scaling.

4.2 Polarization Switching Characteristics

The electrical switching properties depend upon the grain size and shape. A theory was developed in detail by *Ishibashi* and *Orihara* [26]. Generally, the switching involves a nucleation time t_{n} during which the grain forms (heterogeneous nucleation at an electrode surface or a grain boundary) and reaches a critical size necessary for further growth (smaller domains shrink and disappear), a propagation time t_{f} during which the domain wall moves from cathode to anode (or vice versa) in a needle-like one-dimensional shape, and finally, as illustrated in Fig. 4, a sideways spreading-out time t_{s} during which the domains fill the entire volume of the film. The forward-growth time is usually just the film thickness d divided by the speed of sound (the domain walls are not supersonic), which is approximately 900 ps for a 90 nm thick film, assuming that domain walls move quickly through grain boundaries, or that grains are columnar from anode to cathode.

Whether the rate-limiting parameter is the nucleation time or the sideways growth time depends critically upon the size of the capacitor. The domain wall speed limited switching time t_{s} is given theoretically by *Stadler* and *Zachmanidis* [27] as:

$$t_\mathrm{s} = (2Nr\nu^2/9)^{-1/3}, \tag{4}$$

where N is the number of nucleation sites per surface area; ν is the average wall speed; and r is the nucleation rate. Nucleation rates are not known for SBT, but for PZT *Duiker* et al. [28] measured r for surface nucleation. By fitting switching current transients $i(t)$ they derived parameters from which a t_s value of 600 ± 300 ps can be inferred for typical PZT capacitors [29]. This number can be compared with the time of about 200 ps which it takes for the output current from one gate to recharge the interconnect between it and the next gate (this, not the clock frequency, determines the basic speed of a silicon chip).

The nucleation time t_n is given in general by

$$t_\mathrm{n} = (NrA)^{-1}, \tag{5}$$

where A is the capacitor area. Hence for the nucleation time to be negligible, we require, from (4) and (5):

$$NrA \gg (Nr\nu^2)^{1/3} \tag{6}$$

or

$$\nu/(Nr) \ll A^{3/2}. \tag{7}$$

Theoretically, both ν and r vary as $E^{3/2}$, so (7) is valid independent of field E (or voltage V). If we insert realistic numbers for ν, N, and r, we find that nucleation will be the rate-limiting step in the switching of ferroelectric capacitor cells that are smaller than 2 μm × 2 μm. So domain wall propagation speed is not important in the access speed of a 1 Gb ULSI FeRAM, only nucleation time. Ishibashi originally assumed that the total switching time t ($t = t_\mathrm{n} + t_\mathrm{f} + t_\mathrm{s}$) varies as

$$t = BE^{-j}\tau^{k}, \tag{8}$$

and empirically found j and k of order of 1–2, where B is a constant; E is the applied field; and τ is the reduced temperature $(T_\mathrm{C} - T)/T_\mathrm{C}$, where T_C is the Curie temperature. t_n, t_f, and t_s stand for, respectively, the nucleation time, the forward-growth time, and the sideways spreading time for ferroelectric domains. In the low-field regime, the theory of Stadler gives $j = 3/2$ exactly, which is confirmed experimentally for KNO_3 films in Fig. 5. At higher fields, a more accurate description (*Scott* et al. [30]) is given by

$$t = B\exp(E_\alpha/E), \tag{9}$$

where E_α is an activation field that is proportional to τ. When expanded to the first few terms, the exponential in (9) averages to T and E power laws with averaging of the linear and quadratic terms in the expansion.

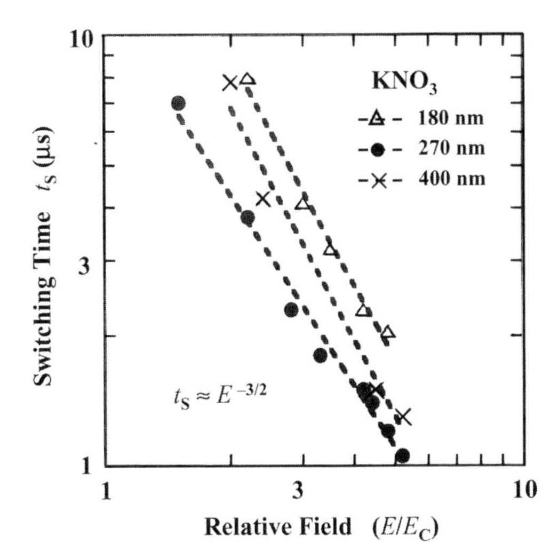

Fig. 5. The switching speed varying as $E^{-3/2}$ in KNO_3 for film thickness of 180, 270, and 400 nm

It is interesting to compare the breakdown fields E_B, the coercive fields E_C, and the activation fields (for switching) E_α in ferroelectric films. There is no expected relationship between the breakdown field and the other two characteristic fields, because the breakdown fields are the same above and below the Curie temperature, in the ferroelectric and paraelectric phases, and hence have nothing to do with domains, whereas the coercive field and the activation field relate to domain creation and domain wall motion. For many oxide ferroelectric films, these values are similar. The breakdown field depends upon thickness, electrodes, temperature, and other factors, but typically it is about 1.3 MV/cm in PZT, 1.0 MV/cm in KNO_3, and 1 MV/cm in SBT. The coercive fields are generally between 1/2 and 1/4 of the activation fields.

For example, E_C in potassium nitrate is 100 kV/cm, while E_α is 180 kV/cm. In PZT the coercive field is typically 65 kV/cm, while E_α is 210 kV/cm. In fact, one might ask why the activation field isn't exactly equal to the coercive field. The probable answer is that the activation field measures the ability to nucleate a domain, and the coercive field to switch an already existing domain. Although these fields should be comparable, it appears that the nucleation field is typically larger by a factor of approximately 3.

4.3 Frequency Dependence

Most laboratory measurements on SBT are in the kHz regime. But FeRAM will operate at very high internal clock-rates – at least 100 MHz and perhaps 300 MHz. The frequency dependence of the coercive field is the most important parameter in this regard. A theory was recently developed [31] which predicts

$$E_C = C f^{D/\alpha} , \tag{10}$$

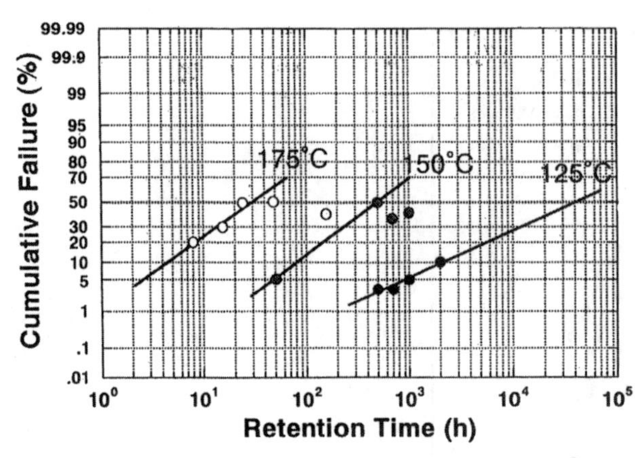

Fig. 6. A log-normal plot of cumulative retention failures in SBT with time at different values of T

where C is a constant, D is the dimensionality of the system ($D = 1$ in highly uniaxial SBT), and α is a constant of about 6.0.

Experimentally, *De Vilbiss* et al. made measurements of this effect in SBT and SBTN [32], with results in agreement with (10). The prediction from extrapolating these data is that the coercive field in SBT at 100 MHz is 118 kV/cm. Hence to achieve complete saturation and consequently good retention, a 2.7 V FeRAM using SBT will have to be 120 nm thick, not 240 nm. The problem becomes worse for the 1.1 V Gbit FeRAMs envisioned. In this case it might be necessary to fabricate 50-nm-thick SBT films. Although this problem is a worry for SBT, it is a nightmare for PZT, where surface layers are usually greater than 20 nm at each side [33], leaving nothing useful in the middle (a "bread sandwich").

4.4 Retention

Retention is the ability of a ferroelectric thin-film memory to retain the polarization sign and magnitude originally written; it is a measure of self-life, as opposed to aging (fatigue), which measures loss of charge on repetitive switching. Lack of retention is usually due to electret-like effects. Consequently, retention is lower at elevated temperatures. Figure 6 shows cumulative retention failures in SBT films at 3 V across about 200 nm at different temperatures [34]. Retention is of only peripheral importance in DRAMs, because they are unipolar devices, but it provides an excellent way of measuring electret-like effects. Electret poling, even unipolar, is slow because it involves real diffusion of charged defects to the surface of the film. In a DRAM capacitor electret-like effects would provide a long tail in the switched charge versus time, as well as undesirable heating.

References

1. T. S. Moise, S. R. Summerfelt, H. McAdams, S. Aggarwal, K. R. Udayakumar, F. G. Celii, J. S. Martin, G. Xing, L. hall, K. J. Taylor, T. Hurd, J. Rodriguez, K. Remack, M. D. Khan, K. Boku, G. Stacey, M. Yao, M. G. Albrecht, E. Zielinski, M. Thakre, S. Kuchimanchi, A. Thomas, B. McKee, J. Rickes, A. Wang, J. Grace, J. Fong, D. Lee, C. Pietrzyk, R. Lanham, S. R. Gilbert, D. Taylor, J. Amano, R. Bailey, F. Chu, G. Fox, S. Sun, T. Davenport: Int. Electron Devices Meeting, Tech. Digest, San Francisco, No.21.1 (2002)
2. Y. Horii, Y. Hikosaka, A. Itoh, K. Matsuura, M. Kurasawa, G. Komuro, K. Maruyama, T. Eshita, S. Kashiwagi: Int. Electron Devices Meeting, Tech. Digest, San Francisco, No.21.2 (2002)
3. M.-K. Choi, B.-G. Jeon, N. Jang, B.-J. Min, Y.-J. Song, S.-Y. Lee, H.-H. Kim, D.-J. Jung, H.-J. Joo, K. Kim: IEEE J. Solid-State Circuits **37**, 1472 (2002)
4. J. F. Scott: IEICE Trans. Electron. **81-C**, 477 (1998)
5. T. Noguchi, T. Hase, Y. Miyasaka: Jpn. J. Appl. Phys. **35**, 4900 (1996)
6. S. K. Dey, P. Alluri, J.-J. Lee, R. Zuleeg: Integr. Ferroelectr. **8**, 715 (1995)
7. D. J. Wouters, G. Willems, G. Groeseneken, H. E. Maes, K. Brooks, R. Klissurska: Proc. EMIF-1, Nijmegen (July 1995)
8. J. F. Scott: Integr. Ferroelectr. **9**, 1 (1995)
9. R. Waser, T. Baiatu, H. Härdtl: J. Am. Ceram. Soc. **73**, 1645 (1990)
10. R. Waser, M. Klee: Integr. Ferroelectr. **2**, 288 (1992)
11. A. Gruverman, H. Tokumoto, A. S. Prakash, S. Aggarwal, B. Yang, M. Wuttig, R. Ramesh, O. Auciello, T. Venkatensan: Appl. Phys. Lett. **71**, 3492 (1997)
12. K. Amanuma, T. Kunio: Jpn. J. Appl. Phys. **35**, 5229 (1996)
13. I. P. Batra, B. D. Silverman: Solid State Commun. **11**, 291 (1972)
14. A. von Hippel: Ergeb. exakt. Naturwiss. **14**, 118 (1935)
15. A. J. Hartmann, J. F. Scott et al.: J. Phys. (Paris) (1998)
16. A. J. Hartmann, J. F. Scott et al.: EMIF-2, Versailles, Sept. 1997
17. K. Watanabe, M. Tanaka, N. Nagel, K. Katori, M. Sugiyama, H. Yamoto, H. Yagi: Integr. Ferroelectr. **14**, 95 (1997)
18. S. M. Sze: *Physics of Semiconductor Devices*, 2nd edn. (Wiley, New York 1981) p. 273
19. T. Mihara, H. Watanabe, H. Yoshimori, C. A. Paz de Araujo, B. Melnick, I. D. MacMillan: Integr. Ferroelectr. **1**, 269 (1992)
20. C. A. Mead, W. G. Spitzer: Phys. Rev. **134**, A713 (1964)
21. V. Joshi, C. P. DaCruz, J. D. Cuchiaro, C. A. Paz de Araujo, R. Zuleeg: Integr. Ferroelectr. **14**, 133 (1997)
22. R. Waser: *Science and Technology of Electroceramic Thin Films* (Kluwer, 1995) p. 223
23. B. M. Melnick, J. F. Scott, C. A. Paz de Araujo, L. D. McMillan: Ferroelectr. **135**, 163 (1992)
24. A. L. Roytburd: Phys. Stat. Sol. **A37**, 329 (1976)
25. B. S. Kwak, A. Erbil, B. J. Wilkens, J. D. Budai, M. F. Chisholm, L. A. Boatner: Phys. Rev. Lett. **68**, 3733 (1992); Phys. Rev. B **49**, 14865 (1994)
26. Y. Ishibashi, Y. Takagi: J. Phys. Soc. Jpn. **31**, 506 (1971)
27. H. L. Stadler, P. J. Zachmanidis: J. Appl. Phys. **34**, 3255 (1963)
28. H. M. Duiker, P. D. Beale, J. F. Scott, C. A. Paz de Araujo, B. M. Melnick, J. D. Cuchiaro, L. D. MacMillan: J. Appl. Phys. **68**, 5783 (1990)

29. J. F. Scott: Ferroelectr. Rev. **1**, 1 (1998)
30. J. F. Scott, L. Kamerdiner, M. Parris, S. Traynor, V. Ottenbacher, A. Shawbkeh, W. F. Oliver: J. Appl. Phys. **64**, 787 (1988)
31. Y. Ishibashi, H. Orihara: Integr. Ferroelectr. **9**, 57 (1995)
32. A. D. DeVilbiss, W. J. Connor, G. F. Derbenwick: Abstracts Int. Symp. Integr. Ferroelectr. No. 18c, Tempe, Arizona (1996)
33. J. F. Scott, D. Galt, J. C. Price, J. A. Beall, R. H. Ono, C. A. Paz de Araujo, L. D. MacMillan: Integr. Ferroelectr. **6**, 189 (1995)
34. Y. Shimada, M. Azuma, K. Nakao, S. Chaya, N. Moriwaki, T. Otsuki: Jpn. J. Appl. Phys. **36**, 5912 (1997)

Novel Si-Substituted Ferroelectric Films

Takeshi Kijima[1] and Hiroshi Ishiwara[2]

[1] Technology Platform Research Center, Seiko Epson Corp.,
281 Fujimi, Fujimi-machi, Nagano, 399–0293, Japan
Kijima.Takeshi@exc.epson.co.jp
[2] Frontier Collabrative Research Center, Tokyo Institute of Technology,
4259 Nagatsuta, Midori-ku, Yokohama 226–8503, Japan
ishiwara@pi.titech.ac.jp

Abstract. In this chapter, properties of novel Si-substituted ferroelectric films are presented. The films are a solid solution between Bi_2SiO_5 (BSO) and conventional ferroelectric materials such as $Bi_4Ti_3O_{12}$, $SrBi_2Ta_2O_9$, and $Pb(Zr,Ti)O_3$, which were formed by a sol-gel spin-coating method. It was found that BSO enhanced crystallization of the ferroelectric materials and finally formed solid solutions with them. As a result, the crystallization temperature of the films decreased by 150–200 °C, and the ferroelectric and leakage current characteristics did not degrade even in an ultra-thin film of 13 nm in thickness. It was also found that the ferroelectric and insulating characteristics of the BSO-added films were dramatically improved by annealing in high-pressure oxygen up to 9.9 atm. A three-orders-of-magnitude improvement of the leakage current density was observed in BSO-added BLT films after annealing at 9.9 atm, while a pronounced increase in the saturation polarization was observed in BSO-added SBT and PZT films.

1 Introduction

Nonvolatile ferroelectric random access memories (FeRAM) have attracted considerable attention with the recent development of portable instruments such as cellular phones, laptop computers, and personal digital assistants (PDA). These instruments handle a large quantity of data such as image and sound, and they require memory that is able to process large data at a high speed with low power consumption. FeRAM is considered most suitable for such requirement. Ferroelectric material retains remanent polarization even after an applied field is removed. It arises from the displacement of anions and cations in relative positions in the crystal. FeRAM utilizes this phenomenon to store the data of "0" and "1". The device structure and operation are the same with dynamic random access memory (DRAM), except that DRAM stores data by charge on the capacitors. Therefore, FeRAM has a large potential to substitute the widely used DRAM.

Currently produced FeRAM has only a small capacity of 256 kb, and has yet to be as highly integrated as DRAM. This is due to the problems of the present ferroelectric materials, in that the surface morphology is too rough to be used in integrated circuits, which operate at a voltage lower than

H. Ishiwara, M. Okuyama, Y. Arimoto (Eds.): Ferroelectric Random Access Memories,
Topics Appl. Phys. **93**, 17–30 (2004)
© Springer-Verlag Berlin Heidelberg 2004

2 V. It is also important to decrease the present crystallization temperature (higher than 600 °C) to fabricate future high-density FeRAMs. In this chapter, we show that these problems can be solved by forming solid solutions between Bi_2SiO_5 (BSO) [1, 2] and conventional ferroelectric materials such as $Bi_4Ti_3O_{12}$ (BIT) [3, 4], $SrBi_2Ta_2O_9$ (SBT) [5] and $Pb(Zr,Ti)O_3$ (PZT). The crystallization temperatures of the new ferroelectric films were lowered by 150–200 °C and the surface of the crystallized films became very flat and smooth. Because of these excellent properties, very thin (13–25 nm thick) ferroelectric capacitors were successfully fabricated, in which the polarization was well saturated at an applied voltage of 0.5 V [6, 7].

2 The Crystallization Process

In the experiment, combinations of BSO and BIT were first attempted, since both have similar Bi-layered structures, and since the lattice mismatch of the basal plane is as small as 0.5% between the two materials. The remanent polarization of a bulk BIT crystal is about 50 μC/cm^2 along the a-axis and it is about 4 μC/cm^2 along the c-axis [4]. On the other hand, BSO has a crystallization temperature as low as 400 °C [1,2], and it is known to act as a solid acid catalyst [8,9]. The film was formed by a simple sol-gel spin-coating method. Commercially available sol-gel solutions of BSO (Bi_2O_3:SiO_2 = 1:1) and BIT (manufactured by Mitsubishi Material Co.) were mixed. Since both of them are dissolved in n-butyl alcohol, they mix well at an arbitrary mol ratio. After several attempts, the optimum mol ratio R of BSO to BIT, at which improvement of the film quality as well as lowering of the crystallization temperature were most clearly demonstrated, was determined to be 0.4.

The crystallization process of the mixed solution was first investigated. Figure 1 shows the results of thermogravimetric-differential thermal analysis (TG-DTA) of sol-gel solutions of BSO, BIT, and the mixed material. The crystallization temperature of BSO is about 330 °C (point A), while the crystallization process of BIT is broad and it has a peak at around 620 °C (point B). For the mixed material, double peaks located at 350 °C (point C) and 400 °C (point D) were observed, as shown in Fig. 1c. Since X-ray peaks were observed at the corresponding temperatures, as discussed later, the peaks C and D are identified as those of BSO and BIT, respectively. It is interesting to note that the crystallization process of BIT is concentrated in a narrow temperature range in the mixed material. It can also be seen from the TG charts of Fig. 1b,c that the mass of the mixed material hardly changes at temperatures higher than 400 °C, while that of BIT gradually decreases in the temperature range from 350 °C to 700 °C.

In order to investigate the origin of the different behavior of the TG analysis, 100-nm-thick films were formed on a Pt/Ti/SiO_2/Si structure by twice repeated spin-coating and pre-annealing processes. The pre-annealing was conducted on a hot plate at 400 °C for 5 min in air. After pre-annealing,

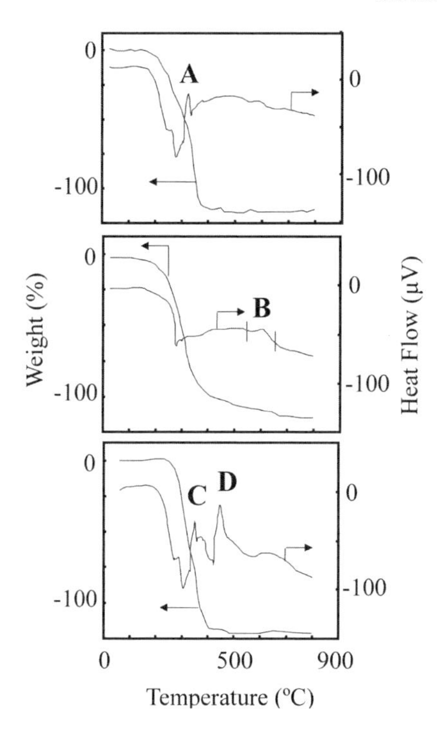

Fig. 1. TG-DTA charts of (**a**) BSO, (**b**) BIT, and (**c**) BSO-added BIT systems. The measurement temperature was increased at a rate of 10 °C/min

quantities of residual carbon were measured by using EPMA (electron-probe micro-analysis), which showed that the carbon quantity in the BSO-added BIT film was less than 1/10 of that in the BIT film, as shown in Fig. 2a,b. It was also found that a large quantity of Si atoms was contained in the BSO-added BIT film, as shown in Fig. 2b. Both films were melted in 1% NH_4OH solution and the pH value of the solutions was measured. The pH value for the BSO-added BIT film was about 5, while the value for the BIT film was about 8. That is, the weak alkaline solution changed to an acid solution after the BSO-added BIT film was melted. We speculate from these results that BSO crystallites in the pre-annealed film act as a solid acid catalyst and enhance the oxidation of residual carbon atoms. The low-temperature crystallization of BIT in the mixed material may be due to both epitaxial growth on BSO crystallites and a low carbon concentration in the material.

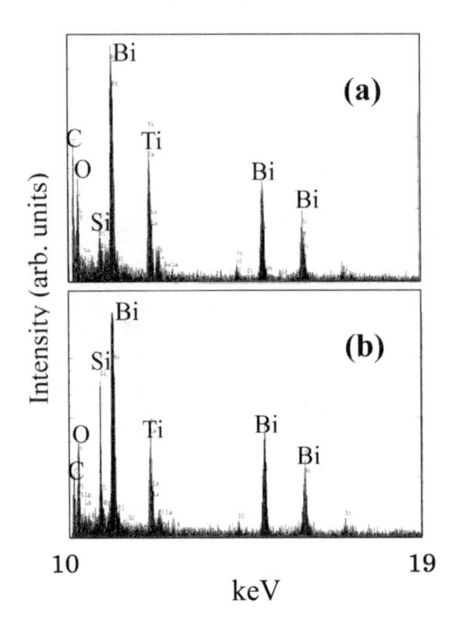

Fig. 2. EPMA spectra for pre-annealed (**a**) BIT and (**b**) BSO-added BIT films

3 The Properties of Bi_2SiO_5-Added $Bi_4Ti_3O_{12}$ Films

In order to further understand the crystallization process, a pre-annealed BSO-added BIT sample was set in a high-temperature X-ray diffractometer (X'Pert-Pro with high-temperature holding attachment; Philips Co.) and X-ray diffraction (XRD) patterns were measured. Figure 3a shows an XRD pattern at 400 °C, in which diffraction peaks only from BSO crystallites appear. The pattern is essentially the same as that measured at room temperature (RT), but the peak positions are shifted because of the thermal expansion of BSO. Figure 3b shows an XRD pattern at 500 °C, in which both peaks of BSO and BIT coexist. This result shows that the BIT phase in the film crystallized at this temperature, as expected from the DTA result in Fig. 1c. Then, the sample temperature was decreased and it was found that the BSO peaks disappeared in the cooling process and only BIT peaks remained at RT, as shown in Fig. 3c. In Fig. 3c, we note that the shift of the diffraction peaks is relatively large, and it was found that these positions were shifted to larger angles by 0.3 ° to 1 ° compared to a pure BIT film.

Next, crystallization annealing of the pre-annealed samples was conducted at 500 °C for 10 min in an O_2 atmosphere using a rapid thermal annealing (RTA) furnace. BIT films were also annealed under the same conditions for comparison. However, no peaks of BIT crystallites were observed in XRD patterns in the BIT films. Scanning electron micrographs of the surfaces of these films (100 nm thick) are shown in Fig. 4a,b. The surface of the BSO-added BIT film is extremely flat and smooth, while that of the BIT film is very rough, although the BIT phase has not been formed yet. The excellent surface

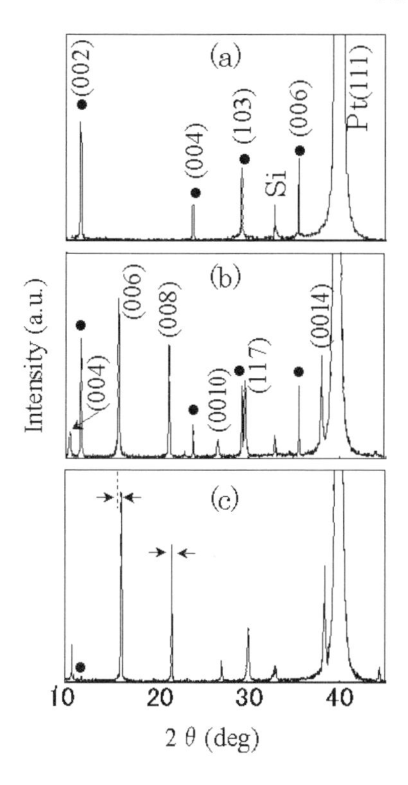

Fig. 3. XRD patterns of a BSO-added BIT film (**a**) at 400 °C, (**b**) at 500 °C, and (**c**) at RT after crystallization. In (**a**) and (**b**), the pattern shift due to thermal expansion was corrected, so that the peak positions of Si (400) were superposed on that in (**c**)

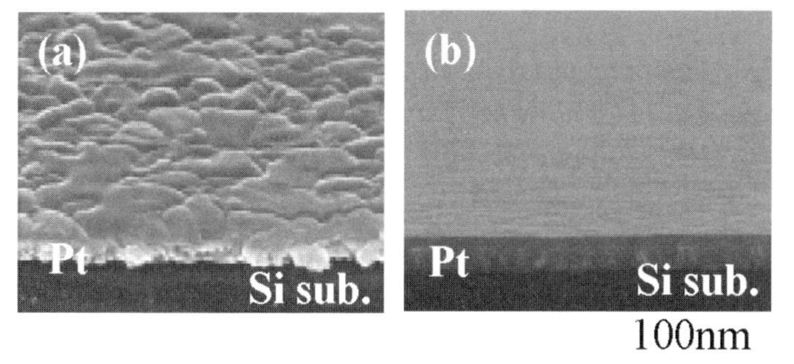

Fig. 4. Surface morphologies of (**a**) BIT and (**b**) BSO-added BIT films crystallized at 500 °C

morphology is considered to originate from the fact that BSO crystallites are uniformly distributed in the pre-annealed film and they act as nucleation sites for the crystallization of BIT.

In the electrical measurement, Pt top electrodes of 100 μm in diameter were deposited and the samples were again annealed at 500 °C for 10 min. The P–V (polarization versus voltage) hysteresis loops and the leakage current density were measured using an RT66 A ferroelectric test system (Radiant

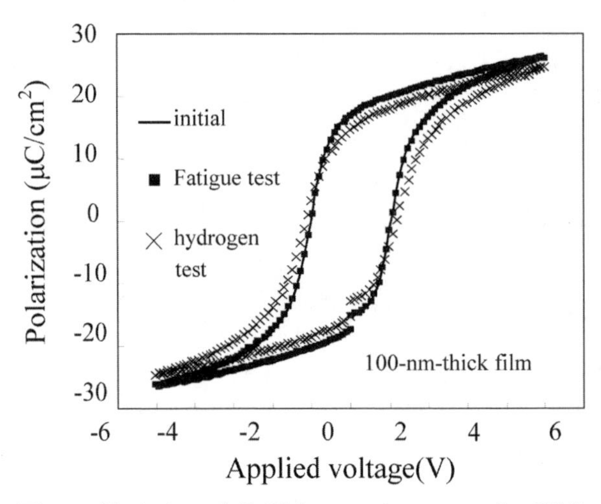

Fig. 5. Variation of P–V hysteresis curves of a BSO-added BIT capacitor. *Solid line*; initial, *closed squares*; after fatigue test of 10^{10} switching cycles at 100 kHz, *cross*; after a hydrogen immunity test at 400 °C

Technologies, Inc.) at a frequency of 1 kHz and a pA meter connected to a DC voltage source, respectively. Figure 5 shows P–V hysteresis loops for the same sample before and after fatigue measurement. As can be seen from the figure, a square-shaped hysteresis loop with the remanent polarization of 18 $\mu C/cm^2$ was obtained and no fatigue (degradation of the hysteresis shape) was observed even after 10^{10} switching cycles with a frequency of 100 kHz. The result for a hydrogen immunity test, in which the sample with top Pt electrodes was annealed in an N_2 atmosphere with 3% H_2 at 400 °C for 10 min, is also shown in the figure. As can be seen from the figure, degradation of the hysteresis loop is not significant in this sample. This result is completely different from the usual case [10], in which the remanent polarization of a ferroelectric film becomes zero by annealing in 3% H_2 at 250 °C. This excellent ferroelectricity is partly due to the fact that a uniformly packed dense film is formed in this method.

So far, we have shown that the BSO peaks in the film disappear during the cooling process to RT and that the ferroelectric properties of the obtained film are excellent. Concerning the stability of a ferroelectric film with a perovskite structure, it is known that the stability is increased when the covalency of the B-site ion is increased [11]. We speculate from these results that the crystal structure of BSO crystallites in the BSO-added BIT film was changed to the Bi-layered perovskite structure during the cooling process of the film, and that a mixed crystal $Bi_4(Ti,Si)_3O_{12}$ with Si concentration as high as 30% was formed. In other words, the SiO_4^{4-} tetragonal structure in a BSO crystal was changed to the SiO_6^{8-} octahedron structure; that is, the coordinate number of the Si atoms was changed from 4 to 6. This is a new hypothesis, because it

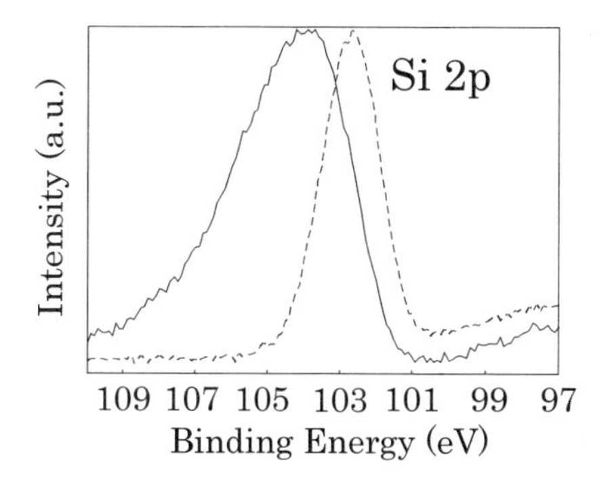

Fig. 6. XPS spectra for Si 2p states in (*solid line*) films

is generally known that the ionic radius of Si is too small to be crystallized in a perovskite structure at atmospheric pressure. Actually, it is reported that $CaSiO_3$ or $MgSiO_3$ with a perovskite structure is formed in the mantle layer of the Earth, where a compressive stress in excess of 20 GPa is applied to the constituent materials [12, 13].

Then, in order to investigate the change of the coordinate number of the Si atoms, X-ray photoelectron spectroscopy (XPS) measurement was conducted for the BSO and BSO-added BIT films. Figure 6 shows Si 2p spectra in both films, in which the peak position in the spectrum shows the binding energy of the Si 2p state. As can be seen from the figure, the shape of the spectra is different between the two samples, and the binding energy of the Si 2p state shifts to the higher energy by about 1.5 eV in the BSO-added BIT film. It is also known that the Si 2p binding energy is about 99 eV in an Si crystal and it is about 103 eV in SiO_2, which is composed of the SiO_4^{4-} tetragonal structure. These results strongly suggest that the coordinate number of the Si atoms was changed from 4 to 6 in the BSO-added BIT film and that a mixed crystal with the Bi-layered perovskite structure was formed.

At present, the precise mechanism on the change of the coordinate number of Si atoms is not well understood. However, we speculate that BSO crystallites in the film are compressed by the surrounding BIT crystals during the cooling stage, because of the difference between the thermal expansion coefficients of the two materials, and the generated stress is high enough to form the SiO_6^{8-} octahedron structure. We can calculate the thermal expansion coefficient of BSO from Fig. 3a and the XRD pattern at RT is about 12×10^{-6} (1/ °C). We can also calculate the coefficient for the BSO–BIT film from Fig. 3b,c. As we mentioned before, the diffraction peaks in Fig. 3c are shifted to a higher-angle direction, compared to a pure BIT crystal. Thus, the thermal expansion coefficient derived from Fig. 3b,c is as large as 90×10^{-6} (1/ °C). In order to calculate the stress generated in this system [14], we further need to know the elastic stiffness constants of BSO and BIT. However, these values are unknown at present. If we assume that these values are equal to that of $PbTiO_3$, the stress value is about 30 GPa, which is high enough to change the coordinate number of the Si atoms.

4 The Formation of Ultra-Thin Films

Based on the results described above, the formation of ultra-thin ferroelectric films was attempted in the BSO-added BIT system. At the same time, the possibility of BSO-added SBT and BSO-added PZT systems was investigated, since the epitaxial alignment effect in the BSO-added BIT system was not found to be essential in obtaining excellent ferroelectricity. It was found from these experiments that BSO was also effective in improving the surface morphology and in lowering the crystallization temperature of the SBT and PZT films. The minimum values of the film thickness were 13 nm, 25 nm, and 25 nm for BSO-added BIT, SBT, and PZT, respectively. The films were crystallized at the lowest temperature for each thickness, below which the films did not exhibit excellent ferroelectricity. Thus, the crystallization temperatures were 500 °C, 550 °C, and 450 °C for the BSO-added BIT, BSO-added SBT, and BSO-added PZT films, respectively. The mole ratios R of BSO to BIT, SBT, and PZT were 0.4, 0.33, and 0.1, respectively.

Figure 7 shows low-magnification and high-resolution cross-sectional TEM images of a BSO-added BIT film. The film thickness is about 13 nm on average and no degradation layer exists at the interfaces to the two electrodes. The upper parts of Fig. 8 show P–V characteristics of (a) 13-nm-thick BSO-added BIT, (b) 25-nm-thick BSO-added SBT, and (c) 25-nm-thick BSO-added PZT films, and the lower parts show the relations between the applied voltage and the remanent polarization P_r of the films with different thicknesses. It can be seen from the figures that the shape of the P–V hysteresis curve and the P_r values are very similar to those of the original ferroelectric materials. It can also be seen that the P_r values are well saturated at an applied voltage of 0.5 V. The leakage current density of the thin BSO-added BIT film was about 10^{-7} A/cm^2 at 500 kV/cm, which was about one order of magnitude higher than that for thicker films (typically 10^{-8} A/cm^2 at 500 kV/cm).

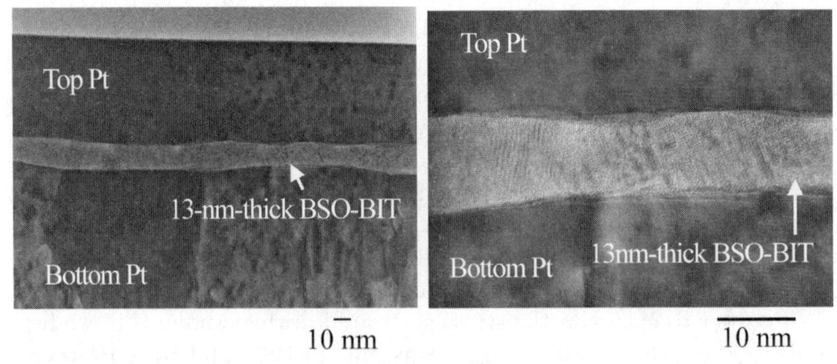

Fig. 7. Cross-sectional TEM micrographs for a Pt/13 nm-thick BSO-added BIT/Pt capacitor

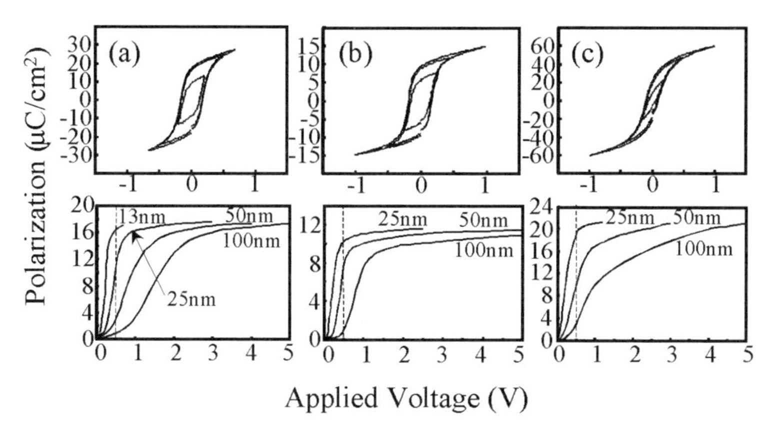

Fig. 8. *Upper row*: P–V hysteresis characteristics and saturation properties for BSO-added (**a**) 13-nm-thick BIT, (**b**) 25-nm-thick SBT, and (**c**) 25-nm-thick PZT films. Three loops in each figure correspond to the applied voltages of 0.3 V, 0.5 V, and 1 V (0.7 V for BIT). *Lower row*: The relation between the applied voltage and the remanent polarization. The crystallization temperatures for BIT, SBT, and PZT films were 500 °C, 550 °C, and 450 °C, respectively

To the best of our knowledge, the thickness of the BSO-added BIT film is about 1/4 of the thinnest polycrystalline perovskite-derived ferroelectric film (48-nm-thick SBT film [15]), in which polarization hysteresis loops were measured electrically using electrodes with a macroscopic size, and it was comparable to the thinnest epitaxial film (12 nm) [16]. Concerning the crystallization temperature, the obtained values for the BSO-added materials were lower by 150–200 °C than the values of the original ferroelectrics, as long as the sol-gel method was used. In the case of PZT or La-doped PZT, however, since there are only a few results in which reasonably good hysteresis loops are obtained at temperatures lower than 450 °C [17, 18], more detailed comparisons, including fatigue and hydrogen-immunity characteristics, are necessary.

5 Annealing Effects in High-Pressure Oxygen

In order to further improve the ferroelectric and leakage current characteristics of these BSO-added ferroelectric films, post-annealing in high-pressure oxygen was carried out [19]. That is, the samples were annealed at 500 °C for 5 min in O_2 ambience at 1.5, 5, and 9.9 atm (atmospheric pressures). Figure 9 shows the XRD patterns of BSO-added BLT ($(Bi,La)_4Ti_3O_{12}$), SBT, and PZT films before high-pressure annealing. The respective annealing temperatures were 600 °C, 650 °C, and 550 °C, and the annealing time was 15 min in O_2. As can be seen from the figure, all films crystallized well at much lower temperatures than the usual crystallization temperatures of the original ma-

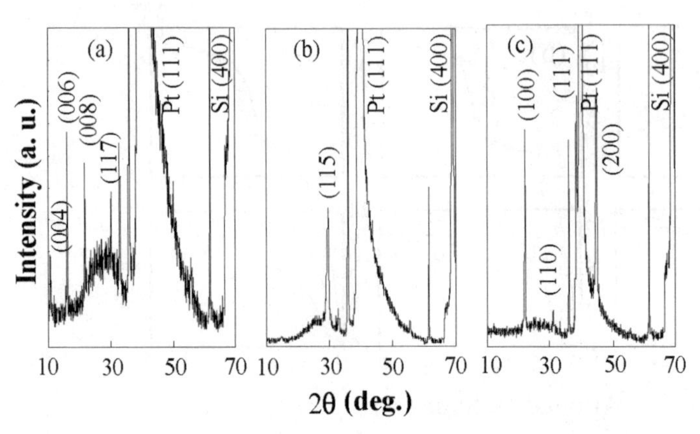

Fig. 9. X-ray diffraction patterns of (**a**) 50-nm-thick BSO-added BLT, (**b**) SBT, and (**c**) PZT films after crystallization annealing

terials. It was also found that the film surfaces became flatter than those of the original materials.

Figure 10 shows variations of the XRD peak intensity and position with oxygen pressure during the high-pressure annealing. The four peaks in each figure correspond to the initial sample and the annealed samples at 1.5, 5, and 9.9 atm from bottom to top. As can be seen from Fig. 10a, the XRD peak position in the BLT film hardly changed, but the peak intensity became somewhat weak as the oxygen pressure increased. On the other hand, in the BSO-added ferroelectric films, the peak intensities became stronger and their positions shifted to higher angles as the oxygen pressure increased. That is, in the BSO-added BLT, SBT, and PZT films annealed at 9.9 atm, the peak intensities became twice, three times, and 1.5 times stronger than the initial intensity, and the peak positions shifted by $0.45\,^{\circ}$, $0.6\,^{\circ}$, and $0.25\,^{\circ}$, respectively. These results clearly show that the crystallinity of BSO-added ferroelectric films is improved by annealing in high-pressure oxygen, which is different from the behavior of a usual ferroelectric film, as shown in Fig. 10a.

Next, the leakage current and *P–V* characteristics of these films were investigated. Typical leakage current characteristics are shown in Fig. 11. As can be seen from the figure, the leakage current densities were dramatically decreased by high-pressure annealing, particularly when the oxygen pressure was 9.9 atm. In a BSO-added BLT film, the leakage current density decreased at least by three orders of magnitude, and in a BSO-added SBT film, the value was less than 1×10^{-8} A/cm^2 at a voltage of 5 V (1 MV/cm).

Figure 12 shows a comparison of the initial *P–V* hysteresis loop with that of the sample annealed at 9.9 atm. As can be seen from Fig. 12a, the hysteresis loop of a BSO-added BLT film was hardly changed by high-pressure annealing. On the contrary, in a BSO-added SBT film, the shape of the hysteresis loop as well as the saturation polarization were improved. Furthermore, in

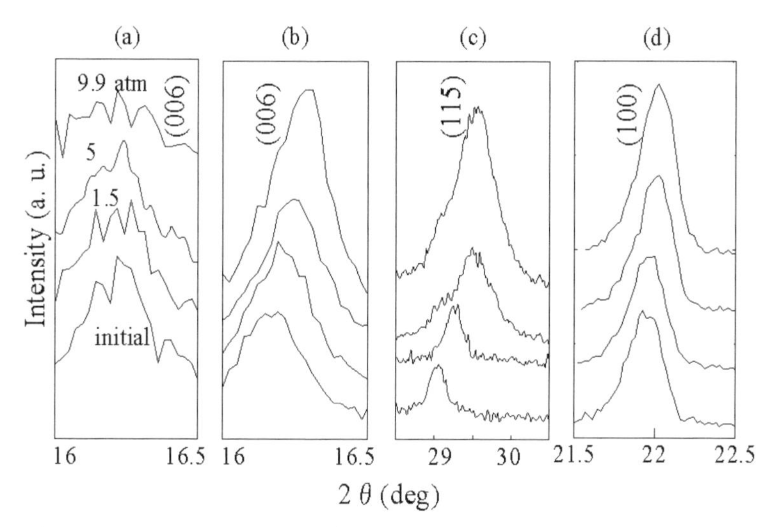

Fig. 10. XRD peak shifts of (**a**) 100-nm-thick BLT, (**b**) 50-nm-thick BSO-added BLT, (**c**) SBT, and (**d**) PZT films after high-pressure annealing at 1.5, 5, and 9.9 atm

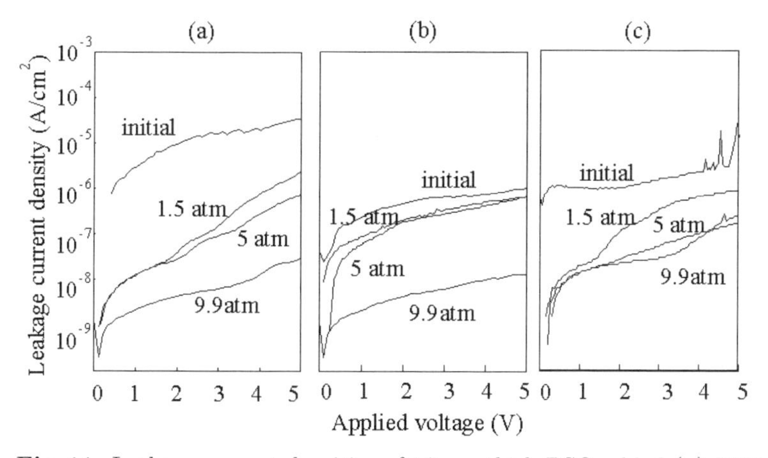

Fig. 11. Leakage current densities of 50-nm-thick BSO-added (**a**) BLT, (**b**) SBT, and (**c**) PZT films after high-pressure annealing at 1.5, 5, and 9.9 atm

a BSO-added PZT film, both the saturation polarization and the coercive voltage were increased with an increase in the oxygen pressure. The reason for the different behavior of the ferroelectric and insulating characteristics is not well understood at present. A difference in the crystal structures, such as between perovskite and Bi-layered structures, may cause the different behavior. Although we need more detailed investigations, we can show evidence of the structural change of the film due to annealing in high-pressure oxygen.

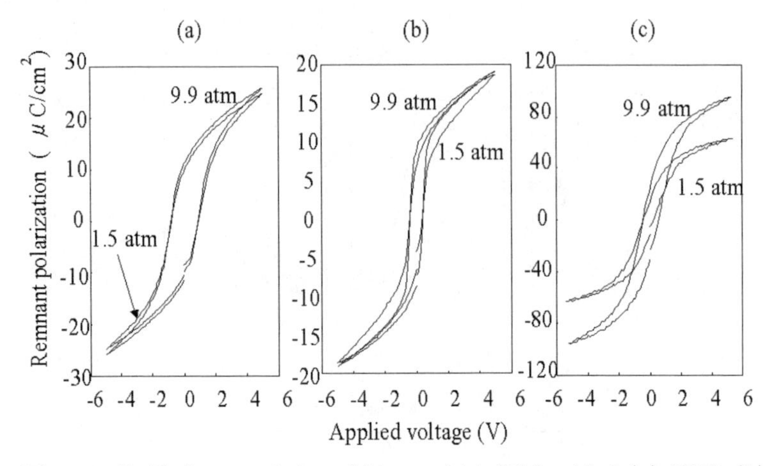

Fig. 12. *P–V* characteristics of 50-nm-thick BSO-added (**a**) BLT, (**b**) SBT, and (**c**) PZT films after high-pressure annealing at 1.5 and 9.9 atm

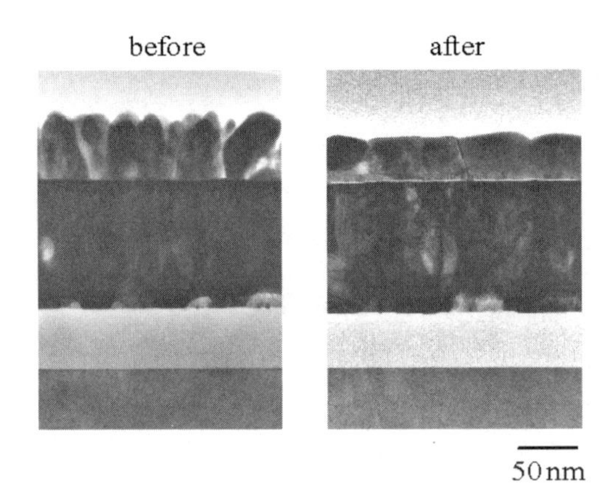

Fig. 13. TEM images of BSO-added SBT films before and after high-pressure annealing at 9.9 atm

Figure 13 shows cross-sectional TEM images of BSO-added SBT films before and after annealing at 9.9 atm. As can be seen from the figure, the film has a columnar structure with a relatively rough surface initially, but it changes to a structure with flat large grains due to the high-pressure annealing. This result suggests that a decrease of the grain-boundary regions, as well as a decrease of the oxygen defects within the grains, are the main reasons for improvement of the ferroelectric and insulating characteristics.

6 Summary

In summary, we succeeded in forming a new class of ferroelectrics by the use of Bi_2SiO_5 in a sol-gel method. Their unique features are summarized as follows. (1) The ferroelectric properties are similar to the original ferroelectrics, such as those of $Bi_4Ti_3O_{12}$, $SrBi_2Ta_2O_9$ and $Pb(Zr,Ti)O_3$, even if the mole ratio of Bi_2SiO_5 is as high as 0.4. (2) The crystallization temperature is lower by 150–200 °C than that of the original ferroelectrics. (3) The morphologies of the surface and the interface with the Pt electrode are excellent, so that a 13-nm-thick ferroelectric capacitor can be formed. Finally, (4) the films are fatigue-free at least up to 10^{10} switching cycles and hydrogen-immune. Moreover, the ferroelectric and insulating characteristics of the BSO-added BLT, SBT, and PZT films were improved by annealing in high-pressure oxygen up to 9.9 atm. It was found in a BSO-added BLT film that the leakage current density decreased by at least three orders of magnitude. It was also found in BSO-added SBT and PZT films that the saturation polarization in the P–V characteristics was increased by the high-pressure annealing. From TEM observation, the origin of these characteristic changes was speculated to be the structural change of the ferroelectric films. At present, we do not know if similar films are formed by a sputtering method or an MOCVD (metal-organic chemical vapor deposition) method. However, since it is certain that these films can be formed by an LSMCD (liquid source misted chemical deposition) method, in which sol-gel solution is used as fine liquid particles, it is considered that these materials will become very important in the fabrication of future high-density FeRAMs.

Acknowledgements

This work was performed under the auspices of the R&D Projects in Cooperation with Academic Institutions (Next-Generation Ferroelectric Memory) supported by NEDO (New Energy and Industrial Technology Development Organization in Japan) and managed by FED (R&D Association for Future Electron Devices).

References

1. T. Kijima, H. Matsunaga: Jpn. J. Appl. Phys. **37**, 5171 (1998)
2. M. Yamaguchi, K. Hiraki, T. Nagatomo, Y. Masuda: Jpn. J. Appl. Phys. **39**, 5512 (2000)
3. E. C. Subbarao: J. Phys. Chem. Solids. **23**, 665 (1962)
4. J. F. Dorrian, R. E. Newnham, D. K. Smith, M. I. Kay: Ferroelectr. **3**, 17 (1971)
5. C. A. Paz de Araujo, J. D. Cuchiaro, L. D. MacMillan, M. C. Scott, J. F. Scott: Nature **374**, 627 (1995)
6. T. Kijima, H. Ishiwara: Jpn. J. Appl. Phys. **41**, L716 (2002)
7. T. Kijima, H. Ishiwara: Ferroelectr. **271**, 289 (2002)
8. K. Mukaida, N. Anbo, N. Iijima: Phosphorus Res. Bull. **1**, 291 (1991)
9. R. Takahashi, S. Sato, T. Sodesawa, M. Yabuki: J. Catal. **200**, 197 (2001)
10. Y. Shimamoto, K. A. Kushida, H. Miki, Y. Fujisaki: Appl. Phys. Lett. **70**, 3096 (1997)
11. H. Miyazawa, E. Natori, S. Miyashita, T. Shimoda, F. Ishii, T. Oguchi: Jpn. J. Appl. Phys. **39**, 5679 (2000)

12. T. Irifune, A. E. Ringwood: Earth Planet. Sci. Lett. **117**, 101 (1993)
13. K. S. Savena, L. S. Dubrovinsky, P. Lasor, Y. Cerenius, P. Haeggkvist, M. Hanfland, J. Hu: Science **274**, 1357 (1996)
14. H. Ishiwara, T. Sato, A. Sawaoka: Appl. Phys. Lett. **61**, 1951 (1992)
15. S. Narayan, V. Joshi, L. McMillan, C. A. Paz de Araujo: Integr. Ferroelectr. **25**, 169 (1999)
16. N. Yanase, K. Abe, N. Fukushima, T. Kawakubo: Jpn. J. Appl. Phys. **38**, 5305 (1999)
17. M. Aratani, K. Nagashima, H. Funakubo: Jpn. J. Appl. Phys. **40**, 4126 (2001)
18. M. Mandeljc, M. Kosec, B. Malic, Z. Samardzija: Integr. Ferroelectr. **30**, 149 (2000)
19. T. Kijima, Y. Kawashima, Y. Idemoto, H. Ishiwara: Jpn. J. Appl. Phys. **41**, L1164 (2002)

Static and Dynamic Properties of Domains

Rainer Waser[1,2], Ulrich Böttger[2], and Michael Grossmann[2]

[1] Institut für Festkörperforschung, Research Center Jülich,
52425 Jülich, Germany
waser@iwe.rwth-aachen.de
[2] Institut für Werkstoffe der Elektrotechnik, RWTH Aachen,
52056 Aachen, Germany

Abstract. The static domain pattern of ferroelectric thin films results from the crystal structure of the film and the misfit strain caused by the substrate. A comparison of conceptual considerations and experimental results of atomic force microscope studies is presented. The motion of the domain walls is analyzed with respect to reversible and irreversible jumps as extracted from small-signal C–V and large-signal P–V measurements. The ferroelectric switching is described in the classical manner for single crystals and compared to the behavior of thin films, in which a superposition of Curie–von Schweidler relaxation processes is observed. The resulting voltage dependence of ferroelectric switching is essential for the speed of FeRAM devices. There are three major types of long-term effects: fatigue, retention, and imprint. The impact of the domain walls and domain wall motion is discussed.

1 Introduction

The electrical behavior of ferroelectric thin films used for FeRAM devices is predominantly determined by the domain configuration within the films and the motion of the domain walls. For this reason, the static and dynamic properties of domains and domain walls are described in this chapter. The presentation starts with a consideration of the static biaxial stress situation of ferroelectric thin films as a function of the substrate-induced misfit strain and the resulting domain configurations. An example of detailed imaging and analysis by atomic force microscopy (AFM) is given. In a subsequent section, the dynamic small- and large-signal behaviour is presented and interpreted in terms of reversible and irreversible domain wall motions. Based on this, ferroelectric switching is introduced. Obviously, the speed of this process ultimately controls the operating frequencies of real FeRAM devices. The empirical Merz equations and their application to ferroelectric single crystals are compared to the results obtained for thin films. The significant differences are traced back to the dielectric relaxation currents, which follow the Curie–von Schweidler law.

The long-term reliability is affected by three processes: the decrease of the ferroelectric polarization upon continuous large-signal cycling is called fatigue, storage at zero bias leads to retention losses due to ferroelectric relaxation, and, in addition, a shift of the hysteresis characteristics along the

H. Ishiwara, M. Okuyama, Y. Arimoto (Eds.): Ferroelectric Random Access Memories,
Topics Appl. Phys. **93**, 31–45 (2004)
© Springer-Verlag Berlin Heidelberg 2004

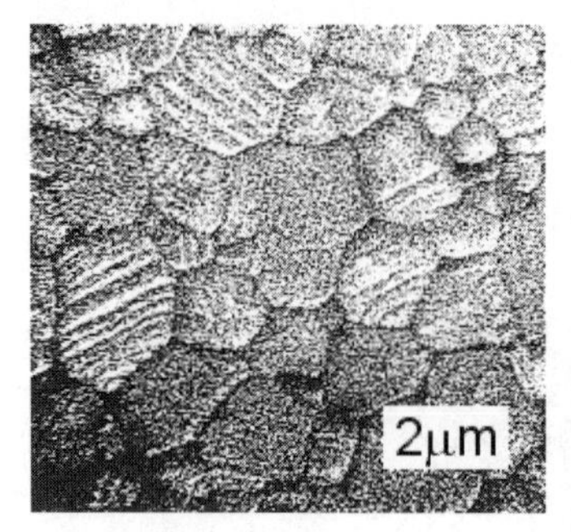

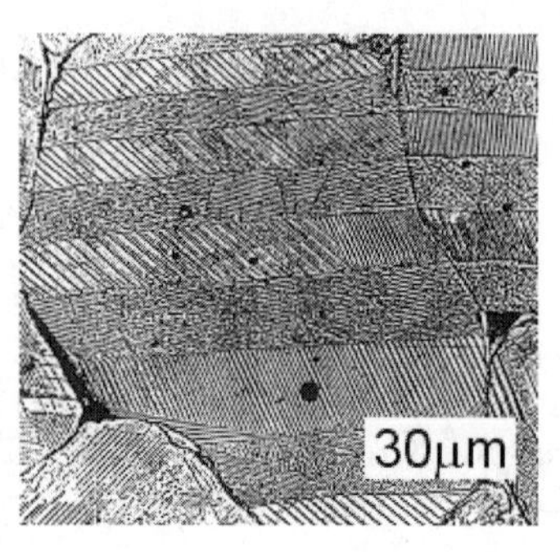

Fig. 1. The domain pattern of etched fine-grained $BaTiO_3$ ceramic by the means of scanning electron microscope (*left*) and that of coarse-grained $BaTiO_3$ ceramic by a light microscope (*right*)

voltage axis, known as imprint, may occur. A brief survey of the possible physical mechanisms is given.

2 Domain Configuration

When a ferroelectric single crystal is cooled below the phase transition temperature, the electrical stray field energy caused by the noncompensated polarization charges is reduced by the formation of ferroelectric domains. The configuration of the domains follows a head-to-tail condition in order to avoid discontinuities in the polarization at the domain boundary, $\nabla \boldsymbol{P} = \sigma$. The buildup of domain walls, elastic stress fields, and free charge carriers counteracts the process of domain formation. In addition, the influence of vacancies, dislocations, and dopants exists.

In polycrystalline bulk ceramics, the pattern of the domains is quite different, because the domain structure of each grain is formed under elastic clamped conditions by its surrounding neighbors, whereas a single crystal is free [1]. It should be noted that only non-180° domains – that is, 90° domains (for tetragonal structures) or 71° and 109° domains (for rhombohedral structures) – have the potential to reduce the elastic energy. Some details of the patterning in ceramics were given by *Cook* [2] and *de Vries* and *Burke* [3]. They observed two types in coarse-grained $BaTiO_3$, called the herringbone and square net patterns. The first one is by far the most common in unpoled ceramics. As shown in Fig. 1, by decreasing the grain size the domain pattern changes from a banded to a laminar structure [4].

In ferroelectric thin films, the process of domain formation is driven by the clamping conditions of the substrate. Ferroelectric thin films are typically crystallized at high temperatures (e.g., $PbZr_{1-x}Ti_xO_3$ (PZT) at about 700 °C [5]). The growth of the film on the substrate is favored in that orientation, which delivers the best match of substrate and thin film. Due to the different thermal expansion coefficients, the subsequent cooling results in

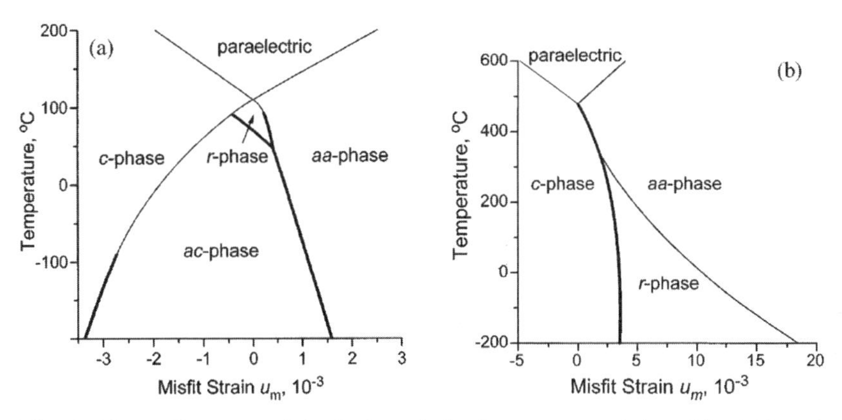

Fig. 2. Phase diagrams of single-domain barium titanate (**a**) and lead titanate (**b**) epitaxial thin films at different misfit strains u_m

two-dimensional elastic stresses on the interface, which are either compressive or tensile. Crossing the ferroelectric phase transition evokes, besides the spontaneous polarization, a spontaneous deformation of the lattice cell that leads to additional elastic stresses on the interface.

Pertsev et al. developed a phenomenological thermodynamic theory for single-domain epitaxial thin films of $BaTiO_3$ and $PbTiO_3$, taking into account the mechanical boundary conditions [6]. They found that the phase transition from the paraelectric to the ferroelectric state is of second order instead of first order in bulk crystals, and that different ferroelectric phases with different orientations of the spontaneous polarization become stable as a function of the misfit strain u_m, as shown in Fig. 2. Assuming an arrangement of the film in the 1–2 plane, the ferroelectric equilibrium states are indicated as being c phase with $\boldsymbol{P} = (0, 0, P_3)$, aa phase with $\boldsymbol{P} = (P_1, P_1, 0)$, r phase with $\boldsymbol{P} = (P_1, P_1, P_3)$, and, in the case of BT, ac phase with $\boldsymbol{P} = (P_1, 0, P_3)$.

It is obvious that integrated ferroelectric thin films as planar capacitors in ferroelectric random access memories (FeRAMs) should suffer compressive rather than tensile stress – that is, negative misfit strains – in order to favor phases with a nonvanishing P_3 component corresponding to the "switchable" part of the polarization.

Real ferroelectric thin films show polydomain structures. In Fig. 3, possible domain patterns of different textures of tetragonal films of $PbZr_{1-X}Ti_XO_3$ with $X > 0.48$ are depicted. In the case of compressive stress, the polarization is predominantly out-of-plane, that is (001), oriented: 90° as well as 180° domains are expected. Such orientation could be realized by the deposition of tetragonal PZT on magnesium oxide substrates [7]. Under the influence of an electric field, the number of 180° domains decreases. The resulting pattern predominantly consists of 90° domains. A (100) orientation – that is, in-plane orientation of the polarization – is caused by tensile stress and is achieved by using a buffer layer of yttrium-stabilized zirconium and an oxide electrode

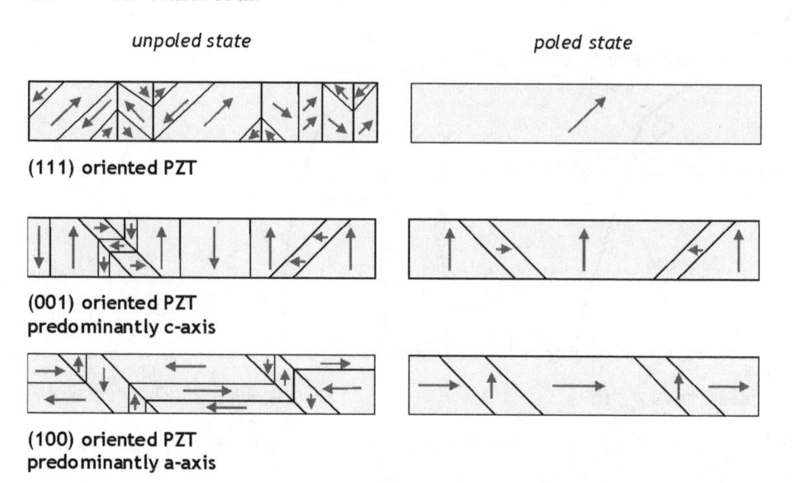

Fig. 3. The domain structure of tetragonal PZT with different orientations

of lanthanum strontium cobaltate, or by depositing on a (100)-$SrTiO_3$ substrate with a $SrRuO_3$ electrode [8]. The change of the domain structure by poling is similar to the (001) orientation, but the a-axis orientation is still preferred. In standard systems for ferroelectric thin films – for example, PZT with platinum electrodes on oxidized silicon wafers – the orientation of the crystallographic axes of PZT is in the (111) direction. Poling should evoke the single-domain state while the "head-to-tail" configuration is required. However, it is still under discussion whether the 90° domain walls are generally mobile.

In epitaxial PZT thin films, a self-polarized polarization mechanism – that is, a remanent polarization without the application of an external field is observed [9]. The three-dimensional piezoresponse force microscope (PFM) enables us to determine different domains and their polarization directions. Investigation of as-grown $PbZr_{0.2}T_{0.8}O_3$ film epitaxially grown on a [001] single-crystalline $SrTiO_3$ substrate by pulsed laser deposition with an intermediate 50-nm-thick $La_{0.5}Sr_{0.5}CoO_3$ oxide layer shows that the out-of-plane polarization in c domains points preferentially toward the bottom electrode [10]. From the asymmetry of the piezoresponse signal across a particular a domain, it can be deduced at which angle (45° or 135°) the boundaries of this a domain are inclined to the substrate. The existing two possibilities are schematically shown in the sketches of Fig. 4, where the left-hand and right-hand side a domains have inclination angles of 45° or 135°, respectively. Inspection of the piezoresponse images shows that the observed configuration is always of the "head-to-tail" type [11].

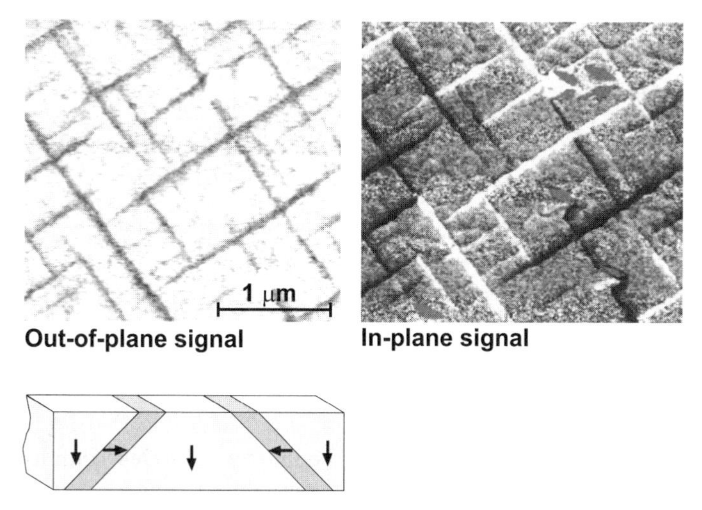

Fig. 4. Out-of-plane and in-plane piezoresponse images of as-grown epitaxial PZT. The contrast corresponds to different orientations of polarizations in the domains

3 Reversible and Irreversible Polarization Contributions

To characterize ferroelectric materials, the dependence of the polarization on the applied voltage is usually measured by means of a Sawyer–Tower circuit or by recording the current response to a voltage step. The $P(V)$ hysteresis curve is used to determine the remanent polarization and coercive field, respectively. These two parameters are of critical importance in the design of external circuits of FeRAMs.

The origin of the ferroelectric hysteresis is the existence of irreversible polarization processes; that is, the irreversible polarization reversal of a single ferroelectric lattice cell as explained by the Landau–Devonshire theory [12]. However, the exact interplay among this fundamental process, domain walls, defects, and the overall appearance of the ferroelectric hysteresis is still not precisely known.

The separation of the total polarization into reversible and irreversible contributions that has long been appreciated in the study of ferromagnetic materials [13] might facilitate the understanding of ferroelectric polarization mechanisms. In particular, the irreversible processes would be important for ferroelectric memory devices, since the reversible processes cannot be used to store information.

For ferroelectrics, mainly two possible mechanisms for irreversible processes exist [14, 15]. First, lattice defects interact with a domain wall and hinder it from returning into its initial position after removal of the electric field that initiated the domain wall motion ("pinning") [16]. Second, there is the nucleation and growth of new domains which do not disappear after the

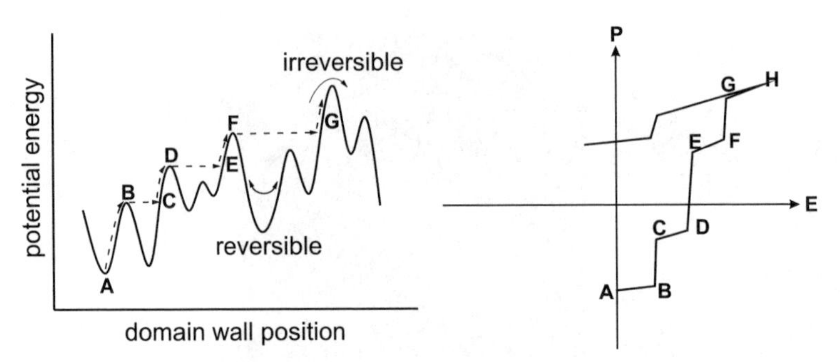

Fig. 5. The motion of a domain wall in the statistical potential (*left*) and the related hysteresis loop (*right*)

field is removed again. In ferroelectric materials the matter is further complicated by defect dipoles and free charges that also contribute to the measured polarization and can also interact with domain walls [17]. Reversible contributions in ferroelectrics are due to ionic and electronic displacements and to domain wall motions with a small amplitude. These mechanisms are very fast. As already mentioned, the reorientation of dipoles and/or defects or free charges also contributes to the total polarization. These mechanisms are usually much slower, but they also might be reversible (relaxation).

The motion of the domain wall under an external electric field happens in a statistical potential generated by their interaction with the lattice, point defects, dislocations, and the neighboring walls. Reversible movement of the wall is regarded as a small displacement around a local minimum. When the driven field is high enough, irreversible jumps above the potential barrier into a neighboring local minimum occur (Fig. 5).

Based on these assumptions, the measurement of the large-signal ferroelectric hysteresis with additional measurements of the small-signal capacitance at different bias voltages are interpreted in terms of reversible and irreversible parts of the polarization. As shown in Fig. 6, the separation is done by subtracting from the total polarization the reversible part; that is, the integrated $C(V)$ curve [18]:

$$P_{\mathrm{irr}}(V) = P_{\mathrm{tot}}(V) - \frac{1}{A}\int_0^V C(V')\,\mathrm{d}V' \,. \tag{1}$$

Typical hysteresis loops are dynamically recorded at certain frequencies. If slow reversible polarization mechanisms such as the above-mentioned relaxations also contribute to the total polarization P_{tot}, the shape of the hysteresis loop, especially the coercive field, becomes frequency dependent (see also the Chapter by *Nagarajan* et al. in this volume). In order to overcome this influence, the measurement should be performed with the lowest frequency possible; that is, quasi-statically.

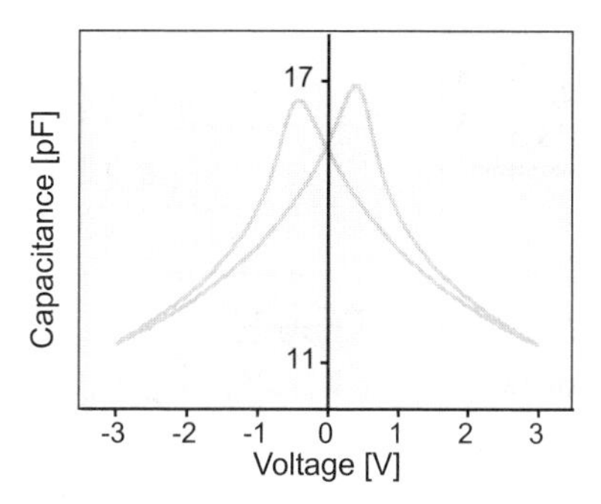

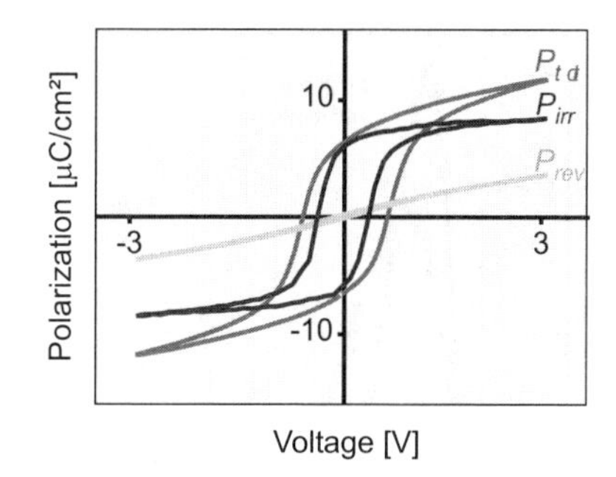

Fig. 6. The $C(V)$ curve (*left*) and reversible and irreversible contributions to the polarization (*right*) of a ferroelectric SBT thin film

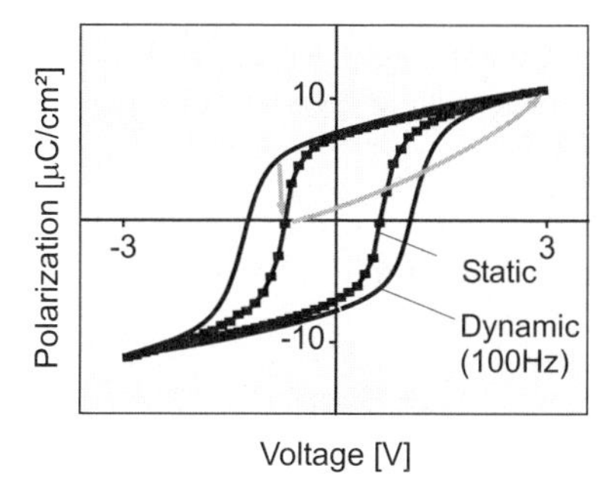

Fig. 7. The dynamic and quasi-static hysteresis loops of a ferroelectric SBT thin film

The procedure of quasi-static hysteresis [19] starts with pre-polarizing by measuring a complete hysteresis curve. Then, the excitation signal is kept constant for one second at a particular voltage V_1. The relaxed quasi-static value of the polarization corresponding to V_1 is determined by measuring and integrating the current response to a subsequent voltage ramp from V_1 to $V_{\max}$.

The values of the coercive voltage and the remanent polarization are significantly smaller in the case of the quasi-static hysteresis loop than those obtained from dynamic hysteresis measurements (Fig. 7).

4 Ferroelectric Switching

The polarization reversal in single crystals has been intensively investigated by direct observation of the formation and the movement of the domain walls. For example, in $BaTiO_3$ single crystals it was found by *Merz* [20], and by *Fousek* and *Brezina* [21], that in response to a voltage step the process happens by the formation of opposing 180° or orthographic 90° domains in the shape of needles and wedges. Both the resulting maximum displacement current $i_{\max}$ as well as the switching time t_s, which is the most significant quantity and describes the duration of the polarization reversal, were mea-

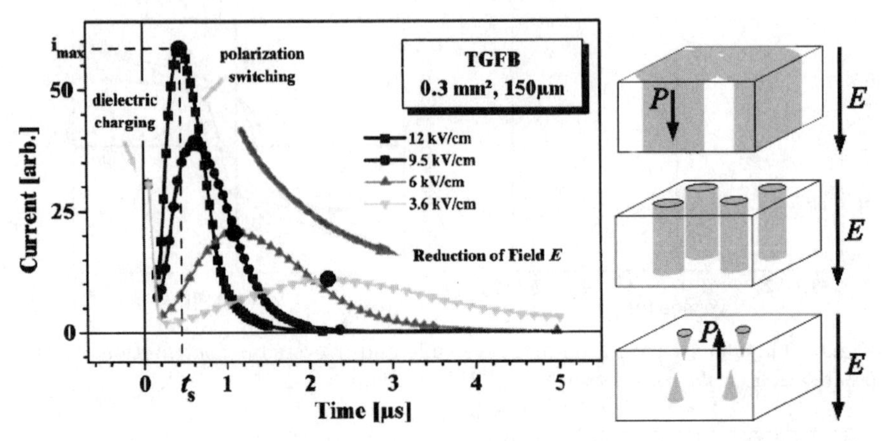

Fig. 8. The polarization switching current of a TGFB single crystal with a decreasing electric field [23] (*left*) and switching by nucleation and growth of domains (*right*)

sured as a function of the applied field E and follow empirical laws:

$$i_{\mathrm{max}} = i_0 \cdot \exp(-E_\alpha/E)\,, \tag{2}$$

$$t_{\mathrm{s}} = t_0 \cdot \exp(E_\alpha/E)\,, \tag{3}$$

where i_0 and t_0 are constants. The constant activation field E_α is the same in both equations [22].

The above-mentioned equations are only applicable when the applied field E is constant during the polarization reversal that is, the time constant of the dielectric charging τ_{RC} must be much smaller than the switching time t_{s}. The dielectric charging is determined by the capacitance of the sample and the inevitable series resistors (source, lines, etc.). The switching time is determined by many factors: the domain structure, the nucleation rate of opposite domains, the mobility of the domain walls, and many others [24]. As shown in Fig. 8, the condition is fullfilled and dielectric charging is clearly separated from the polarization switching hump.

Another approach to analyzing the polarization reversal process in ferroelectrics is the hysteresis loop. The problem of an undetermined excitation voltage step is avoided by the constant increasing and decreasing of the electric field, which can be much more easily electrically controlled. However, the interpretation of the loop is much more difficult and needs an integral calculation method. This has been evolved successfully by the group of *Ishibashi* and *Orihara* [25]. Based on the Kolmogorov–Avrami model [26] that was originally developed for phase transition kinetics, they have included the field dependence of the polarization switching and modeled the D–E hysteresis loops, which depend on the excitation frequency. They found good agreement with the experimental data for tri-glycine fluoroberyllic (TGFB) single crystals, especially the pronounced frequency dependence of the coer-

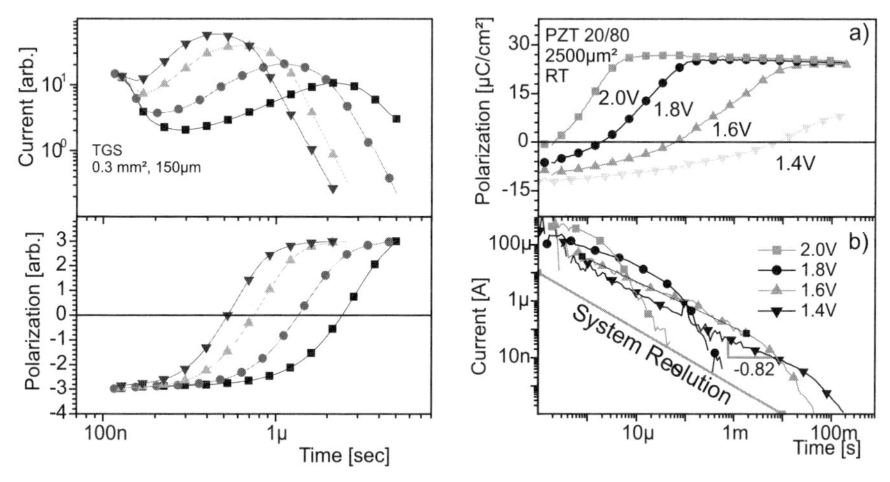

Fig. 9. The polarization switching current and calculated polarization at different electric fields for a TGFB single crystal [23] (*left*) and PZT thin film [32] (*right*)

cive field E_C:

$$E_C \propto f^{-\beta} . \tag{4}$$

The frequency dependence of the coercive field in ferroelectric thin films does not yet give a clear picture: it depends on the ferroelectric itself and the electrode material [27]. In $SrBi_2Ta_2O_9$ with Pt electrodes, the exponential factor β of (4) is about 0.1 in a raw approximation, while in $PbZr_{1-X}Ti_XO_3$ with Pt electrodes it is about an order of magnitude less.

Applying voltage steps on a PZT capacitor up to voltages above the coercive voltage, the current response shows no maximum i_{max}, even if the switching is complete (Fig. 9c,d). This is in contrast to the typical behavior of a single crystal, as shown in Fig. 9a,b. From the application point of view for memory cells, a film with such characteristics could lead to a read error because the collected charges are not distinguished between the two switching states $-P_r$ and $+P_r$; that is, the two logic states.

During polarization switching, the current response shows a typical Curie–von Schweidler behavior and can be described by the following law:

$$J \propto t^{-\kappa} , \tag{5}$$

where κ is a constant that has a value less than or equal to one. For PZT thin films, a value of $\kappa = 0.82$ was observed.

The detection of this kind of current response leads to a different view of polarization inversion, since the Curie–von Schweidler behavior is found in all thin films; in dielectric [28] as well as in ferroelectric thin films [29]. Two main models are considered to be responsible for the Curie–von Schweidler relaxation. The first is a many-body interaction model that is based on the

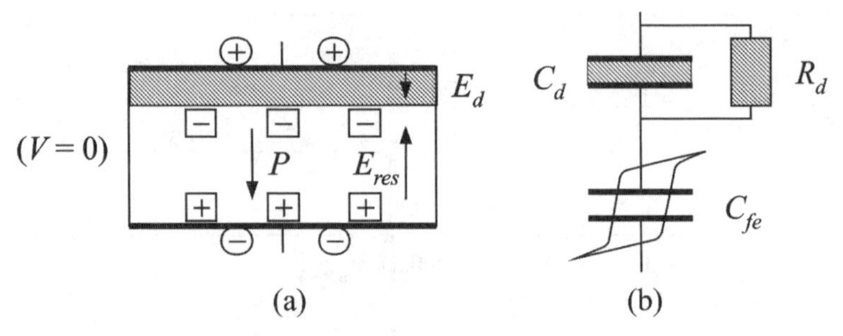

Fig. 10a,b. The simplified scheme for a real ferroelectric capacitor and a equivalent circuit

fact that the hopping motion of a charged particle always affects the motion of the neighboring charges, which can lead to a Curie–von Schweidler law [30]. And, secondly, the distribution of relaxation times (DRT) model is based on a superposition of Debye-type relaxations with a large distribution of relaxation times, which may be caused by variation in the charge transport barrier; for example, at the grain boundaries [31]. The physical origin is not discussed further, but all evidence suggests that dielectric polarization processes are responsible.

The Curie–von Schweidler law in the behavior of nonferroelectric materials leads to a linear dispersion in the frequency regime, which can be expressed as

$$\epsilon = \epsilon_\infty + k_0 \cdot \omega^{\alpha-1} , \tag{6}$$

where ϵ_∞ is the permittivity at very high frequencies, k_0 is a constant, and α is the same constant as in (5).

Using a simplified picture of a real ferrelectric capacitor, the switching behavior is modeled by a simulation tool for integrated circuits such as, for example, SPICE. The capacitor constists of an ideal ferroelectric and a nonferroelectric interface layer. The interface layer implies defects in the ferroelectric and amorphous grain boundaries, as well as the real interface between the electrode and the ceramic film. It is well known that these regions have a strong dielectric dispersion; that is, dielectric losses. The behavior of the dielectric interface is described by a parallel circuit of linear *RC* terms, as shown in Fig. 10. *Lohse* added to the ideal nondispersive ferroelectric capacitor the dielectric *RC* behavior in the above-mentioned concept of the distribution of relaxation times (DRT) model, on a superposition of Debye-type relaxations [32]. The switching is implemented by the Kolmogorov–Avrami process of nucleation of domains and the domain wall motion is depicted in Fig. 10b.

As a result, it can be shown that the dielectric interface has a strong influence on the switching process. The delay of the polarization reversal is determined by dielectric dispersive polarization mechanisms rather than by

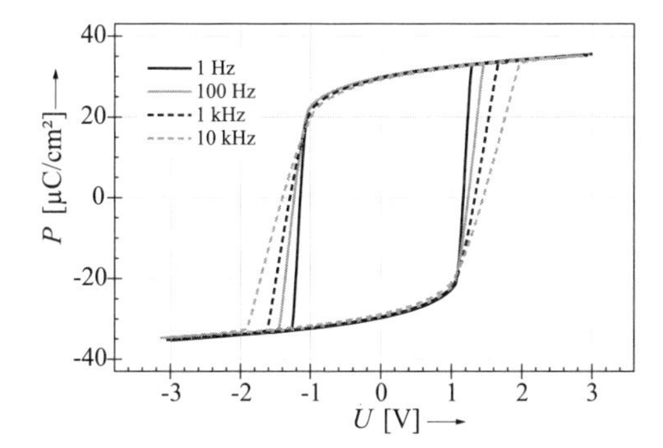

Fig. 11. The simulated frequency behavior of a ferroelectric thin film

real ferroelectric switching. Frequency-dependent hysteresis loops calculated using the above model are shown in Fig. 11.

5 Long-Term Effects

Long-term effects describe the deterioration of the ferroelectric properties with time. These deteriorations can lead to a failure of the ferroelectric memory cell. Generally, three different long-term effects are considered as major failure mechanisms for the ferroelectric memory cell: fatigue, retention loss, and imprint (Fig. 12). In what follows, these major failure mechanisms will be briefly introduced, as well as the physical mechanisms that might cause these deteriorations.

5.1 Fatigue

Fatigue describes the flattening of the ferroelectric P–V hysteresis loop upon continuous cycling (Fig. 12a). Due to the decrease of the remanent polarization, the charge difference between logic "0" and "1" becomes smaller. This effect can lead to a failure of the device if this difference falls short of the safety margin of the sensing amplifier. Usually, different mechanisms are suggested to explain this phenomenon, which can be divided into bulk- and interface-related effects.

The bulk-related effects assume the pinning of domain walls by charged defects within the bulk of the film. As a result, the pinning leads to a reduction of switchable polarization and, thus, explains the polarization decrease [33].

Another approach assumes the interface between the ferroelectric and the electrode to play the dominant role in the fatigue scenario [42]. Some authors consider domain wall pinning at the interface due to the electromigration of charged defects such as oxygen vacancies and the subsequent formation of defect planes parallel to the interface [34]. Also, an electronic mechanism is considered by another group of authors. Due to a large electric field during the switching of the polarization, electronic charges are injected from the

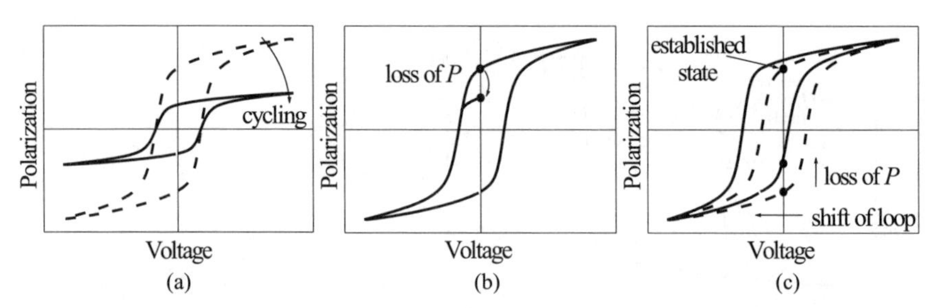

Fig. 12.a–c Three major failure mechanisms of ferroelectric thin films in view of memory applications

electrode into the film. These charges then become trapped in the vicinity of a thin surface layer and, hence, lead to the inhibition of domain nucleation [35]. Another approach assumes the existence of a thin surface layer with suppressed ferroelectric properties and a lower permittivity compared to the ferroelectric bulk. In the course of fatigue, the extension of the surface layer increases, resulting in an increased voltage drop across this layer [36]. This effect can also explain the polarization decrease during fatigue.

5.2 Retention Loss

After establishing a remanent polarization, the ferroelectric cannot ideally retain its remanent polarization. The slight polarization decrease with time is referred to as retention loss (Fig. 12b). Similarly to fatigue, the difference between switching and nonswitching becomes smaller in the case of retention loss.

In the case of retention loss, the interface between the ferroelectric and the electrode also seems to be the important parameter. *Gruvermann* and *Tanaka* [37] assume the presence of a polarization-independent built-in bias at the bottom electrode interface, caused, for example, by the formation of interfacial depletion layers. This built-in bias triggers the back-switching of domains and, hence, is responsible for the retention loss. Other authors suggest a residual internal field pointing in the opposite direction to the spontaneous polarization to be the cause of retention loss. This residual field arises due to insufficient screening caused by a thin surface layer [38].

5.3 Imprint

Imprint describes the preference of one polarization state over the other or the inability to distinguish between two different polarization states. Imprint leads to a shift of the hysteresis loop on the voltage axis as well as to a loss of the remanent polarization state opposite to the established one (Fig. 12c). Hence imprint can lead to either a read or a write failure of the memory cell.

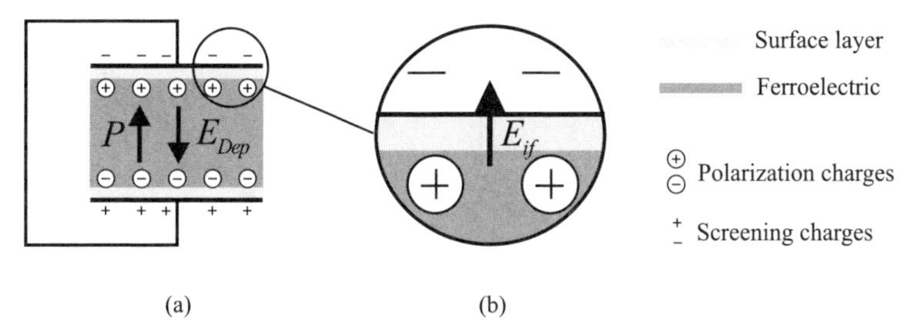

Fig. 13. (**a**) A sketch of a ferroelectric capacitor with a thin surface layer. (**b**) An enlargement of the interfacial region, showing the driving force $E_{\rm if}$ of imprint according to the interface screening model

For the explanation of imprint, two different types of mechanism are generally discussed. The defect dipole alignment model [39], on the one hand, involves the motion of ions, whereas the recently suggested interface screening model [40] assumes imprint to be of electronic origin.

In ferroelectric bulk materials, such as ceramics and single crystals, an effect similar to imprint is observed, which is called *internal bias*. In these bulk materials, a significant enhancement of the internal bias is observed upon doping with acceptor ions. A qualitative and quantitative model has been proposed that explains the internal bias effect as being due to the gradual alignment of defect dipoles consisting of acceptor ions and oxygen vacancies in the direction of the spontaneous polarization [39]. When the polarization is switched by an external field, these aligned defect dipoles remain in their positions and hence cause a shift of the hysteresis loop. The defect dipole alignment model has also often been discussed as the origin of imprint in ferroelectric thin films. In ferroelectric thin films, however, the imprint effect seems to be independent of dopant additions [40]. Furthermore, a significant enhancement of imprint is observed due to illumination [41]. Both observations indicate that imprint in ferroelectric thin films is in contrast to the defect dipole alignment model of electronic origin. Recently, with the interface screening model, an imprint model has been proposed that explains, both qualitatively and quantitatively, the imprint behavior as a function of various experimental parameters [40]. The interface screening model also assumes a thin surface layer with suppressed ferroelectric properties. As a result, a large electric field $E_{\rm if}$ arises within the surface layer, pointing in the direction of the polarization (see Fig. 13). The field $E_{\rm if}$ is responsible for charge separation within the surface layer according to a Frenkel–Poole effect (see Fig. 14a). In Fig. 13b, experimental results are compared to simulation results based on the interface screening model. Good agreement is obtained between experiment and simulation. With the interface screening model, the impact of other experimental parameters, such as time, temperature, film thickness, and the improvement due to the use of oxide electrodes [40], can also be successfully simulated.

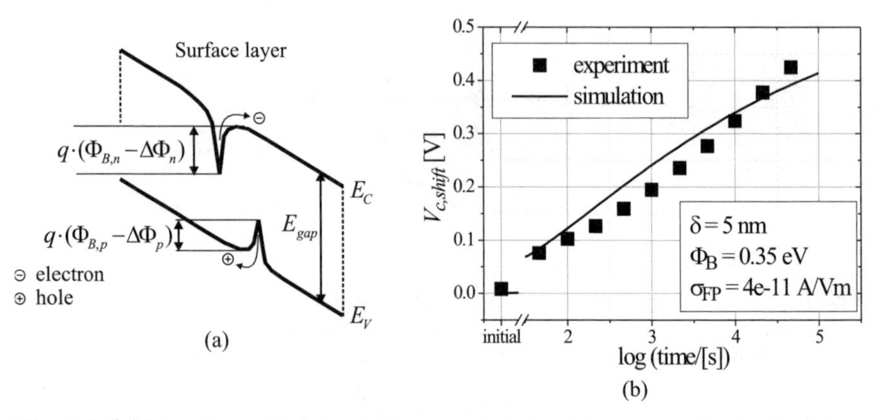

Fig. 14. (**a**) Experimental data and a numerical simulation according to the interface screening model. (**b**) A sketch of the Frenkel–Poole emission within the surface layer

References

1. G. Arlt, D. Hennings, G. de-With: J. Appl. Phys. **58**, 1619 (1985)
2. W.R. Cook: J. Am. Ceram. Soc. **39**, 17 (1956)
3. R.C. de Vries, J.E. Bruke: J. Am. Ceram. Soc. **40**, 200 (1957)
4. G. Arlt: J. Mater. Sci. **25**, 2655 (1990)
5. C.D. Lakeman, D.A. Payne: J. Am. Ceram. Soc. **75**, 3091 (1992)
6. N.A. Pertsev, A.G. Zembilgotov, A.K. Tagantsev: Phys. Rev. Lett. **80**, 1988 (1998)
7. K. Nashimoto, D.K. Fork, G.B. Anderson: Appl. Phys. Lett. **66**, 822 (1995)
8. K. Nagashima, M. Aratani, H. Funakubo: J. Appl. Phys. **89**, 4517 (2001)
9. A. Roelofs, N.A. Pertsev, R. Waser, F. Schlaphof, L.M. Eng, C.S. Ganpule, V. Nagarajan, R. Ramesh: Appl. Phys. Lett. **80**, 1 (2002)
10. C.S. Ganpule, V. Nagarajan, H. Li, A.S. Ogale, D.E. Steinhauer, S. Aggarwall, E. Williams, R. Ramesh: Appl. Phys. Lett. **77**, 292 (2000)
11. C.S. Ganpule, V. Nagarajan, B.K. Hill, A.L. Roytburd, E.D. Williams, R. Ramesh, S.P. Alpay, A. Roelofs, R. Waser, L.M. Eng: J. Appl. Phys. **91**, 1477 (2002)
12. M.E. Lines, A.M. Glass: *Principles and Applications of Ferroelectrics and Related Materials* (Clarendon, Oxford 1977)
13. S. Chikazumi, S.H. Charap: *Physics of Magnetism* (Wiley, New York 1964)
14. D. Damjanovic, M. Demartin: J. Phys. Condens. Matter **9**, 4943 (1997)
15. D. Damjanovic: Phys. Rev. B **55** (2), R649 (1997)
16. T.J. Yang, V. Gopalan, P.J. Swart, U. Mohideen: Phys. Rev. Lett. **82**, 4106 (1999)
17. O. Boser, D.N. Beshers: Mater. Res. Soc. Symp. Proc. **82**, 441 (1987)
18. O. Lohse, D. Bolten, M. Grossmann, R. Waser, W. Hartner, G. Schindler: Ferroelectric Thin Films VI, Symp. Mater. Res. Soc. **493**, 267 (1998)
19. D. Bolten, O. Lohse, M. Grossmann, R. Waser: Ferroelectr. **221**, 251 (1999)
20. W.J. Merz: Phys. Rev. **95**, 690 (1954)
21. J. Fousek, B. Brezina: Czech. J. Phys. B **10**, 511 (1960)

22. C. F. Pulvari, W. Kuebler: J. Appl. Phys. **29**, 1742 (1958)
23. O. Lohse, M. Grossmann, U. Böttger, D. Bolten, R. Waser: J. Appl. Phys. **89**, 2332 (2001)
24. V. Shur, E. Rumyantsev, S. Makarov: J. Appl. Phys. **84**, 445 (1998)
25. Y. Ishibashi, H. Orihara: Integr. Ferroelectr. **9**, 57 (1995)
26. M. Avrami: J. Chem. Phys. **7**, 1103 (1939)
27. J. F. Scott, F. M. Ross, C. A. Paz de Araujo, M. C. Scott, M. Huffmann: MRS Bull. **21**, 33 (1996)
28. M. Schumacher, G. W. Dietz, R. Waser: Integr. Ferroelectr. **10**, 231 (1995)
29. X. Chen, A. I. Kingon, L. Mantese, O. Auciello, K. Y. Hsieh: Integr. Ferroelectr. **3**, 355 (1993)
30. L. A. Dissado, R. M. Hill: J. Mater. Sci. **16**, 1410 (1981)
31. H. Kliem: IEEE Trans. Electr. Insul. **24**, 185 (1989)
32. O. Lohse: Dissertation, RWTH Aachen (2001)
33. E. L. Colla, D. V. Taylor, A. K. Taganstev, N. Setter: Appl. Phys. Lett. **72**, 2478 (1998)
34. M. Dawber, J. F. Scott: Appl. Phys. Lett. **76**, 1060 (2000)
35. I. Stolichnov, A. K. Tagantsev, E. L. Colla, N. Setter: Appl. Phys. Lett. **73**, 1361 (1998)
36. A. M. Bratkovsky, A. P. Levanyuk: Phys. Rev. Lett. **84**, 3177 (2000)
37. A. Gruvermann, M. Tanaka: J. Appl. Phys. **89**, 1836 (2001)
38. A. K. Tagantsev, M. Landivar, E. Colla, N. Setter: J. Appl. Phys. **78**, 2623 (1995)
39. G. Arlt, H. Neumann: Ferroelectr. **87**, 109 (1988)
40. M. Grossmann: Dissertation, RWTH Aachen (2001)
41. M. Grossmann, S. Hoffmann, S. Gusowski, R. Waser, S. K. Streiffer, C. Basceri, C. B. Parker, S. E. Lash, A. I. Kingon: Integr. Ferroelectr. **22**, 83 (1998)
42. R. Ramesh, A. Inam, W. K. Chan, B. Wilkens, K. Myers, K. Remschnig, D. L. Hart, J. M. Tarascon: Science **252**, 944 (1991)

Nanoscale Phenomena in Ferroelectric Thin Films

Valanoor Nagarajan[1], Chandan S. Ganpule, and Ramamoorthy Ramesh[2]

[1] Department IFF, Forschungzentrum Juelich, 52425 Juelich, Germany
valanoor@iwe.rwth-aachen.de

[2] Department of Materials Science and Engineering and Department of Physics, 219 Hearst Mining Building,
University of California, Berkeley, CA 94720, USA
rramesh@uclink.berkeley.edu

Abstract. In this chapter, recent progress in our group in the area of thin-film ferroelectrics is reviewed. The specific focus is on nanoscale phenomena such as polarization relaxation dynamics and piezoelectric characterization in model thin films and nanostructures, using voltage-modulated scanning force microscopy. Using this technique, we show the three-dimensional reconstruction of the polarization vector in lead zirconate titanate (PZT) thin films. Secondly, the time-dependent relaxation of remanent polarization in epitaxial PZT ferroelectric thin films, containing a uniform two-dimensional grid of 90° domains (*c*-axis in the plane of the film), has been investigated extensively. The 90° domain walls preferentially nucleate the 180° reverse domains during relaxation. Relaxation occurs through the nucleation and growth of reverse 180° domains, which subsequently coalesce and consume the entire region as a function of relaxation time. The kinetics of relaxation is modeled through the Johnson–Mehl–Avrami–Kolmogorov approach, with a decreasing driving force. In addition, we also present results on investigation of the relaxation phenomenon on a very local scale, where pinning and bowing of domain walls has been observed. We also show how this technique is used for obtaining quantitative information on piezoelectric constants and, by engineering special structures, how we realize ultra-high values of piezoconstants. It is shown that the piezo-constant of free-standing nanoscale capacitors is much higher than clamped values, and that there is a complex interplay between the substrate-induced constraint and the piezoresponse. Lastly we also show direct hysteresis loops on nanoscale capacitors, where there is no observable loss of polarization in capacitors as small as 0.16 μm^2 in area.

1 Introduction

At present, there is considerable interest in ferroelectric thin films as a medium for nonvolatile data storage [1]. In particular, much attention is focused on investigating high-density giga-bit data storage using scanning probe techniques [2]. As ferroelectric random access memory (FeRAM) devices scale down to the Gbit density regimes, it becomes imperative to understand physical phenomena that affect device performance at the nanoscale. For example, the ferroelectric element in these storage media is subject to progressive loss

H. Ishiwara, M. Okuyama, Y. Arimoto (Eds.): Ferroelectric Random Access Memories,
Topics Appl. Phys. **93**, 47–68 (2004)
© Springer-Verlag Berlin Heidelberg 2004

of polarization, commonly referred to as "polarization relaxation". Polarization relaxation shows a very complex interplay between various factors, such as defects, the domain wall dynamics, the crystallographic structure, etc., and hence it has been a subject of intense scientific research. All of the above-mentioned factors have characteristic dimensions in the range of a few hundred to a few thousand angstroms, which really means that in order to fully appreciate their role, one must have an approach to the study of phenomena at those scales.

With the advent of Piezoresponse Scanning Force Microscopy (Piezoresponse SFM), the investigation of nanoscale ferroelectric phenomena has not only become viable but also a necessity. It is a unique nondestructive characterization tool coupled with a relatively easy method of sample preparation, which can provide critical information on key nanoscale processes. In this chapter, we show in the first part how this tool has been so effective in observing domain relaxation. In this section, we focus on imaging nanoscale phenomena, particularly the phenomena of relaxation. It will be clear how "defects" such as domain walls act as nucleation centers and how nanoscopic sites act as pinning centers to inhibit complete reversal. The physical nature of this problem and early studies on the microscopic origins of polarization reversal using scanning force microscope techniques [3] show that they have the potential to quantify the physical origin of the relaxation kinetics. The Johnson–Mehl–Avrami–Kolmorogov (JMAK) theory is a useful approach for systems exhibiting nucleation and growth-based phase change kinetics [4]. We review in detail a study to apply this theory to the nanoscale investigation of polarization reversal in ferroelectric thin films via the atomic force microscope. In addition, the same technique can be used to image the domain structure of epitaxial ferroelectric thin films. We also show, by using careful interpretation of domain images, one can reconstruct the three-dimensional polarization picture in these thin films.

The second section on our observations will concentrate on obtaining quantitative information from such nanoscale capacitors. In particular, the dependence of the capacitor on orientation, size, and constraint has been investigated as far as the d_{33} goes. Since the probe itself measures only several tens of nanometers, it gives us the advantage of probing piezoelectric and ferroelectric properties of sub-micron capacitors. Consequently, we have been able to show piezoelectricity in capacitors as small as 70 nm × 70 nm and direct hysteresis loops in sub-micron capacitors. We summarize this review with some results of direct hysteresis and fatigue measurements on nanoscale capacitors. Finally, some areas for future research are also highlighted.

2 Experimental Details

In order to create films with systematically varying crystalline quality, three different types of substrates were used: (1) conventional platinized silicon (Pt/

$TiO_2/SiO_2/Si$, abbreviated to Pt–Si); (2) $(La_{0.5}Sr_{0.5})CoO_3$-buffered Pt–Si (LSCO/Pt–Si); and (3) LSCO/STO/Si. The thickness of STO is 15 nm. Details of the STO deposition process are described elsewhere. About 70 nm thick LSCO top and bottom electrodes were deposited by RF sputtering and crystallized in oxygen for 1 h at 600–650 °C. The PZT thin films (with Zr : Ti ratio = 40 : 60 (40/60), if not specified) were prepared by sol-gel and spin-coating techniques. The annealing of the PZT was done at 600–650 °C for 1 h and at 450 °C for the samples prepared with the new precursor. X-ray diffraction (XRD) and transmission electron microscopy (TEM) were employed to probe the phase formation and microstructure of the samples. The top electrodes (Pt and LSCO) were made by photolithography and sputtering, so that the capacitors for electrical measurements show a symmetric configuration – Pt/PZT/Pt–Si, Pt/LSCO/PZT/LSCO/Pt–Si, and Pt/LSCO/PZT/LSCO/STO/Si, respectively. The ferroelectric measurement was done by the use of an RT-66A tester.

Our experimental studies typically involve two distinct types of samples. A significant fraction of our domain imaging studies use epitaxial PZT films as model systems. Our piezoresponse measurements involve films with systematically different microstructures (epitaxial to polycrystalline) as well as a variety of substrates. Films are grown by either pulsed laser deposition [5] and/or by sol-gel processing. The epitaxial PZT films of high tetragonality (e.g., 20/80) contain a uniform two-dimensional grid of 90° domains (i.e., *c*-axis in the plane of the film). Such 90° domains have been shown to have a strong impact on the hysteretic polarization behavior of ferroelectric films [6]. Due to the absence of a top electrode, the polarization and piezoresponse hysteresis loops show a marked asymmetry, thereby indicating a built-in internal field.

Voltage-modulated scanning force microscopy [7] has been employed to study the piezoelectric effect and to image and modify the domain structure in thin-film samples. The experimental setup is shown in Fig. 1. Piezo-SFM is based on the detection of the local electromechanical vibration of the ferroelectric sample caused by an external AC voltage. The voltage is applied through the probing tip, which is used as a movable top electrode. The external driving voltage with frequency ω generates a sample surface vibration with the same frequency due to the converse piezoelectric effect. This surface displacement can be approximated by

$$\Delta z(t) = \int_0^{T_\mathrm{f}} d_{33}^{\text{thin film}} E_\mathrm{AC}(t,z)\,\mathrm{d}z\,, \tag{1}$$

where E_AC is the AC electric field as a function of time t and displacement z, and T_f is the thickness of the ferroelectric layer. The modulated deflection signal from the cantilever, which oscillates together with the sample, is detected using the lock-in technique. The amplitude of the first harmonic signal from the lock-in amplifier is a function of the piezoelectric displacement

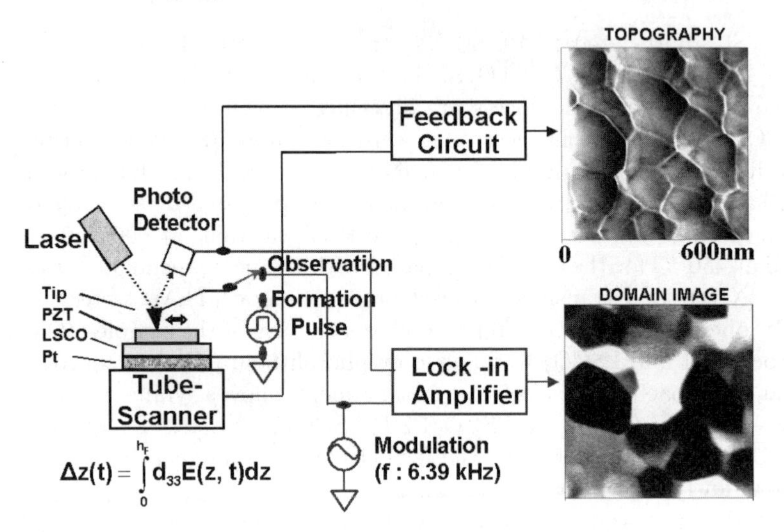

Fig. 1. The experimental setup for piezoresponse imaging

and the phase shift between the AC field and the cantilever displacement. Mathematically, it is given as

$$H_{\varpi} = k_{\mathrm{d}} d_{33}^{\text{thin film}} V_{\mathrm{AC}} \cos\theta \,. \tag{2}$$

The first two terms in the above equation represent the cantilever displacement, which is coupled to magnitude of the polarization. The last term ($\cos\theta$) represents the phase of the lock-in, which (through some lock-in techniques) can be made to assume values of 1, 0, or -1. This means that regions with opposite orientations of polarization, vibrating in counterphase with respect to each other under the applied AC field, should appear as regions of opposite contrast in the piezoresponse image. Since this technique depends on the piezoresponse of the ferroelectric, it is also termed piezoresponse microscopy.

For our experiments, we used a commercially available Digital Instruments Nanoscope IIIA Multimode scanning probe microscope, equipped with standard silicon tips, coated with Pt/Ir alloy for electrical conduction. The typical force constant of these tips was 5 N/ m and the apex radius was approximately 20 nm. The contact force was estimated to be around 70–100 nN.

3 Nanoscale Domain Imaging in Ferroelectric Thin Films

Figure 2 shows AFM images of a typical epitaxial PZT (20/80) film with (a) the topography and (b) the piezoresponse image. The topography clearly shows the two-dimensional 90° domain structure. The topographical relief in the structure originates from the spontaneous strain in the ferroelectric and

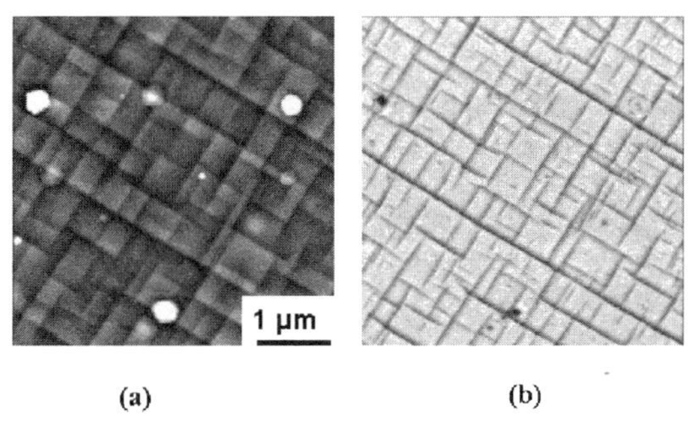

Fig. 2. (**a**) The topography of epitaxial $PbZr_{0.2}Ti_{0.8}O_3$ film grown via pulsed laser deposition. (**b**) Piezoresponse images of the same film. The 90° domain structure is clearly seen in both images as a two-dimensional grid

the tilting of a-domains away from the normal, which has been discussed in detail elsewhere [8, 9, 10]. The piezoresponse image compares very well with the topographical image. Piezoresponse images of the film surface show the presence of long needle-like orthogonal structures that have been identified to be 90° twinned domains. These domains form to relax the strains associated with lattice mismatch and phase transformation from the cubic to the tetragonal (ferroelectric) phase [8]. The image contrast in between the needle-like regions is uniformly bright, suggesting that the entire region is pre-poled in a specific direction. This was confirmed by measurements of the piezo-hysteresis loop from a local region [11], which showed a marked asymmetry indicating a built-in internal field. This as-grown sample was switched into the opposite state by scanning the surface of the film with the AFM tip biased at $-10\,\mathrm{V}$ (scan speed 1 Hz), leading to a strong change in the image contrast from bright to dark, as will be shown later. We observed that this switched state (in the c-axis regions of the film) begins to relax back to the original bright state.

We first discuss the use of this technique to map out the polarization orientation in a given ferroelectric thin film. This technique was first applied by *Eng* et al. [12] to image the three-dimensional polarization map in a $BaTiO_3$ single crystal, and later *Roelofs* et al. [13] demonstrated 3D polarization imaging on polycrystalline PZT thin films. In order to map the polarization vectors in the a-domains, the lateral differential signal from the photodiode was tracked. This signal effectively represents the torsion deformation of the AFM cantilever. As the tip scans over the a-domain regions, the electrical field through the tip produces the shearing deformation mentioned earlier, giving rise to the torsion signal in the photodiode. This is schematically illustrated in Fig. 3a,b for oppositely oriented a-domains. The phase of this torsion motion of the tip with respect to the input sinusoidal electric field provides information about the orientation of polarization in the domain. It should be pointed out that the scan direction of the tip is chosen to be such

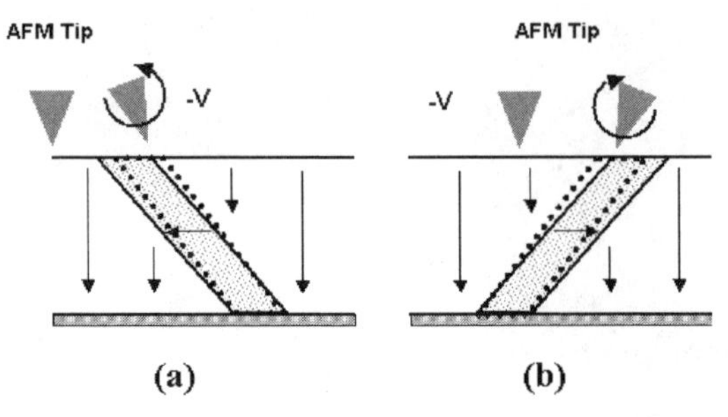

Fig. 3. The interaction of the field with the tip to give the torsion signal. Depending upon the output (**a**) or (**b**), the orientation of the in-plane image can be mapped out

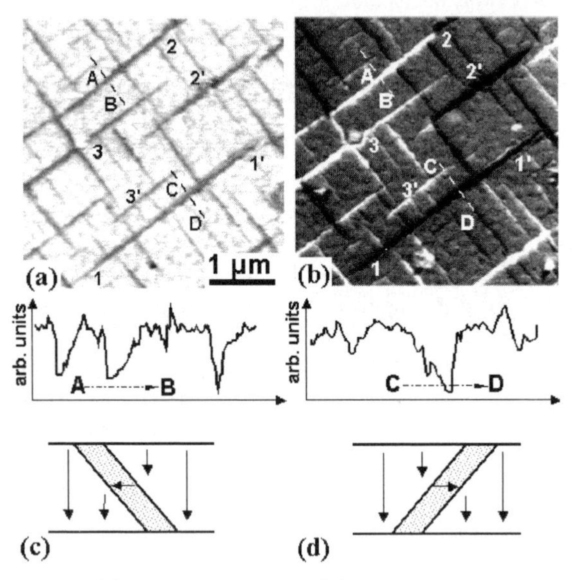

Fig. 4. (**a**) Out-of-plane and (**b**) in-plane piezoresponse images of the same region in a PZT film. The sections A–B and C–D are shown in the line intensity plots below the images. (**c**) and (**d**) show the schematic of the in-plane polarizations based on the line scans A–B and C–D

that the orthogonal a-domains contribute equally to the tip motion. The tip, therefore, scans at 45° to the a_1- and a_2-domains.

Figure 4a,b shows typical out-of-plane and in-plane piezoresponse images of the same region in the PZT film. In an earlier publication, we showed that it was possible to find out the inclination of the a-domain with respect to the substrate by obtaining a line scan of the piezoresponse signal across the domain [11]. By combining this information with the knowledge about the polarization direction within the a-domain as seen in Fig. 4b, it is possible to see that the head-to-tail configuration of the polarization vectors is maintained across the 90° domain wall. Thus, domains aligned at 45° and 135° to the substrate are phase shifted by 180° and therefore lead to different contrast levels in the in-plane image. Figure 4c plots a line scan of the vertical differential signal across A–B in Fig. 4a, thereby implying that the domain is oriented as illustrated schematically in Fig. 4c. Figure 4d plots a similar line scan across C–D in Figure 4a, thus showing that this *a*-domain is oppositely oriented with respect to the substrate. As expected, the image contrast within these domains is reversed, as noted in the in-plane image of Fig. 4b.

By suitably biasing the scanning AFM tip, it is possible to reverse the domain state within the film. In Fig. 5a (out-of-plane piezoresponse image), the regions of dark contrast were poled into the opposite polarity by scanning the tip with a DC bias of $-8\,\mathrm{V}$. Figure 5b shows a corresponding in-plane piezoresponse image. Thus, we have effectively created 180° domain walls. If the tip is directly on top of a 180° domain wall and the AFM cantilever is aligned along the wall, the electric field penetrates regions of oppositely poled domains on either side of the cantilever. The differential piezoelectric activity in these domains causes the AFM cantilever to torque, as illustrated in Fig. 5c. Therefore, it is in principle possible to image a single 180° domain wall. However, the nonuniform field distribution below the tip and the geometry of the tip leads us to believe that the width of the 180° domain wall (20 nm) as seen in the line scan (Fig. 5d) is clearly an upper bound.

As mentioned earlier, the absence of a top electrode in our samples leads to a favorable state of polarization along the *c*-axis, as seen in Figs. 2b and 4a. This stable state of polarization can be switched into the opposite unstable state, as shown in Fig. 6a. This unstable state decays over time, and through nucleation and growth of the original stable state converts the switched region back into the as-grown state of polarization.

For the relaxation experiments, a $5\,\mu\mathrm{m} \times 5\,\mu\mathrm{m}$ area of the sample was switched from its original state (bright contrast) to the reversed state (dark contrast) by scanning the film with the AFM tip biased at $-5\,\mathrm{V}$. Piezoresponse images were recorded from the same region, as a function of time from a few minutes to several days. The polarization spontaneously reversed its direction, as illustrated in the sequence of piezoresponse images in Fig. 6b–d. However, it is evident in Fig. 6d that complete reversal of the polarization does not occur; rather, there are some small regions of dark contrast, which seem to be pinned.

The key aspects of the reversal experiments are as follows. The internal field present due to the asymmetry of the electrode structure drives the reversal. Nucleation of the reverse domains was found to occur preferentially at the twin boundaries, as illustrated schematically in Fig. 7a, and was followed by the growth of the reversed regions to cover an increasing area of the film. These observations were found to be very repeatable in several experiments conducted over various parts of the sample. The thermodynamic driving force for relaxation is a result of the built-in field due to charge re-

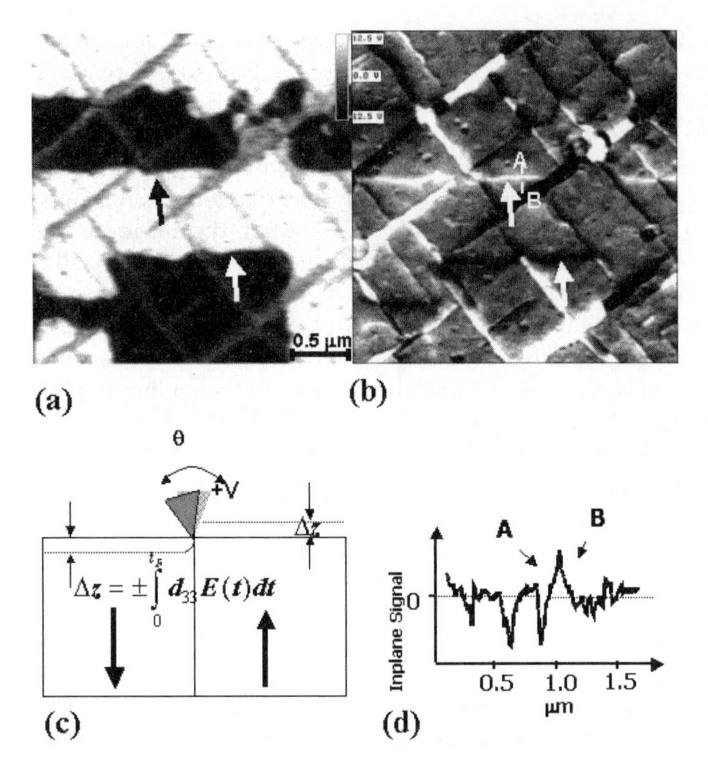

Fig. 5. (**a**) Out-of-plane (vertical differential signal) and (**b**) inplane (lateral differential signal) images of PZT (20/80). Regions with dark contrast in (**a**) have been poled into the opposite direction by scanning the surface of the film selectively with −8 V. (**c**) shows a schematic of the tip deflection on top of a 180° domain wall, with the AFM cantilever aligned along the wall. Differential vertical displacements on either side of the wall give rise to a torque on the cantilever, which can be seen by tracking the lateral differential signal from the photo detector. The presence of 180° domain walls can therefore be seen in (**b**) as *thin lines.* (**d**) shows a line scan across one such domain wall, marked AB in (**b**)

distribution within the electrode. In the as-grown state, this built-in field is compensated by the film polarization, so that the net electrostatic field in the film is close to zero. The polarization charge is considerably less than P_s due to the effect of internal screening near the interface. Since this poled state can exist for a long time and can be repeatedly obtained after reheating the sample, it is to be naturally assumed that this is close to the "stable" state of the film and that the electrostatic field in the film is absent (or close to zero) for this state. After "writing" or switching the polarization vector under an external DC field, an unstable state of polarization is created, which then relaxes toward the as-grown stable state. (*Gruverman* and *Tanaka* similarly observed suppression of relaxation in a recent publication [14]).

Fig. 6. (**a**) A 5 μm × 5 μm region is switched into opposite polarity by applying a DC volatge through the tip. (**b**)–(**d**) show the relaxation of the switched state after times of 39 040, 75 520, and 322 740 s, respectively

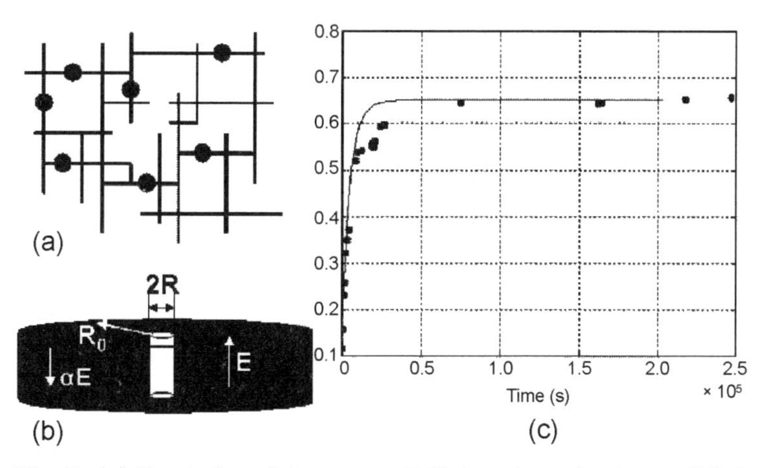

Fig. 7. (**a**) Simulation of the reverse 180° domain nucleation on 90° domain walls. (**b**) A schematic of the nuclei shape used for modeling the relaxation, based on the JMAK theory. (**c**) Relaxation of the black region as function of time; the *line* shows the relaxation of the black regions to white regions, fitted to equation [9]

Our original interpretation of these relaxation results followed an empirical fit of the experimental data, which yielded a stretched exponential-type behavior. Such a stretched exponential dependence, of course, is very well known in condensed-matter physics, and is a consequence of either a broad distribution of pinning energies and/or activation times for pinning sites with the same pinning energy.

An alternative approach to the interpretation of this data uses a classical nucleation and growth argument. In such a model, we use the Johnson–Mel–Avrami–Kolmogorov (JMAK) model to understand this time dependence of the fraction of the reversed regions. Consider the mesh of $c/a/c/a$-domains as shown schematically in Fig. 7a. The spacing between the a-domains is random. Nucleation is modeled to occur midway along each of the a-domain lines at time $t = 0$.

From the JMAK theory, we can relate the fractional volume transformed to the extended volume through

$$f = 1 - \exp\left(-V_{\mathrm{ext}}\right) , \tag{3}$$

where f is the actual transformed volume and V_{ext} is the extended volume. To calculate V_{ext}, consider the growth of a cylindrical stable (white) domain surrounded by an unstable (black) region (Fig. 7b). For a small change in the radius dR, we can write the change in the free energy $d\Delta F$:

$$\mathrm{d}\Delta F = -2\pi R h_{\mathrm{f}} P_{\mathrm{s}} E(1-f)\mathrm{d}R + 2\pi R h_{\mathrm{f}} P_{\mathrm{s}} \alpha E \mathrm{d}R + 2\pi \Gamma h_{\mathrm{f}} \mathrm{d}R , \tag{4}$$

where ΔF is the change in free energy, h_{f} is the film thickness, R is the instantaneous radius of the growing nucleus, P_{s} is the spontaneous polarization of the ferroelectric thin film ($0.6\,\mu\mathrm{C/cm^2}$), f is the fraction of reversed domains, E is the imprint driving field in the sample, αE is a reverse driving field, and Γ is the surface energy of the 180° domain wall. From the piezoresponse images at early times ($t < 10^4\,\mathrm{s}$), it can be seen that domain wall is rough. This implies that we can preclude 2D-nucleation-assisted domain wall motion. Therefore, we can write the rate equation as follows:

$$\frac{\mathrm{d}R}{\mathrm{d}t} = -\frac{L}{2\pi R}\frac{\mathrm{d}\Delta F}{\mathrm{d}R} , \tag{5}$$

where the rate constant L can be expressed through [15, 16]

$$L = \upsilon \nu_0 \frac{\exp\left(-U_{\mathrm{a}}/kT\right)}{kT} , \tag{6}$$

where U_{a} is the activation energy, h_{f} is the film thickness, υ is the volume of the critical nucleus, and ν_0 is the jump frequency ($10^{13}\,\mathrm{Hz}$).

The critical radius can be estimated from

$$\left.\frac{\mathrm{d}\Delta F}{\mathrm{d}R}\right|_{R=R_{\mathrm{crit}}} = 0 : \tag{7}$$

$$R_{\mathrm{crit}} = \left.\frac{\Gamma}{P_{\mathrm{s}} E(1-f-\alpha)}\right|_{f\to 0} = \frac{\Gamma}{P_{\mathrm{s}} E(1-\alpha)} . \tag{8}$$

Therefore,

$$\frac{\mathrm{d}R}{\mathrm{d}t} = -\frac{L}{2\pi R}\frac{\mathrm{d}\Delta F}{\mathrm{d}R} = P_{\mathrm{s}}ELh_{\mathrm{f}}(1-f-\alpha)\left(1-\frac{R_{\mathrm{crit}}}{R}\right) . \tag{9}$$

R_{crit} has been calculated to be of the order of 40 Å, which is much smaller than the resolution limit of our imaging technique. Thus we can ignore the term involving R_{crit} in (9). This leads to an estimate for V_{ext} as

$$\begin{aligned} V_{\mathrm{ext}} &= \int_0^t N(\xi)\pi\left[R(t-\xi)\right]^2 h_{\mathrm{f}}\mathrm{d}\xi \\ &= N_0\pi h_{\mathrm{f}}^3 P_{\mathrm{s}}^2 E^2 L^2 \left[\int_0^t (1-f-\alpha)\mathrm{d}t\right]^2 , \end{aligned} \tag{10}$$

where $N(\xi)$ is the nucleation density. We assume that the number of nucleation sites does not change with time and therefore it is a constant, N_0. Finally, we can write the reversed fraction f as

$$\begin{aligned} f &= 1-\exp(-V_{\mathrm{ext}}) \\ &= 1-\exp\left\{-N_0\pi h_{\mathrm{f}}^3 P_{\mathrm{s}}^2 E^2 L^2\left[\int_0^t (1-f-\alpha)\mathrm{d}t\right]^2\right\} . \end{aligned} \tag{11}$$

This equation can be plotted numerically to fit the pre-exponent parameters

Table 1. Values of N_0, α, and U_{a} obtained after fitting the data to (11) for various DC write voltages

Voltage (V)	N_0	α	L ($\mathrm{m^3/Js}$)	U_{a} (eV)
−8.0	8.89×10^{12}	0.35	1.428×10^{-17}	0.9076
−10.0	7.77×10^{12}	0.40	2.381×10^{-18}	0.9522
−12.0	1.89×10^{12}	0.82	7.142×10^{-18}	0.9249

$C = N_0\pi h_{\mathrm{f}}^3 P_{\mathrm{s}}^2 E^2 L^2$ and α. Figure 7c plots the numerical solution to (11) fitted to the experimental data. The values of L and α are $5.95\times10^{-18}\,\mathrm{m^3/Js}$ and 0.1, respectively. This gives an estimate of 0.962 eV for U_{a}. It is enticing to correlate this activation energy with the activation energy for the diffusion of specific defects, such as oxygen vacancies. However, detailed understanding of the atomic-scale origins of this activation process needs considerable further study. An important question that arises is this: Is this polarization reversal accompanied by reversal within the a-domain?

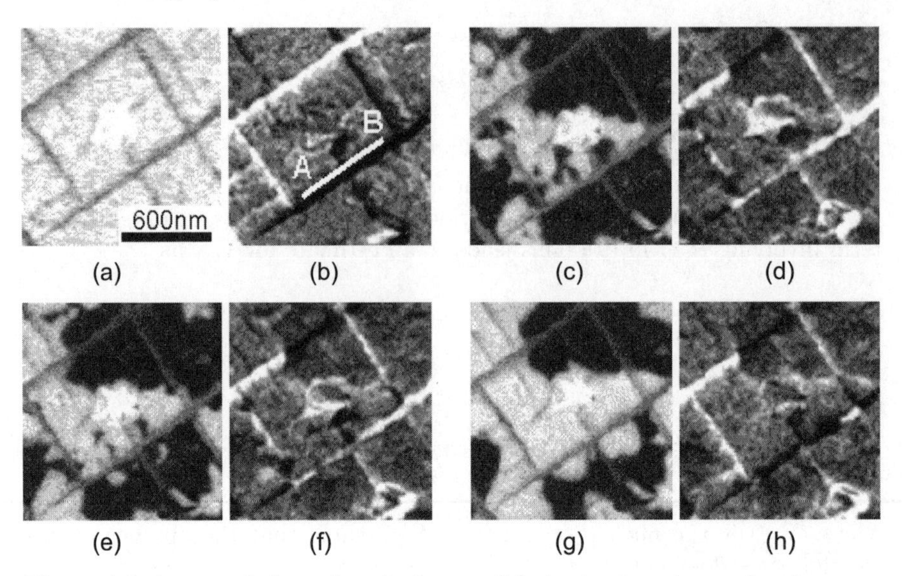

Fig. 8. (**a**) An out-of-plane domain image. (**b**) An in-plane image for the same region. (**c**) The same region, after being poled into opposite polarity, shows relaxation of the unstable black region into stable white regions. (**d**) The accompanying in-plane images. The region marked AB shows that in-plane polarization has also switched from the original black contrast to white contrast to maintain uncharged domain walls. (**e**) and (**g**) show out-of-plane images after longer waiting periods, (**f**) and (**h**) are the corresponding in-plane images. Note that in the in-plane images 180° domains are present inside the 90° domains

Figure 8a,b is in-plane and out-of-plane domain images of the PZT film before writing (as grown). The out-of-plane polarization state within the region scanned was switched into the opposite state by scanning with the tip biased at -8 V. The in-plane polarization within the a-domains was seen to switch at the same time, such that no charged interfaces could be identified. Images (c) and (d), (e) and (f), and (g) and (h) were acquired after wait times of 1.8×10^4 s, 2.6×10^4 s, and 5.9×10^4 s, respectively. The reversed c-domains in Fig. 8c are seen to reach the c/a domain interface AB, but no reversal of the in-plane polarization within the a-domain is observed. This interface is therefore likely to be charged, as it represents a head-to-head arrangement of dipoles. The polarization within the a-domain is seen to reverse only after stable c-domains have nucleated on the opposite side of AB, as seen in Fig. 8e, and have grown to meet the interface AB (in Fig. 8g), thereby causing electrostatic pressure to be applied on both sides of the a-domain. It is possible that the sluggish nature of in-plane reversal under zero external field (relaxation) could be attributed to the small dimension (width) of the a-domain.

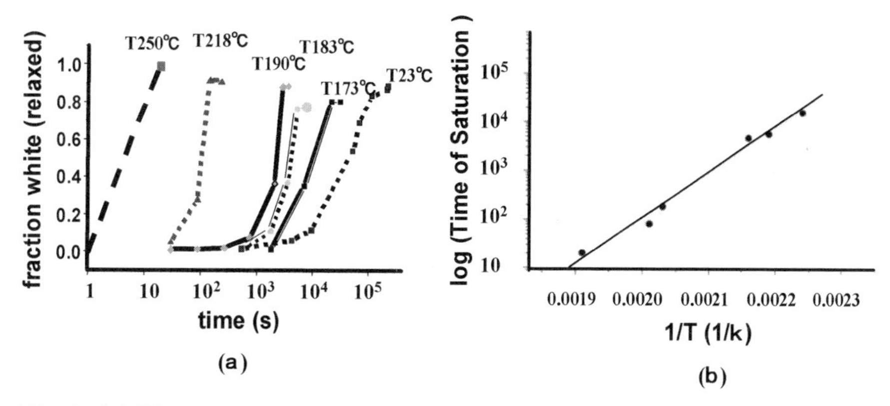

Fig. 9. (**a**) The temperature dependence of polarization relaxation. It is seen that as the temperature is increased the time for achieving full relaxation decreases. (**b**) An Arrhenius plot, showing the time for saturation (on log scales) versus $1/T$. The slope of the plot yields an activation energy of 0.8 eV

An obvious question that comes to mind in the discussion of such kinetic phenomenon is the role of temperature. To investigate the temperature dependence of the relaxation process, a 10 μm × 10 μm region of the sample was switched into the oppositely poled state and the sample was incrementally heated to the following temperatures: 175 °C, 185 °C, 190 °C, 220 °C, 250 °C, and 300 °C. The sample was held at the temperature of interest for various time periods, ranging from a few seconds to a day, depending on the temperature, and then cooled to ambient temperature. For each such time interval, a domain image was captured until saturation to the white region was achieved. Figure 9a shows the fraction of white domains plotted as heating times for each temperature. Clearly, the saturation time shows a strong dependence on the heating temperature. The higher the temperature, the shorter is the time for saturation; that is, there is a faster rate of relaxation. In other words, as one would expect, the kinetics of polarization reversal is accelerated with increasing temperature. Since the saturation time for each temperature is known, an Arrhenius plot of the saturation time versus the inverse of the temperature is shown in Fig. 9b. This analysis yields an activation energy of 0.8 eV, which is obtained from the slope of a straight-line fit to the data in Fig 9b.

Until now, we have discussed the polarization phenomenon more or less on the macroscale. It becomes pertinent to investigate this physical phenomenon on a very local scale. Figure 10a–d shows the transformation in a single "cell" that is 500 nm wide. The domain wall is seen to meander across from the nucleation site inside one of the sides toward the opposite a-domain wall until it gets pinned by defect sites, as observed in Fig. 10b. In the early stages there is enough driving force for the domain wall to overcome the pinning sites, as can be seen in the series of images shown in Fig. 10b,c.

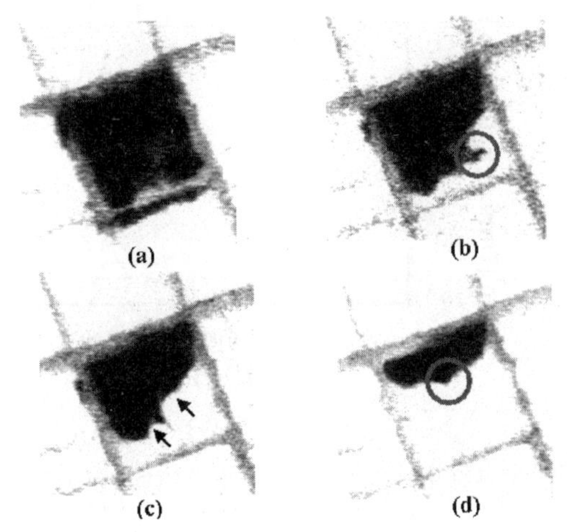

Fig. 10. (**a**)–(**c**) Relaxation in a "single cell". The data illustrate the role of pinning in governing the local velocity of the 180° domain wall. In (**b**) pinning is clearly observed inside the *circle*. The moving domain wall breaks away from the site in (**c**), moving in the direction shown by the *arrows*, and gets pinned at another site (**d**)

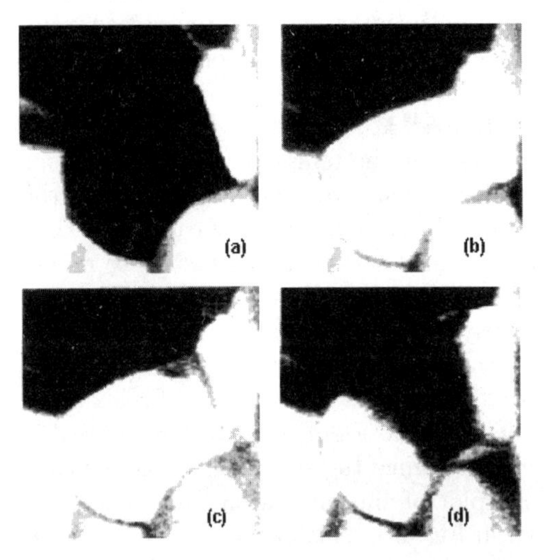

Fig. 11. Relaxation in a "single grain". The data illustrate the role of grain boundaries in the relaxation process. In (**c**) polarization reversal is clearly observed adjacent to the grain boundary

However, as the relaxation proceeds to later stages, the driving force is exhausted and the wall, unable to break free from the pinning sites, bows out. The difficulty of estimating this small driving field at the end of the relaxation prevents an accurate estimation of the wall energy. The direct observation of pinned walls at small fields and their noncrystallographic orientation allows us to suggest that the main controlling mechanism of domain wall mobility is the overcoming of pinning points in the field. The domain wall movement should then require 2D nucleation and this diminishes the mobility dramatically. In addition to these features, we have also reported faceting of the domain walls, which results in metastable shapes with clear crystallographic orientations [16].

Figure 11a–c shows the relaxation observed for a polycrystalline film. It is clear that grain boundaries play a crucial role in the relaxation behavior. Figure 11a is the virgin grain, which shows that the polarization has a dark contrast. Figure 11b is the image of the same grain after a DC bias of 6 V. It shows that the DC bias has switched the grain into opposite polarity.

Figure 11c,d is images taken after 60 s and 10^6 s, respectively. In Fig. 11c we observe that the relaxation process proceeds from the grain boundaries. In this way, it is analogous to the role of 90° domain boundaries in the nucleation of reverse polarization nuclei. Secondly, it is seen that even after 10^6 s the polarization is not completely reversed back. In fact, it is pinned by almost 50 % of its original value. This set of images clearly reveals the role of grain boundaries on polarization relaxation.

4 Nanoscale Piezoelectric and Hysteresis Behavior

This section focuses in detail on our recent observations and findings on the quantification of ferroelectric and piezoelectric phenomena in sub-micron capacitors. The SFM probe tip is used to make contact with the ferroelectric through a top electrode, typically LSCO/Pt, and hence the SFM can now be used as a "nano-probe" station. We begin this section with a brief description of processing approaches to control the crystalline quality as well as the orientation in the ferroelectric layer.

As shown in Fig. 12, the PZT layer on STO/Si is not only pure perovskite but is also found to be epitaxial. Because of the highly oriented nature, the (00l) peak intensities of epitaxial PZT are much higher than those in the polycrystalline samples. The inset of Fig. 12a is the ϕ-scan of the epitaxial PZT, indicating its four-fold symmetry. TEM observation (not shown) has revealed that the interfaces between layers are clean and flat; no obvious diffusion between adjacent layers was observed. Figure 12b compares the hysteresis loops for four types of films. The (001)-textured PZT (on LSCO/Pt–Si) has slightly higher polarization than the polycrystalline PZT (on Pt–Si); a larger polarization, however, was obtained on the epitaxial PZT. This polarization difference between polycrystalline and epitaxial PZT is a primary consequence

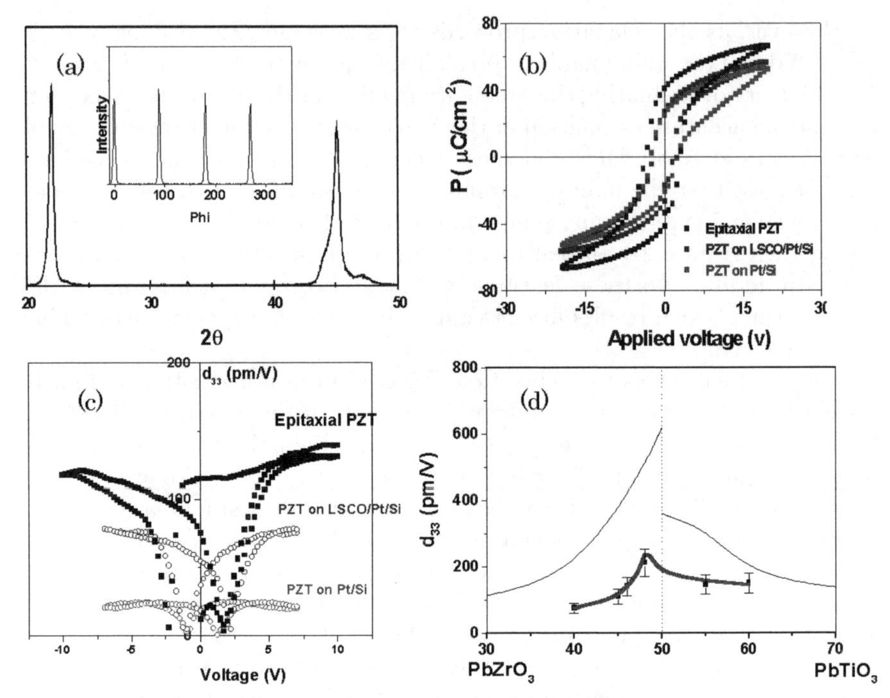

Fig. 12. Orientation effects on piezoelectric properties. The data illustrate the role of crystalline quality on d_{33}. It is seen clearly in (**c**) that d_{33} increases with a commensurate improvement in the crystalline properties. (**d**) shows the measured composition dependence of d_{33}, while the *black line* shows theoretical values predicted by *Haun* and *Cross* [17]

of the higher degree of crystallographic orientation in the epitaxial films. As expected, we found that bipolar fatigue of the Pt/PZT/Pt films was quite strong, while the films with LSCO contacts did not show any appreciable fatigue.

Piezoelectric measurements, as shown in Fig. 12c, were made using piezoresponse force microscopy. For the PZT on Pt–Si, the saturated piezoelectric coefficient is about 25 pm/V. The (001)-textured PZT shows a saturated d_{33} of about 75 pm/V. The epitaxial PZT exhibits a d_{33} of 125 pm/V, the highest among the three samples. We have further investigated the compositional dependence of d_{33}. In the range of PZT 60/40 to PZT 40/60, we found that all thin films grown on LSCO/STO/Si are epitaxial (annealing temperature 600–650 °C). The d_{33} measurement results are shown in Fig. 12d. The piezoelectric coefficient versus Zr : Ti ratio shows a trend that is similar to the theoretical prediction in the literature [13]; however, it is clear that our measured saturation values are far lower than the single domain values predicted by *Haun* et al. [17]. This is a direct result of substrate clamping on the ferroelectric. This suggests that although we have been able to improve the

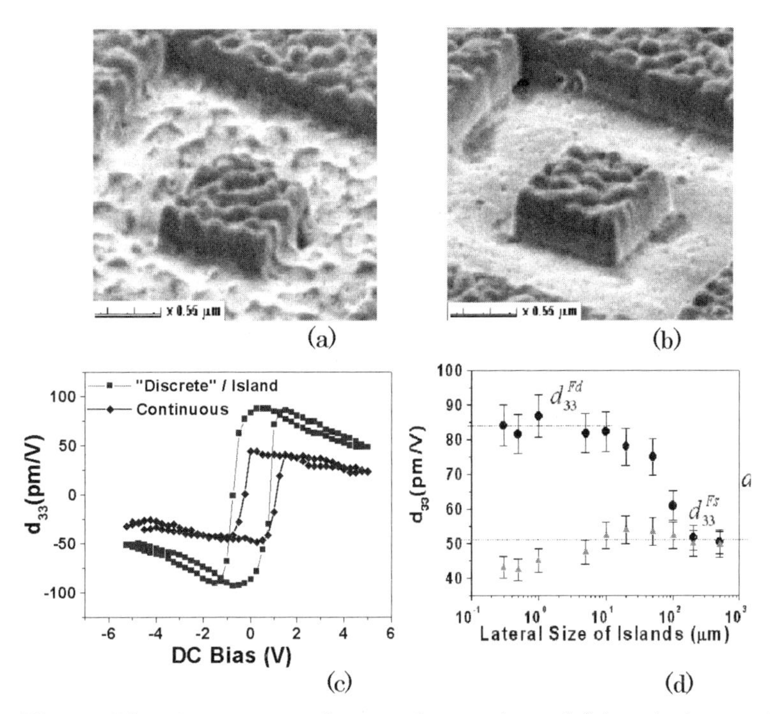

Fig. 13. The piezoresponse of nanoscale capacitors. (**a**) is a single capacitor, where only the top Pt and LSCO has been milled via FIB. (**b**) is a nanoscale capacitor, where the ferroelectric has also been milled out. (**c**) shows that there is a marked increase in the measured d_{33} after cutting the ferroelectric free. (**d**) shows the complete size dependence, for both free ferroelectric and the case in which only the top electrode has been milled

piezoresponse significantly by improving the crystallographic orientation, the strong clamping effect through the silicon substrate is still dominant. Such a clamping effect comes from the mismatch of lattice parameters and the mismatch of thermal properties between layers in the device. *Lefki* and *Dormans* predicted this loss due to clamping to be approximately 30–40 % of the theoretical single domain value [18]. Thus, to further improve the piezoelectric performance, it is necessary to modify the interface between the substrate and the ferroelectric.

One such approach to modifying the interface is to change the mechanical boundary conditions on the film by "nanostructuring" it into small dimensions (of the order of 1 μm or smaller). This has been successfully achieved via focused ion beam milling. Figure 13a,b shows two such capacitors, each 0.55 μm × 0.55 μm, that have been fabricated via focused ion beam (FIB) milling. The only difference between the two capacitors is that while for the capacitor in Fig. 13a only the top platinum has been milled, the capacitor in Fig. 13b has also its ferroelectric layer milled. As a result, on the nanoscale,

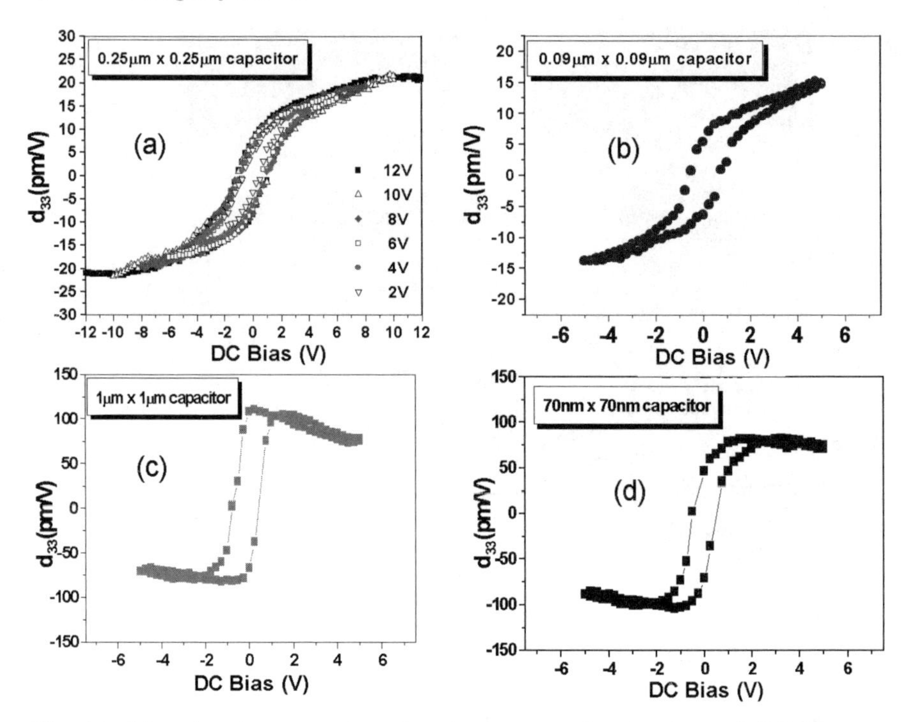

Fig. 14. Piezoelectric properties of various nanoscale capacitors: (**a**) 0.25 μm × 0.25 μm SBT cap; (**b**) 0.09 μm × 0.09 μm SBT cap; (**c**) 1 μm × 1 μm PZT cap; and (**d**) 0.07 μm × 0.07 μm PZT cap

we now have effectively free-standing ferroelectric thin film. Figure 13c plots the measured d_{33} for a constrained capacitor and a free-standing capacitor, which is 0.01 μm^2 in area. It is clear that the free-standing capacitor shows a higher value of d_{33} than the continuous film. The measured value of d_{33} is 50 % more than the clamped value and hence in good agreement with the Lefki and Dormans model. Figure 13d plots the piezoresponse from various such capacitors that have been fabricated by FIB from 100 μm × 100 μm down to 0.1 μm × 0.1 μm. A clear increase in d_{33} is seen when the capacitors are milled to 10 μm or below. However, when one mills only the top platinum, it is seen that the piezoresponse is clamped and does not change much with capacitor size; this is plotted as solid triangles in Fig. 13d.

Figure 14 is the piezoresponse from a set of nanoscale capacitors down to 70 nm × 70 nm in size. Figure 14a,b shows the piezoresponse from SBT capacitors, while Fig. 14c,d is the response from PZT capacitors. All the capacitors show clearly that they are piezoelectric, and hence ferroelectric. To date, this is the smallest capacitor stack whose piezoelectric coefficients have been reported.

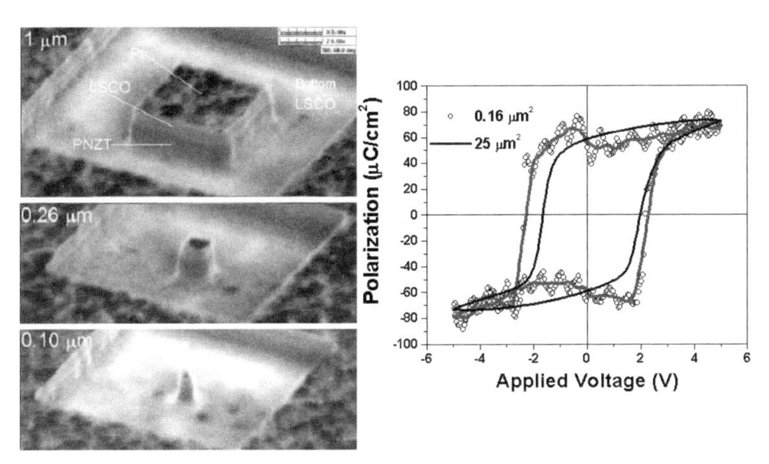

Fig. 15. (**a**) AFM images of sub-micron capacitors fabricated via FIB milling. (**b**) Static polarization hysteresis loops obtained for 25 μm^2 and 0.16 μm^2 pads via the SFM/Aixacct setup

Figure 15a shows various capacitors with different pad sizes. Figure 15b is the static hysteresis loops that are measured directly via the scanning force microscope with a conductive Pt/Ir coated tip as the probe for a 25 μm^2 pad and a 0.16 μm^2 pad. The cantilever was connected to a commercial TF Analyzer 2000 (from Aixacct Systems) to apply the excitation signal and record the current response of the device [19]. The response obtained from the 0.16 μm^2 capacitor is shown after subtracting the contribution from the parasitic capacitances and is compared with the response from the 25 μm^2 cap. It can be clearly seen that there is no loss of polarization even when scaled down to the sub-micron sized ferroelectric capacitors. Another question of significant technological interest, is as follows: What fatigue behavior do sub-micron capacitors show? We have performed fatigue on LSCO/PZT/LSCO and Pt/PZT/Pt capacitors before and it is well known that oxide electrodes prevent fatigue. However, in nanoscale regimes, such an investigation again becomes imperative. Figure 16 plots fatigue for two nanoscale capacitors, one with LSCO electrodes and the other with Pt electrodes. It is clear that for the capacitor with LSCO electrodes no loss of polarization is observed, while the capacitor with Pt electrodes shows a marked decrease in switched polarization.

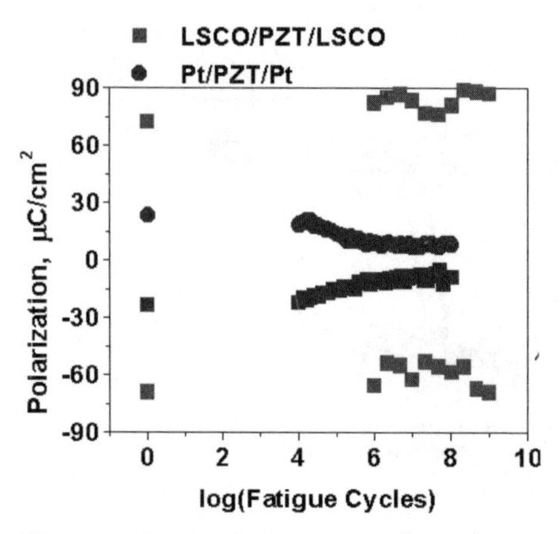

Fig. 16. Nanoscale fatigue: Pt/PZT/Pt capacitors show polarization fatigue while LSCO/PZT/LSCO capacitors are fatigue free

5 Summary

Scanning piezoforce microscopy is evolving to be a powerful tool that has enabled investigations of a variety of phenomena in ferroelectric thin films at the nanoscale. Our work has focused on size scaling issues, the dynamics of both 180° and 90° domain walls in constrained as well as constraint-relieved nanostructures, and a fundamental understanding of polarization relaxation behavior. This surface probe, in conjunction with nano-fabrication techniques such as focused ion beam milling or e-beam lithography, provides an ideal "playground" for nanoscale science of ferroelectric materials.

Over the past few years, our studies on polarization relaxation behavior indicate that relaxation occurs through the nucleation of reverse domains at heterogeneities. The JMAK model for phase change kinetics has been applied to describe the observed transformation-time data for 180° domain reversal under a constantly decaying internal driving field. The effect of local curvature on driving the domain relaxation kinetics has been studied and it has been observed that the curvature of growing domains decreases. Pinning and bowing of 180° domain walls is observed in the late stages of relaxation.

Acknowledgements

We have been fortunate to be able to collaborate with several colleagues both within Maryland as well as around the world. We wish to gratefully acknowledge their on-going contributions to the central theme of our research effort; specifically, the work of Dr. A. Stanishevsky on focused ion beam milling, that of Professor A. Roytburd on theoretical modeling, that of Professor Ellen Williams on fundamental studies of dynamics, and that of Dr. Baoting Liu on thin film processing. This work is supported by the NSF under grants

DMR 00-80008 and DMR 99-03279, and by DOE grants from PNNL and the Argonne National Laboratory.

References

1. J.F. Scott: *Ferroelectric Memories*, Springer Ser. Adv. Microelectron. **3** (Springer, Berlin, Heidelberg 2000)
D. Damjanovic: Rep. Prog. Phys. **61**, 1267 (1998)
2. T. Hidaka, T. Mayurama, M. Saitoh, N. Mikoshiba, M. Shimizu, T. Shiosaki, L.A. Wills, R. Hiskes, S.A. Dicarolis, J. Amano: Appl. Phys. Lett. **68**, 2358 (1996)
C.H. Ahn, T. Tybell, L. Antognazza, K. Char, R.H. Hammond, M.R. Beasley, Ø. Fisher, J.-M. Triscone: Science **276**, 1100 (1997)
3. A. Gruverman, H. Tokumoto, A.S. Prakash, S. Aggarwal, B. Yang, M. Wuttig, R. Ramesh, O. Auciello, T. Venkatesan: Appl. Phys. Lett **71**, 3492 (1997)
4. J.D. Axe, Y. Yamada: Phys. Rev. B **34**, 1 599 (1986)
K. Sekimoto: Physica A **125**, 261 (1984), Phys. Rev. B **128**, 1 32 (1984), Phys. Rev. B **135**, 328 (1986)
S. Ohta, T. Ohta, K. Kawasaki: Physica A **140**, 478 (1987)
R.M. Bradly, P.N. Strenski: Phys. Rev. B **40**, 8967 (1989)
K. Sekimoto: Int. J. Mod. Phys. B **5**, 1843 (1991)
Y.A. Andrienko, N.V. Brilliantov, P.V. Krapivsky: Phys. Rev. A **45**, 2263 (1992)
5. G.E. Pike, W.L. Warren, D. Dimos, B.A. Tuttle, R. Ramesh, J. Lee, V.G. Keramidas, J.T. Evans, Jr.: Appl. Phys. Lett. **66**, 484 (1995)
6. V. Nagarajan, I.G. Jenkins, S.P. Alpay, H. Li, S. Aggarwal, L. Salamanca-Riba, A.L. Roytburd, R. Ramesh: J. Appl. Phys. **86**, 595 (1999)
7. P. Güthner, K. Dransfeld: Appl. Phys. Lett. **61**, 1137 (1992)
A. Gruverman, O. Auciello, H. Tokumoto: Appl. Phys. Lett. **69**, 3191 (1996)
T. Tybell, C.H. Ahn, J.-M. Triscone: Appl. Phys. Lett. **72**, 1454 (1998)
O. Auciello, A. Gruverman, H. Tokumoto, S.A. Prakash, S. Aggarwal, R. Ramesh: Mater. Res. Bull. **23**, 33 (1998)
8. A.L. Roitburd: Phys. Status Solidi A **37**, 329 (1976)
9. B.S. Kwak, A. Erbil, J.D. Budai, M.F. Chrisholm, L.A. Boatner, B.J. Wilkens: Phys. Rev. B **49**, 14865 (1994)
10. C.M. Foster, Z. Li, M. Buckett, D. Miller, P.M. Baldo, L.E. Rehn, G.R. Bai, D. Guo, H. You, K.L. Merkle: J. Appl. Phys. **78**, 2607 (1995)
11. C.S. Ganpule, V. Nagarajan, H. Li, A.S. Ogale, D.E. Steinhauer, S. Aggarwal, E. Williams, R. Ramesh: Appl. Phys. Lett. **77**, 292 (2000)
12. L.M. Eng, M. Abplanalp, P. Günter: Appl. Phys. A **66**, S679 (1998)
13. A. Roelofs, U. Böttger, R. Waser, F. Schlaphof, S. Trogisch, L.M. Eng: Appl. Phys. Lett. **77**, 3444 (2000)
14. A. Gruverman, M. Tanaka: J. Appl. Phys. **89**, 1836 (2001)
15. P. G. Shewmon: *Transformations in Metals* (McGraw-Hill, New York 1969)
B.Y. Lyubov: *Kinetic Theory of Phase Transformation* (National Bureau of Standards, New Delhi 1978)
16. C.S. Ganpule, A.L. Roytburd, V. Nagarajan, B.K. Hill, S.B. Ogale, E.D. Williams, R. Ramesh, J.F. Scott: Phys. Rev. B **65**, 14101 (2002)

17. M. J. Huan, L. E. Cross: Ferroelectr. **99**, 45 (1989)
18. K. Lefki, G. J. M. Dormans: J. Appl. Phys. **76**, 1764 (1994)
19. S. Tiedke, T. Schmitz, K. Prume, A. Roelofs, T. Schneller, U. Kall, R. Waser, C. S. Ganpule, V. Nagarajan, A. Stanishevsky, R. Ramesh: Appl. Phys. Lett. **79**, 3678 (2001)

Part II

Deposition and Characterization Methods

The Sputtering Technique

Koukou Suu

ULVAC Inc., 1220-1 Suyama, Susono, Shizuoka 410-1231, Japan
koukou_suu@ulvac.com

Abstract. In this chapter, the formation of ferroelectric films using the sputtering technique is described. A main topic is a mass production system for Pb(Zr,Ti)O_3 (PZT) films, which includes discussions on the concept of the system, optimization of the sputtering conditions, the ferroelectric properties of sputtered PZT films, and so on. It is concluded from the switching charge data of PZT films sputtered on 1000 substrates that the uniformity and stability of the system are excellent. Sputtering data for related materials, such as $SrBi_2Ta_2O_9$ films and hydrogen barrier layers, are also presented in the last part of this chapter.

1 Introduction

Ferroelectric random access memories (FeRAMs) have interesting features such as a high-speed rewriting capacity and nonvolatility characteristics, and thus they are referred to as the ultimate memory. In general, ferroelectric thin films for FeRAMs are required to have the following characteristics: 1. a large amount of polarization reversal charge ($P_r > 10\,\mu C/cm^2$); 2. a small coercive electric field (low-voltage operation); 3. a strong resistance to polarization fatigue; 4. a high polarization reversal speed; 5. a small leakage current (not exceeding $10^{-6}\,A/cm^2$); 6. excellent data retention characteristics (10 years and over); and 7. a small dielectric constant (large S/N ratio).

Pb(Zr,Ti)O_3 (PZT) and $SrBi_2Ta_2O_9$ (SBT) films are considered to be the most promising candidates as ferroelectric materials that satisfy the above characteristics for FeRAM. We have been developing mass production technology for PZT [1, 2, 3, 4, 5], since we consider PZT as the most promising candidate for the ferroelectric material used in FeRAMs, due to its longer period of research, the existence of actual production, its manufacturing capability within the tolerable temperature range of general Si LSI processes, and the recent reports on the dramatic improvement of the fatigue characteristics and imprint/retention characteristics [6, 7]. In addition, with an eye toward the potentiality for larger integration in the future, we are simultaneously developing mass production technology for SBT ferroelectric thin films [8].

The characteristics of major PZT thin-film formation methods are shown in Table 1. We selected the sputtering method for the mass production of ferroelectric thin films from the following factors: 1. good compatibility with conventional Si LSI processes; 2. superb controllability of film quality (e.g.,

H. Ishiwara, M. Okuyama, Y. Arimoto (Eds.): Ferroelectric Random Access Memories,
Topics Appl. Phys. **93**, 71–83 (2004)
© Springer-Verlag Berlin Heidelberg 2004

film composition), which enables relatively easy thin-film deposition; 3. better possibilities of obtaining uniform surfaces on large-diameter substrates (e.g., ∅ 6–8"); 4. plasma-assisted lowering of deposition and crystallization temperatures; 5. the feasibility of high-speed deposition; 6. the same deposition method as electrodes (Pt, Ir, Ru, etc.), which will facilitate in-situ integration; and 7. the present difficulties and lack of future potentialities in other methods (e.g., impurity residues, lack of good sources).

Table 1. A comparison of formation methods for PZT thin films

Method	Advantages	Disadvantages
Sputtering	• Uniformity over a large area • Repeatability • High-speed deposition • No residual impurity	• Composition fluctuation (Pb content) • Poor step coverage
MOCVD	• Low-temperature process • Conformal step coverage	• Poor uniformity • Poor repeatability • Slower deposition • Less adaptive source • Residual organic impurities
Sol-gel	• Uniformity over a large area • Repeatability • Easy composition control	• Poor step coverage • Higher processing temperature • Residual organic impurities

2 A Sputtering System for Mass Production

With an emphasis on both the mass production capability and the advanced process capacity of ferroelectric PZT thin-film sputtering, we consider the following as important for development.

1. Throughput – realization of high-speed deposition: Compared to the other processes (e.g., electrode deposition), the deposition speed of ferroelectric sputtering is considerably slower and thereby limits the throughput. Between the two methods for improving the throughput – high-speed deposition and thinner film deposition – high-speed deposition is more promising for the improvement of throughput, since thinner films are apt to cause deterioration of the film characteristics.

2. Control of film composition: The film composition determines other film qualities (crystal structure, electric characteristics). Since volatile contents are included in the materials, such as PZT or SBT, they are sensitive and they fluctuate easily according to the temperature or plasma status. Film composition control is the fundamental factor in this process.
3. Uniformity on large-diameter substrates: Large-diameter substrates of ∅ 6–8" are expected to be used for mass production. Film thickness and uniformity of the film quality are the keys to mass production.
4. Process stability/reliability: Under circumstances in which there are many unknown factors, such as new materials, ceramic targets, insulator sputtering, and so on, process stability/reliability is the more pertinent factor. In fact, there has been a problem in film composition that has changed over time.
5. Measures against particles: While characteristics such as ceramic targets or insulating thin film increase the mechanical factors (e.g., adherence, thermal expansion) for particle occurrence, they also produce electrical factors (e.g., dielectric breakdown due to charge-up) and make the measures difficult to obtain.
6. Others (to be discussed in Sect. 6 "Future Development Tasks")

Based on these concepts, we designed and set up a multi-chamber type mass production sputtering system, equipped with an exclusive sputtering module for ferroelectric materials, CERAUS ZX-1000 from ULVAC (Fig. 1). This system has the following features, in addition to features such as easy maintenance, short exhausting time, short down time, etc. 1. It can mount ∅ 300 mm (∅ 12") targets and process large-diameter substrates of ∅ 200 mm. At present, deposition of PZT and SBT thin films on ∅ 6" and ∅ 8" substrates and of $(Ba,Sr)TiO_3$ (BST) thin films on ∅ 8" substrates is performed using ∅ 12" single ceramic targets. 2. Including the heat chamber, this system has five process chambers, thereby achieving high flexibility. The system is presently executing the following in-situ processes as standard: pre-heating of substrate → substrate sputtering (e.g., Ti, TiN, Pt) → ferroelectric material sputtering. 3. As a substrate heating mechanism, this system is capable of precise and rapid heating over a wide range from low to high temperatures, with the aid of an electrostatic chuck type hot plate, in addition to lamp heating. 4. This system uses RF sputtering for ferroelectric deposition and counters RF noises.

We adapted the RF magnetron sputtering method for sputtering of PZT targets. PZT sputtering methods include high-temperature sputtering [9,10], where film deposition is carried out at substrate temperatures of 500 °C or more, and low-temperature deposition, where the films are deposited at around room temperature and then crystallized by heat treatment. We chose the latter, where the crystallization was performed after deposition through heat treatment, because of the high volatility of the Pb component, which made composition control difficult in high-temperature deposition. In what

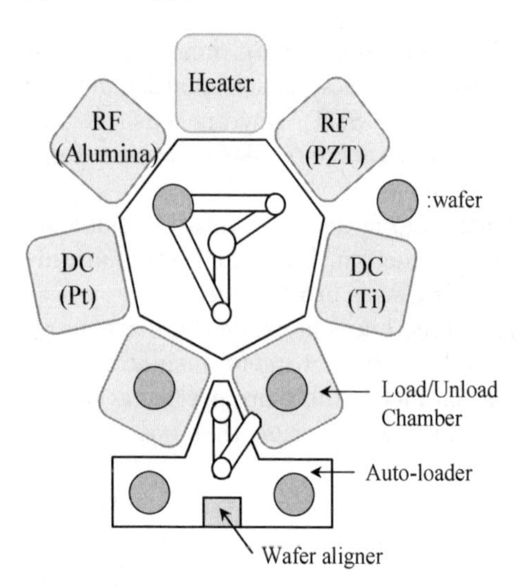

Fig. 1. The multi-chamber type mass production sputtering system

follows, the improvement in high-speed deposition, film composition control, and stability of sputtering processes are discussed.

The representative PZT sputtering conditions are shown in Table 2. Since a ceramic target with an inferior thermal conductivity was used, the application of high power led to the destruction of the target. Therefore, we used a backing plate with an elevated cooling efficiency and a higher-density target, and achieved further rate improvement by an increase in the sputtering rate and high power application. Presently, the necessary film thickness on a device is considered not to exceed 200 nm, and with this sputtering method, we made it possible to create a PZT film in approximately 3 min, 7 min shorter than the conventional deposition method. This will lead to a considerable improvement in throughput, which is presently limited by the PZT film deposition process.

Table 2. Typical PZT sputtering conditions

Sputtering method	RF magnetron sputtering
Target	Sintered PZT ceramic target (12")
Substrate	Pt(100 nm)/TiO_x(20 nm)/SiO_2/Si
Base pressure	1.0×10^{-5} Pa
Gas (Ar) flow rate	10–50 sccm
RF power	1.0–2.0 kW
Substrate temperature	Room temperature
Film thickness	< 200 nm

3 The Optimization of Sputtering Conditions

The control of the film composition is translated to control of the Pb content within a film. Various factors are believed to exist that influence the Pb content within the film (Fig. 2) and we investigated the influence of the sputtering conditions (e.g., RF power, Ar gas pressure), the strength of the magnetic field, and the electrical potential of the substrate. As shown in Fig. 3, the Pb content in the film is sensitively influenced by the sputtering power, the Ar gas pressure, and so on. Therefore, one can dynamically and precisely control the Pb content by utilizing a combination of these factors. Furthermore, as shown in Fig. 4, control of the substrate's electrical potential is effective in controlling the Pb content. Therefore, in cases in which the sputtering conditions cannot be changed, largely for fear of damaging the film uniformity, film composite control by this method is very effective for the process.

In addition to the known characteristic of PZT sputtering that the Pb content within the film fluctuates easily, it has been confirmed that the Pb content within the film is unstable even when the deposition is performed under identical conditions. The conceivable causes for that phenomenon are the instability of the target, fluctuation of the temperature in the sputtering chamber, and variation of the plasma status over time. In the case in which a new target is put to use, sufficient burn-in treatment is necessary. Although the input power for stabilizing the Pb content depends on the type or density

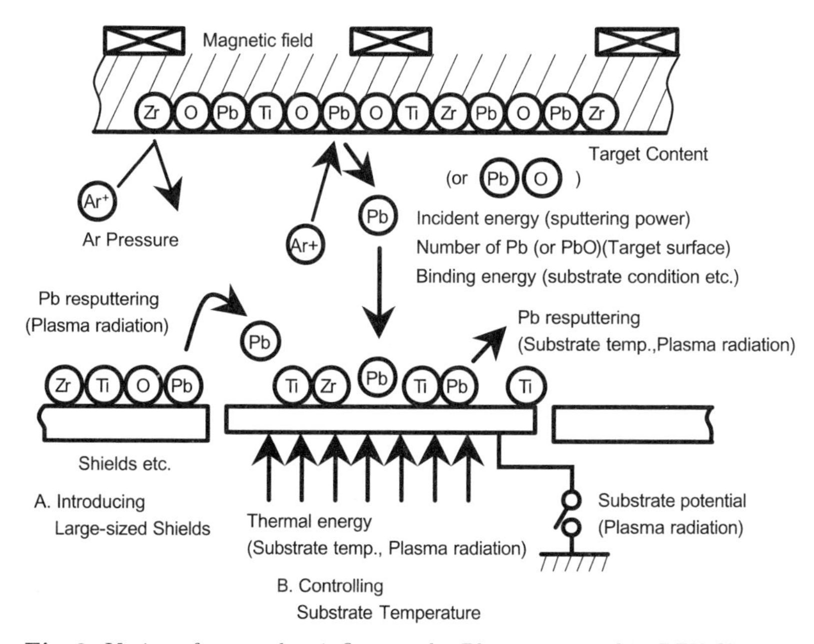

Fig. 2. Various factors that influence the Pb content within PZT film

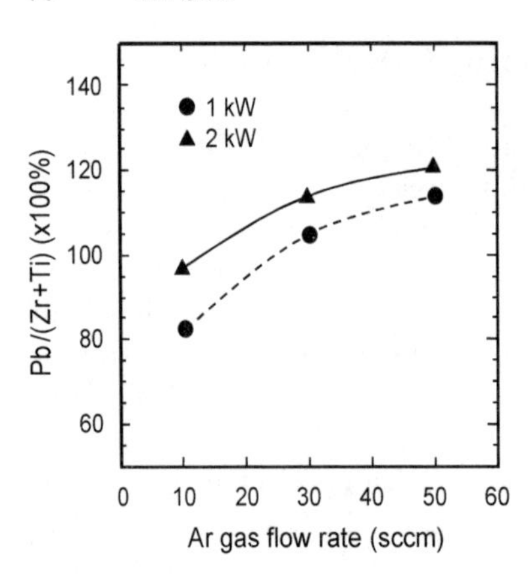

Fig. 3. The relation between the Pb content and the Ar gas flow rate

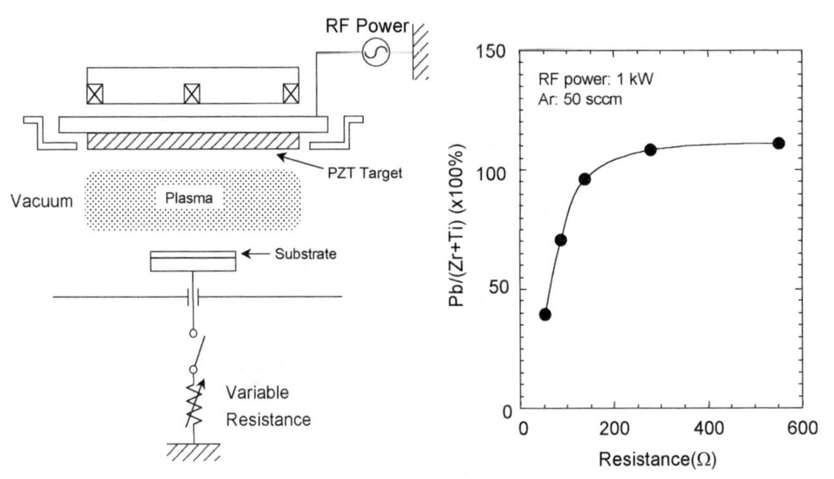

Fig. 4. Control of the substrate's electrical potential in controlling the Pb content

of the targets, our accumulated experience demonstrates that it is safer to take 10 kWh sputtering as the burn-in treatment for targets.

Another factor for film composition fluctuation is fluctuation of the temperature. Temperature increases in the sputtering system mean an increase in the substrate temperature or the substrate stage temperature, which then has an influence on the temperatures in the shields or other chamber components around the substrate that occur with the progress of sputtering. In general, the Pb content is reduced by the temperature increase and the behavior of the Pb content becomes unstable when the substrate temperature exceeds 150 °C in PZT sputtering. Therefore, it is necessary to control the substrate temperature below 150 °C so that fluctuation of the film composition does not occur. To counter this fluctuation of the Pb content due to temperature increases, we used a method for pre-sputtering that absorbed the increase in temperature and started the full-fledged sputtering. In the

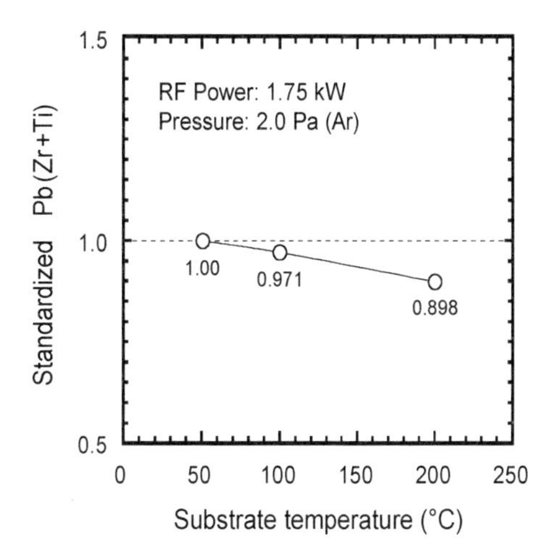

Fig. 5. The dependence of the Pb content on the substrate temperature

case of sputtering at 1.5 kW, approximately one hour of pre-sputtering was found to be sufficient to avoid instability of the Pb content within the film in the early stages of sputtering.

On the other hand, sudden increases in the substrate temperature, which have the largest influence on the Pb content within the film from room temperature to 200 °C or over, lead, in a few minutes of deposition, to fluctuation of the film composition and deterioration of uniformity. As shown in Fig. 5, temperature increases up to 100 °C do not cause large fluctuations in composition, but when the temperature rises to 200 °C, the Pb content is reduced by approximately 10%. Because of this, we considered that it was necessary to limit the substrate temperature in deposition to less than 100 °C, or preferably around 50 °C, and we introduced the electrostatic chuck (ESC) type low-temperature plate. Using this plate, the rise in substrate temperature was successfully limited to less than 50 °C, when sputtering at 1.5 kW was continued for 10 min (a PZT film thickness of approx. 450 nm).

Even after solving the temperature problem, however, a change in the Pb content was observed with the passage of sputtering time. For example, continuous sputtering of 2.0 kW for 10 h showed an approximately 30% decline in the Pb content compared to that at the beginning. We found that the reason for this was a change in the plasma status. As shown in Fig. 6, when insulating PZT films adhere to the shields of the ground potential, charge-up occurs, the impedance of the system changes, plasma is pushed to the center of the chamber, exposure to plasma is enhanced, and, as a consequence, the Pb content within the film is reduced. In order to stabilize the status of the plasma, we installed a stable anode; that is, an anode that avoided charge-up due to the adhesion of PZT film and maintained its role as an anode. Consequently, as can be seen in Fig. 7, stable transition of the Pb content within the film in continuous sputtering has been confirmed. Figure 8 shows an example of the thickness and the Pb content uniformity in a PZT film

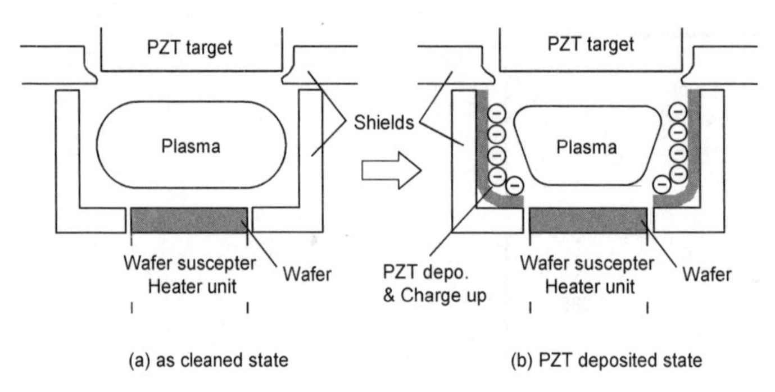

Fig. 6. Change in the plasma status, which is responsible for Pb content variation

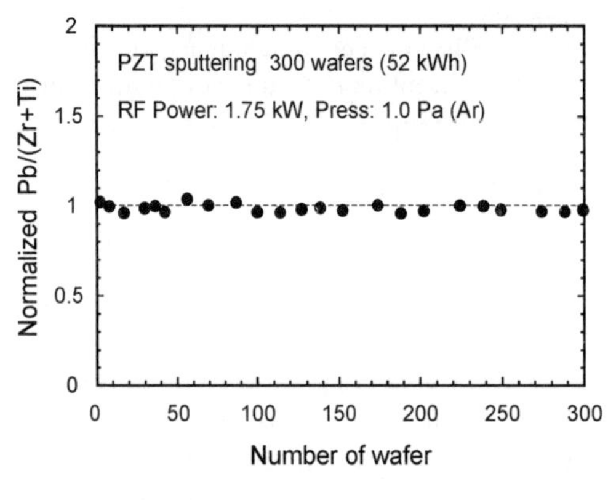

Fig. 7. The stable transition of the Pb content within film in continuous sputtering

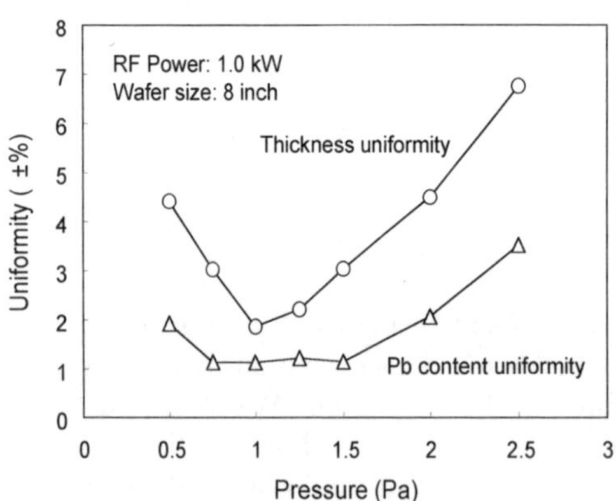

Fig. 8. Uniformity in the ⌀8" area

on an 8 " substrate. Both the thickness and the Pb content uniformity varied according to the deposition conditions and they were minimized around a sputtering pressure of 1.0 Pa. Uniformity in the film thickness and the Pb content represents a good result, as low as ±1.9% and ±1.1%, respectively. These uniformity values were confirmed as being good and stable enough to satisfy the requirements of mass production.

4 The Ferroelectric Characteristics of Sputtered PZT Capacitors

In this section, the ferroelectric characteristics of PZT and La-doped PZT (PLZT) capacitors are presented. Figure 9 shows the transition of the switching charge ($Q_{\rm sw}$) and saturation characteristics of a Pt/PLZT(200 nm)/Pt capacitor. $Q_{\rm sw}$ with 5 V applied was approximately 34 μC/cm^2 and the 90% saturation voltage ($V_{90\%}$) was 3.1–3.2 V. The composition of the PLZT film, which also contained Ca and Sr for improvement of the retention and imprint characteristics, was excellent. Figure 10 shows the stability of $Q_{\rm sw}$ in continuous sputtering of 1000 substrates. The result shows high stability, sufficient for mass production. Also, the reference datum is that before chamber cleaning. The fact that it is equivalent to the data after chamber cleaning demonstrates the reliability of the system process. In the fatigue measurements, decreases in $Q_{\rm sw}$ of 10% and 26% were observed by 3 V and 5 V pulses, respectively, after 10^9 switching cycles. The retention and imprint characteristics were also far better than usual [11, 12].

Further improvement of the ferroelectric performance is needed, because scaled thinner capacitors (sub-100 nm) will be demanded in the near future. A Pb-deficient surface layer was confirmed to be responsible for the degradation of the ferroelectric performance. Thus, the bottom electrode and the PZT deposition process were modified to further improve the ferroelectric performance and achieve a thinner capacitor with good performance [13]. Figure 11 shows that the $Q_{\rm sw}$ transition curve of an IrO_x/PZT(80 nm)/Pt capacitor measured at 3 V. $Q_{\rm sw}$ with 1.8 V applied was approximately 24 μC/cm^2 and $V_{90\%}$ was 1.4 V. In this sample, $Q_{\rm sw}$ did not decrease even after 10^9 switching cycles. From these results, the sputter-derived PZT capacitor was proved to be suitable for the 0.18 μm technology node in the near future.

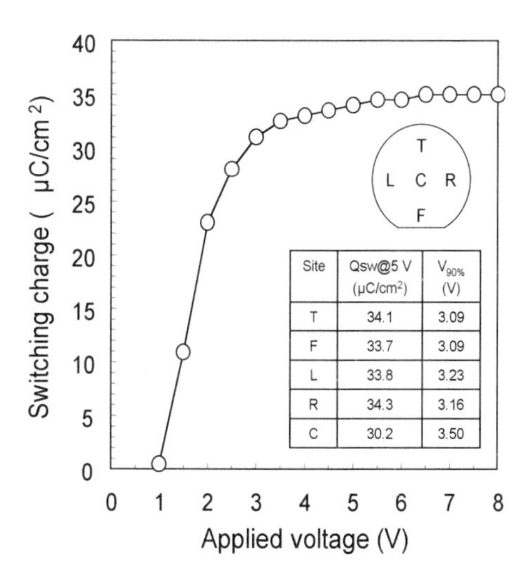

Site	Qsw@5 V (μC/cm²)	$V_{90\%}$ (V)
T	34.1	3.09
F	33.7	3.09
L	33.8	3.23
R	34.3	3.16
C	30.2	3.50

Fig. 9. The transition of the switching charge of a Pt/PLZT(200 nm)/Pt capacitor

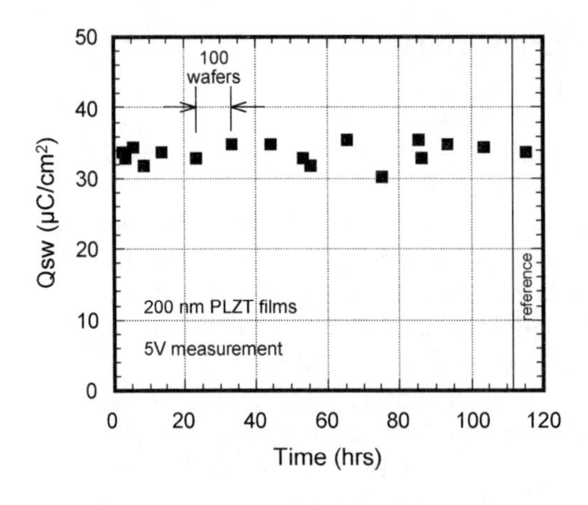

Fig. 10. The stability of Q_{sw} in the continuous sputtering of 1000 substrates

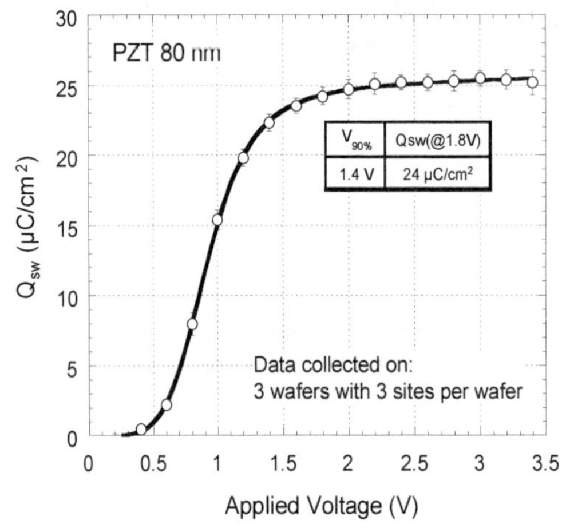

Fig. 11. The transition of the switching charge of an IrO_x/PZT(80 nm)/Pt capacitor

5 Sputtering of SBT Films and Hydrogen Barrier Layers

We have also been trying sputtering of SBT films. In addition to PZT sputtering, we have had difficulty in SBT film sputtering due to the high volatility of Bi, which is an obstacle to composition control. We have also had difficulty with narrow processing margins, such as the requirement for high-temperature processing, and so on [14]. However, recent experiments have shown that superb ferroelectric characteristics can be obtained through the optimization of the film composition and the conditions for crystallization. Therefore, sputtering is expected to become a powerful means of mass-producing carbon-free SBT films. Figure 12 shows the Q_{sw} and fatigue characteristics of a 200 nm-thick Nb-doped SBT film (SBTN). In 5 V application, Q_{sw} is as high as 12 μC/cm^2 and it does not degrade until at least after 10^9 cycles [15, 16]. In addition, the temperature required for the crystallization

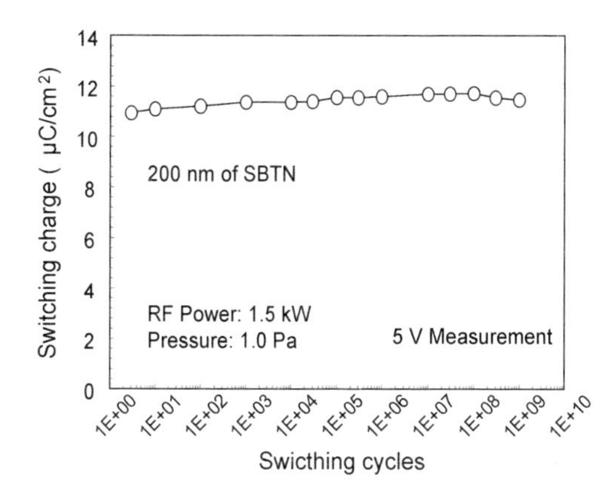

Fig. 12. The fatigue characteristics of Pt/SBTN(200 nm)/Pt capacitor

process of SBTN films was 730 °C, which was lower than usual. This is a great step toward mass production.

In the fabrication of FeRAMs, there is a serious issue in that ferroelectric layers are damaged by hydrogen during the process associated with the integration of ferroelectric capacitors. In order to solve this problem, it is proposed that the ferroelectric capacitors should be covered with a hydrogen barrier layer. Alumina (Al_2O_3) thin film is one of the most promising candidates as a hydrogen barrier layer, because of its high chemical stability [17, 18].

Alumina thin films as a hydrogen barrier layer were prepared by a RF magnetron sputtering method using a sintered alumina ceramic target, and the fundamental properties of these films, such as the refractive index, were primarily investigated. Ferroelectric capacitors coated with alumina barrier layers were annealed in forming gas and further characterized in terms of ferroelectric properties. Alumina films 100 nm in thickness were deposited by RF reactive sputtering. The oxygen partial pressure was varied from 10% to 80% under 2.0 kW of RF power and 1.0 Pa of sputtering pressure at room temperature (R.T.). In this experiment, sufficiently oxidized alumina, which is supposed to be one of the essential conditions for a hydrogen barrier layer, was obtained under the condition of a higher oxygen partial pressure.

Next, the hydrogen barrier effect was investigated by annealing PZT capacitors in forming gas. The effects of the oxygen partial pressure on alumina deposition were investigated first, as shown in Fig. 13. The RF power and sputtering pressure were maintained at 2.0 kW and 1.0 Pa, respectively. Forming gas (3% $H_2 + N_2$) annealing was performed at a temperature of 450 °C, at atmospheric pressure. As can be seen in this figure, ferroelectric capacitors without an alumina layer were obviously degraded, while those with an alumina layer were not so affected. Within the samples covered with alumina, the barrier performance of the sample deposited with oxygen as a sputtering gas was relatively higher. The dependency on the alumina film thickness

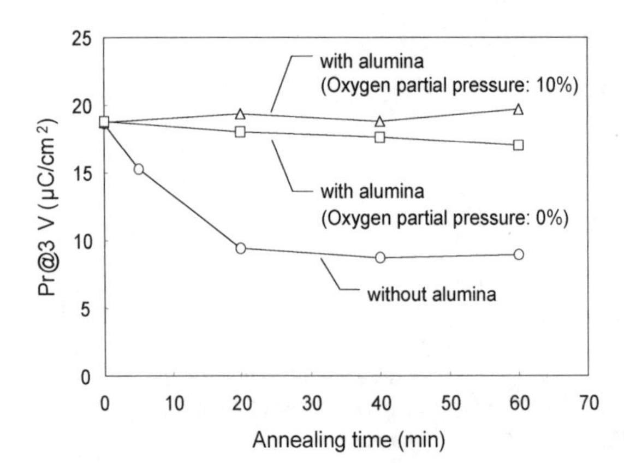

Fig. 13. The forming gas annealing effect for a PZT capacitor

was also investigated. As a result, the barrier performance of not less than 50-nm-thick alumina could be confirmed [19].

6 Future Development Tasks

As we have listed in Sect. 1, throughput, control of film composition, uniformity on large-diameter substrates, process stability/reliability, and measures against particles are important to be solved. The first priority is to grapple with these tasks and keep improving their degrees of completion, in order to succeed in mass production. Also, the deposition and characteristics of ferroelectric films are largely dependent on the conditions of the substrate electrodes, or on heat treatment such as post-annealing. In particular, in order to achieve the deposition of thinner films and high-performance devices, it is important to understand the correlation of the processes. In other words, optimization of the process integration is desired.

At present, we are performing experiments on the sputtering of Pt and oxide electrode materials, in order to optimize the conditions for laminating with ferroelectric films. In addition, the correlation with crystallization or thermal processing, such as the annealing process, is being examined. In the future, we will commit ourselves to expanding this optimization of process integration to other series of processes, and we will create an IPS (Integrated Production Support) for ferroelectric-related memory production technology.

References

1. K. Suu, A. Osawa, N. Tani, M. Ishikawa, K. Nakamura, T. Ozawa, K. Sameshima, A. Kamisawa, H. Takasu: Jpn. J. Appl. Phys. **35**, 4967 (1996)
2. K. Suu, A. Osawa, N. Tani, M. Ishikawa, K. Nakamura, T. Ozawa, K. Sameshima, A. Kamisawa, H. Takasu: Integr. Ferroelectr. **14**, 59 (1997)
3. K. Suu, A. Osawa, Y. Nishioka, N. Tani: Jpn. J. Appl. Phys. **36**, 5789 (1997)

4. K. Suu, Y. Nishioka, A. Osawa, N. Tani: Oyo Buturi **65**, 1248 (1996) (in Japanese)
5. K. Suu: Proc. Semicon. Korea Tech. Symp. (1998) p. 255
6. T. Nakamura, Y. Nakao, A. Kamisawa, H. Takasu: Appl. Phys. Lett. **65**, 1522 (1994)
7. T. D. Hadnagy, T. Davenport: Integr. Ferroelectr. **22**, 183 (1998)
8. K. Suu, T. Masuda, Y. Nishioka, N. Tani: Proc. Int. Symp. Appl. Ferroelectr. (ISAF XI) (1998) p. 332
9. N. Inoue, Y. Maejima, Y. Hayashi: Tech. Dig. Int. Electron Devices Meeting (1997) p. 605
10. N. Inoue, T. Takeuchi, Y. Hayashi: Tech. Dig. Int. Electron Devices Meeting (1998) p. 819
11. F. Chu, G. Hickert, T. D. Hadnagy, K. Suu: Integr. Ferroelectr. **26**, 47 (1999)
12. K. Suu, N. Tani, F. Chu, G. Hickert, T. D. Hadnagy, T. Davenport: Integr. Ferroelectr. **26**, 9 (1999)
13. F. Chu, G. Fox, T. Davenport, Y. Miyaguchi, K. Suu: Integr. Ferroelectr. **48**, 161 (2002)
14. K. Suu, T. Masuda, Y. Nishioka, N. Tani: Integr. Ferroelectr. **21**, 407 (1998)
15. S. Sun, G. R. Fox, T. D. Hadnagy: Integr. Ferroelectr. **26**, 31 (1999)
16. T. Masuda, Y. Miyaguchi, K. Suu, S. Sun: Integr. Ferroelectr. **31**, 23 (2000)
17. I. S. Park, Y. K. Kim, S. M. Lee, J. H. Chung, S. B. Kang, C. S. Park, C. Y. Yoo, S. I. Lee, M. Y. Lee: Tech. Dig. Int. Electron Devices Meeting (1997) p. 617
18. D. J. Jung, B. G. Jeon, H. H. Kim, Y. J. Song, B. J. Koo, S. Y. Lee, S. O. Park, Y. W. Park, K. Kim: Tech. Dig. Int. Electron Devices Meeting (1999) p. 279
19. T. Jimbo, Y. Miyaguchi, S. Kikuchi, I. Kimura, M. Tanimura, K. Suu, M. Ishikawa: Integr. Ferroelectr. **45**, 105 (2002)

A Chemical Approach Using Liquid Sources Tailored to Bi-Based Layer-Structured Perovskite Thin Films

Kazumi Kato[1,2]

[1] Ceramics Research Institute, National Institute of Advanced Industrial Science and Technology, 2266-98 Anagahora, Shimoshidami, Moriyama-ku, Nagoya 463-8560, Japan
[2] Frontier Collaborative Research Center, Tokyo Institute of Technology, 4259 Nagatsuda-cho, Midori-ku, Yokohama 226-8503, Japan
kzm.kato@aist.go.jp

Abstract. The electrical properties of Bi-based layer-structured perovskite compounds strongly depend on their anisotropic crystal structure. Because of this dependence, the properties are expected to be designed or controlled through two kinds of crystallographic approaches: the choice of cations in both the A and B sites of the perovskite layers between the Bi-O layers; and adjustment of the number, n, of oxygen octahedra along the c-axis. Chemical solution deposition is superior to other deposition techniques in precise compositional control and a low processing temperature for multi-component oxide thin films. In particular, it would be a promising technique for control of the crystal structure when the molecules in the precursor solutions are tailored to have optimum structures for easy transformation to the functional oxides. In this chapter, discussions are focused on the effects of the choice of Ca^{2+} as the A site cation and adjustments of n in the range of 2 to 5 on the properties of thin films. Namely, thin films of $CaBi_2Ta_2O_9$ ($n = 2$), $CaBi_3Ti_3O_{12-\delta}$ ($n = 3$), $CaBi_4Ti_4O_{15}$ ($n = 4$), and $Ca_2Bi_4Ti_5O_{18}$ ($n = 5$) were deposited using precursor solutions which were synthesized by chemical reactions of metal alkoxides on the basis of the original concept. The properties of the novel thin films were primarily investigated. The relationship between the crystallographic and electronic properties is fundamentally considered and then the potential for application to FeRAMs is addressed.

1 Introduction

Chemical processing is a promising technique for the synthesis of thin films of multi-component electronic ceramics such as ferroelectrics. The electrical properties of the ceramic thin films are predominantly determined by their crystal structure, composition, and microstructure. Precise control of the three fundamental factors is the key to developing high performance. In particular, one should focus on chemical processing via liquid sources, because there are many attractive advantages. By designing the molecular structure of the liquid source and optimizing its reactivity for the hydrolysis reaction, photoreaction, and combustion, low-temperature crystallization

H. Ishiwara, M. Okuyama, Y. Arimoto (Eds.): Ferroelectric Random Access Memories, Topics Appl. Phys. **93**, 85–94 (2004)
© Springer-Verlag Berlin Heidelberg 2004

and a uniform microstructure would be achieved in the ceramic thin film. Furthermore, the use of liquid sources with designed structures is essential for integration of the ceramic thin films into silicon, because of the capability of low-temperature crystallization. In this chapter, the synthesis of a series of traditional and novel Bi-based layer-structured perovskite thin films by using liquid sources containing structure-designed molecules is described, primarily addressing their electrical properties.

2 $SrBi_2(Ta, Nb)_2O_9$ and $CaBi_2Ta_2O_9$ Thin Films Deposited via Triple Alkoxides of Sr–Bi–Ta/Nb and Ca–Bi–Ta

Thin films of layer-structured perovskite compounds, such as $SrBi_2Ta_2O_9$ (SBT) and $SrBi_2(Ta,Nb)_2O_9$, have been considered to be promising materials for nonvolatile ferroelectric memory applications because of their excellent ferroelectric properties, especially with respect to fatigue performance [1]. The solid solutions, $SrBi_2M_2O_9$ (where M is Ta or Nb), known as Aurivillius compounds, consist of a stack of alternating layers of $Bi_2O_2^{2+}$ with two pseudoperovskite layers of oxygen octahedra, $(SrM_2O_7)^{2-}$, in the c-direction. The high spontaneous polarization is parallel to the plane of the layers (in the a-direction), because this direction contains the O–M–O chains that are known to have high polarizibilities, as in the perovskite ferroelectrics. It is imperative to lower the processing temperatures of these compounds to achieve compatibility with the complementary metal-oxide–semiconductor (CMOS) technology. Another problem is that the bottom electrode thin film interface reactions at relatively high temperatures cause degradation of the electrical properties.

Precursors for layer-structured perovskite thin films of SBT and $SrBi_2Nb_2O_9$ (SBN) have been prepared by the reactions of a strontium–bismuth double methoxyethoxide and tantalum or niobium methoxyethoxide in methoxyethanol, followed by partial hydrolysis. Several spectroscopic techniques, such as ^{1}H-, ^{13}C-, and ^{93}Nb-NMR (nuclear magnetic resonance), and Fourier transform infrared spectroscopy, can be used to analyze the arrangement of the metals and oxygen in the precursor molecules. The precursors contain Sr–O–M (where M is Ta or Nb) bonds (i.e., a strontium is connected to two MO_6 octahedra) and Sr–O–Bi bonds, with a bismuth atom bonded to the oxygens of the MO_6 octahedron. The arrangement of metals and oxygen is considered to be similar to the layer-structured perovskite crystal sublattice (Fig. 1).

As a result, the alkoxy-derived SBT thin films can be crystallized by rapid thermal annealing in an oxygen atmosphere below 550 °C, and they exhibit a preferred (115) orientation. The crystallinity is improved and the crystallite size increases with temperature up to 700 °C. In the case of SBN thin films, a low heating rate (2 °C/min) is necessary for control of the crystallographic

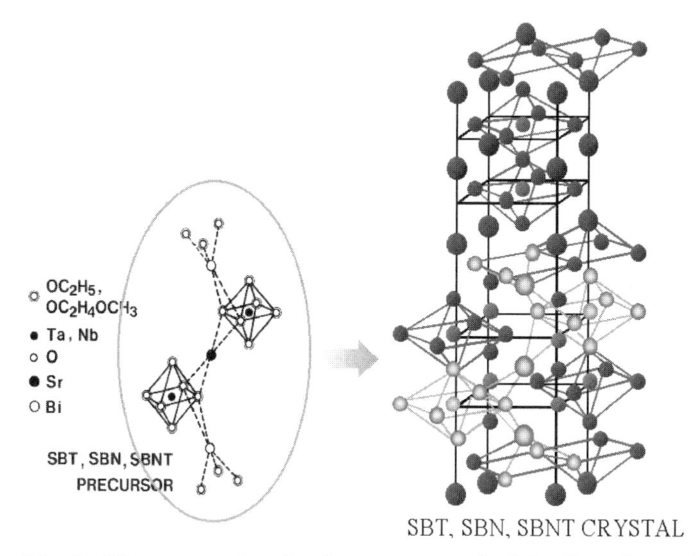

Fig. 1. The proposed molecular structure of triple alkoxides and the crystal structures of $SrBi_2Ta_2O_9$, $SrBi_2(Ta,Nb)_2O_9$, and $SrBi_2Nb_2O_9$

(115) orientation, whereas a rate of 200 °C/s (rapid thermal annealing) produces films that exhibit a *c*-axis orientation. The (115) SBT thin film, heated to 700 °C, exhibits improved ferroelectric properties [2, 3, 4, 5]. The 700 °C-annealed SBT thin film does not show fatigue after 10^{10} cycles of switching.

The film, deposited from the complex ethoxide solution and afterwards calcined at 250 °C in the mixture of water vapor and an oxygen flow, contains a relatively smaller amount of the remaining alkoxy groups. Therefore, it crystallized to SBT via rapid thermal annealing at a temperature below 500 °C in an oxygen flow. The crystallinity had improved significantly and the microstructure had developed up to 650 °C. Remarkable grain growth was observed, particularly in the SBT thin film deposited from the ethanolic solution of the complex ethoxide. Films annealed at 650 °C exhibited saturated P–E hysteresis loop and improved endurance behavior [6, 7, 8].

As described above, the ferroelectric properties of Bi-based layer-structured perovskite originate in the unique and anisotropic crystal structure. On the basis of crystallographic aspects, there are two important approaches to controling the ferroelectric properties: precise substitution of the A-site and B-site cations and adjustment of the number of oxygen octahedra along the *c*-axis in the perovskite layer. Substitution of the Sr^{2+} ion by the Ca^{2+} ion in SBT is a typical approach of the former method, which leads to the formation of $CaBi_2Ta_2O_9$ (CBT).

CBT thin films were prepared on Pt-coated quartz glass substrates by means of the solution of a Ca–Bi–Ta triple alkoxide which was synthesized by a method similar to that used for Sr–Bi–Ta triple alkoxide, except for the

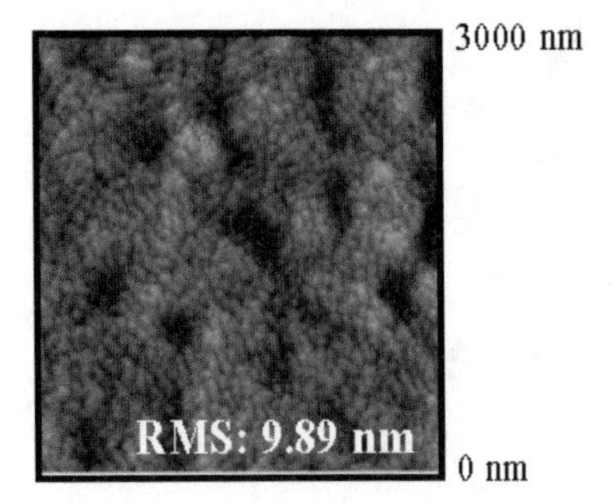

Fig. 2. An AFM image of the surface morphology of $CaBi_2Ta_2O_9$ thin film

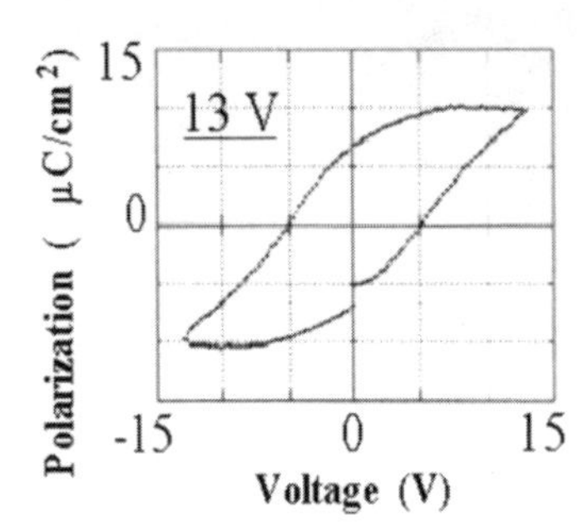

Fig. 3. The ferroelectric hysteresis loop of $CaBi_2Ta_2O_9$ thin film [10]

use of Ca instead of Sr. The 750 °C-crystallized thin film was a single phase of layer-structured perovskite CBT and showed random orientation. The CBT thin film consisted of fine and isotropic grains (Fig. 2) and exhibited a P–E hysteresis loop at a relatively high voltage of 13 V (Fig. 3) and good fatigue properties up to 10^{11} cycles at 5 V (Fig. 4). The uniform surface morphology indicates that the thin film is to be preferred to SBT thin films for integration at the nano-size level. The dielectric constant of the randomly crystallized CBT thin film was about 130 and the loss factor was sufficiently low in the frequency range from 10 kHz to 1 MHz [9, 10, 11].

Non-c-axis-oriented CBT thin films were successfully deposited via the triple alkoxide solutions on Pt-coated Si substrates by inserting thin buffer layers of $BaBi_2Ta_2O_9$ (BBT). The BBT thin buffer layers, which were prepared on Pt-coated Si, were a key material for suppression of the nonpolar c-axis orientation and the promotion of a ferroelectric structure perpendicular to the in-plane direction of the CBT thin films. The details of the crystallographic and microstructural development of the CBT thin films depended on the annealing temperature and the thickness of the BBT buffer layers. The resultant 650 °C-crystallized CBT/BBT (30 nm) thin film on Pt-coated Si was a single phase of layer-structured perovskite and showed the lowest degree of c-axis orientation with high intensities of (115), (200), and (020) diffraction lines (Fig. 5). The dielectric constant and loss factor of the stacked CBT/BBT thin films were 140 and 0.024, respectively, at 100 kHz and were

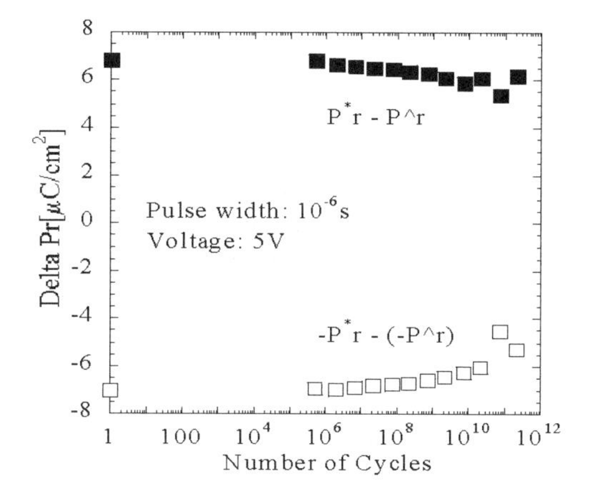

Fig. 4. The fatigue behavior of $CaBi_2Ta_2O_9$ thin film [9, 13]

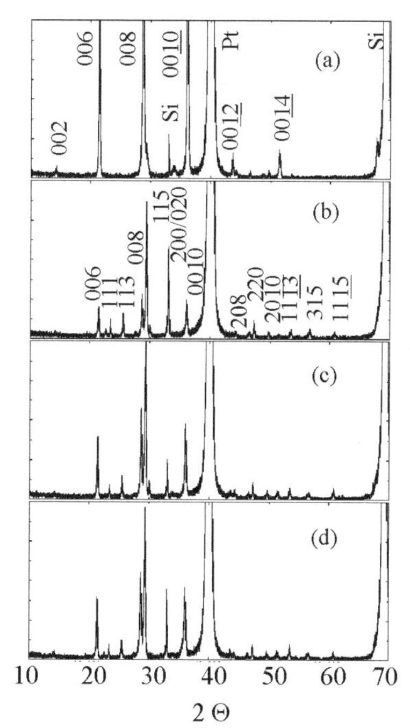

Fig. 5. XRD profiles of $CaBi_2Ta_2O_9$ and stacked $CaBi_2Ta_2O_9/BaBi_2Ta_2O_9$ thin films crystallized at 650 °C on Pt-passivated Si with BBT buffer thicknesses of: (**a**) 0, (**b**) 30, (**c**) 60, and (**d**) 90 nm [11]

almost constant in the frequency range from 10 kHz to 1 MHz. The stacked CBT/BBT thin films exhibited P–E hysteresis loops at the relatively high voltage of 11 V and exhibited little fatigue, which depended on the applied voltage. The dielectric, ferroelectric, and fatigue properties of the stacked CBT/BBT thin films depended on the annealing temperature and the thickness of the BBT buffer layer. The 650 °C-annealed CBT/BBT (30 nm) thin film exhibited the best properties [12, 13].

3 $CaBi_3Ti_3O_{12-X}$, $CaBi_4Ti_4O_{15}$, and $Ca_2Bi_4Ti_5O_{18}$ Thin Films Deposited via Mixtures of Ca–Bi and Bi–Ti Double Alkoxides

Another crystallographic approach to controling the ferroelectric properties is adjustment of the number of oxygen octahedra along the *c*-axis in the perovskite layer. The displacement of B-site cations from the center of the octahedron and the rotation and bending at the connections of the octahedra strongly relate to the ferroelectricity of the Bi-based layer-structured perovskite. Therefore, the number is imperiative to determine the properties. A series of layer-structured perovskite with various numbers of oxygen octahedra can be obtained in the system of $(Bi_2O_2)^{2+}(M_{m-1}Ti_mO_{3m+1})^{2-}$ (M: Ca or Bi). $Ca[Bi(OR)_4]_2$ (R: C_2H_5, $C_2H_4OCH_3$) double alkoxide is again available for thin films of the series, as used for the synthesis of Ca–Bi–Ta triple alkoxide as a precursor of CBT thin films. A Bi–Ti double alkoxide has been prepared by the reaction of $Bi(OCOCH_3)_3$ and $Ti_2(OC_2H_4OCH_3)_8$ for synthesis of $Bi_4Ti_3O_{12}$ (BIT) thin films [14]. By the use of the double alkoxide, a pyrochlore-free perovskite phase began to form at 670 °C. Homogeneity in mixtures of the Ca–Bi and Bi–Ti complexes is a key to giving the series of layer-structured perovskite thin films with various numbers of oxygen octahedra along the *c*-axis in the perovskite layer.

Thin films with three oxygen octahedra along the *c*-axis between the Bi–O layers, $CaBi_3Ti_3O_{12-x}$ (CBTi133), were prepared using a mixture solution of Ca–Bi and Bi–Ti double alkoxides with an atomic ratio of Ca : Bi : Ti = 1 : 3 : 3. These films crystallized below 550 °C. The crystal structure and surface morphology of these films changed between 600 °C and 650 °C (Fig. 6). The 650 °C-annealed thin film, with a thickness of about 100 nm, consisted of well-developed grains and exhibited *P–E* hysteresis loops. The remanent polarization and the coercive electric field were 8.5 μC/cm^2 and 124 kV/cm, respectively, at 7 V. The dielectric constant and loss factor were about 250 and 0.048, respectively, at 100 kHz. The primarily evaluated fatigue behavior was unfortunately rather poor and similar to that of typical $Pb(Zr, Ti)O_3$ (PZT) thin films on Pt. The polarization began to decrease after 2×10^6 switching cycles and the change was independent of the applied voltage. The behavior was considered to relate to oxygen vacancies which were generated to compensate for charge neutralization [15].

$CaBi_4Ti_4O_{15}$ (CBTi144) thin films with four oxygen octahedra were prepared on Pt-coated Si substrates by using a mixture solution of Ca–Bi and Bi–Ti double alkoxides with an atomic ratio of Ca : Bi : Ti = 1 : 4 : 4. As-deposited thin films initiated crystallization below 550 °C and enhanced the crystallinity of a single phase of layer-structured perovskite at 650 °C via rapid thermal annealing in oxygen. The 650 °C-annealed CBTi144 thin films with a thickness of about 130 nm consisted of uniform and isotropic grains (Fig. 7), and had a closely packed columnar structure (Fig. 8). The character-

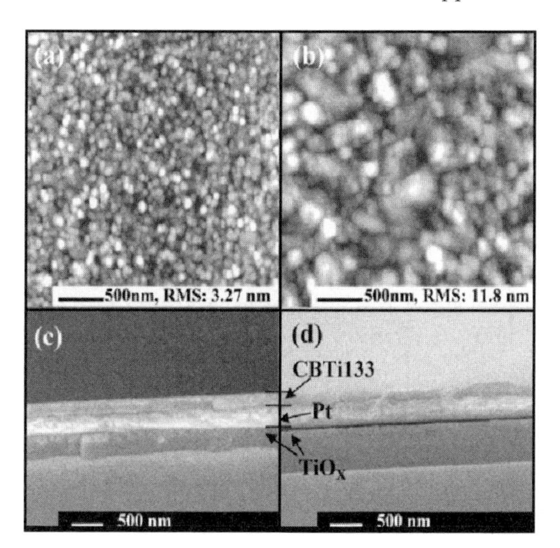

Fig. 6. AFM images and SEM cross-sectional profiles of (**a**), (**c**) 600 °C and (**b**), (**d**) 650 °C-annealed $CaBi_3Ti_3O_{12-\delta}$ thin films [15]

istic uniform structure is preferable for integration. These thin films showed random orientation and exhibited P–E hysteresis loops. The remanent polarization and the coercive electric field were 9.4 μC/cm^2 and 106 kV/cm, respectively, at 11 V. The dielectric constant and the loss factor were 300 and 0.033, respectively, at 100 kHz. The change of the polarization with the number of switching cycles depended on the applied voltage and was constant up to 2×10^{10} cycles at 7 V (Fig. 9) [16, 17]. SBTi144 thin films were also prepared on Pt-coated silicon substrate. Compared with CBTi144 thin films, the SBTi144 thin films crystallized to the perovskite phase at relatively low temperatures and consisted of small and nonuniform grains. Plate-shaped grains developed randomly in heating up to 750 °C. The dielectric constant and the loss factor of the 650 °C-annealed SBTi144 thin film were 330 and 0.04, respectively, at 100 kHz. P_r and E_c were 3.3 μC/cm^2 and 66 kV/cm, respectively, at 10 V (Fig. 10). By comparing the properties of CBT and SBT thin films, it was clarified that the dielectric and ferroelectric properties depend on the cation in the A site, the size of which determines how far it can be displaced in the site, and affects the microstructural development. Annealing at 750 °C resulted in grain growth and an interface reaction that caused anisotropic ferroelectric properties in the case of the SBTi144 thin film. As superior properties were obtained for CBTi144 thin films annealed at lower temperatures, CBTi144 thin film would be preferable to SBTi144 thin film for application to integrated systems [18].

$Ca_2Bi_4Ti_5O_{18}$ (CBTi245) thin films with five oxygen octahedra were prepared by spin-coating a mixture solution of of Ca–Bi and Bi–Ti double alkoxides with an atomic ratio of Ca : Bi : Ti = 2 : 4 : 5. The onset of crystal-

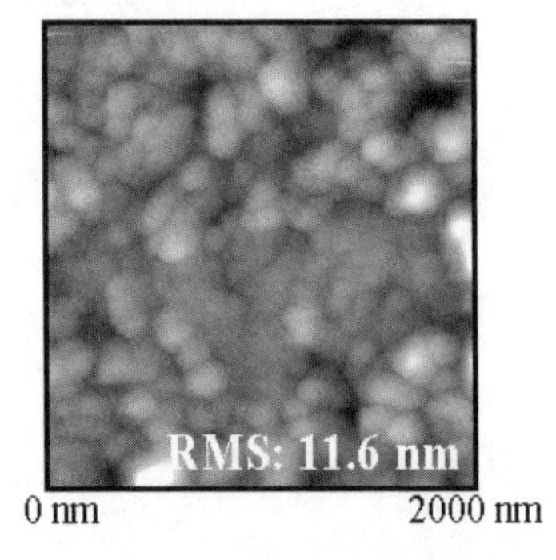

Fig. 7. An AFM image of the surface morphology of 650 °C-annealed $CaBi_4Ti_4O_{15}$ thin film [16, 17, 18]

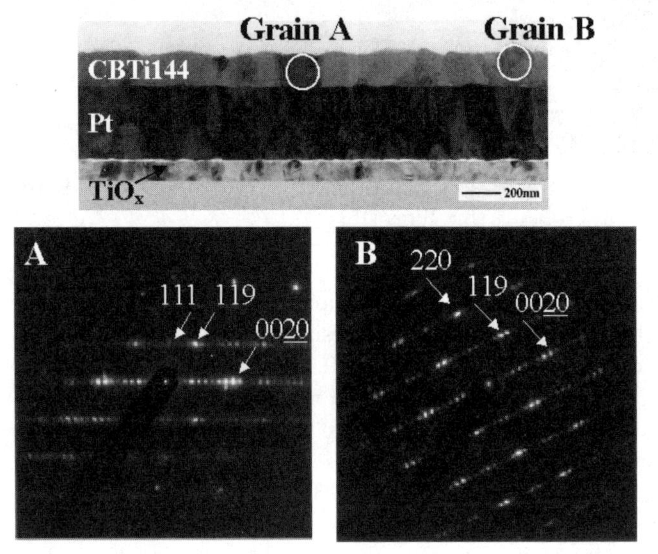

Fig. 8. A TEM cross-sectional profile and electron diffraction patterns of 650 °C-annealed $CaBi_4Ti_4O_{15}$ thin film [17, 18]

lization of the thin films to a pyrochlore phase was below 550 °C, via rapid thermal annealing in oxygen. Further annealing at temperatures of 650 °C or higher resulted in the development of a perovskite layer. The CBTi245 thin films crystallized on Pt-passivated Si substrates showed random orientation, a columnar structure, and P–E hysteresis loops. The remanent polarization and coercive electric field of the 650 °C-annealed CBTi245 thin film were 4.7 μC/cm^2 and 111 kV/cm, respectively, at 12 V. The dielectric constant and the loss factor were 330 and 0.028, respectively, at 100 kHz. The polarization did not change after 10^{11} switching cycles at 7 V [19, 20].

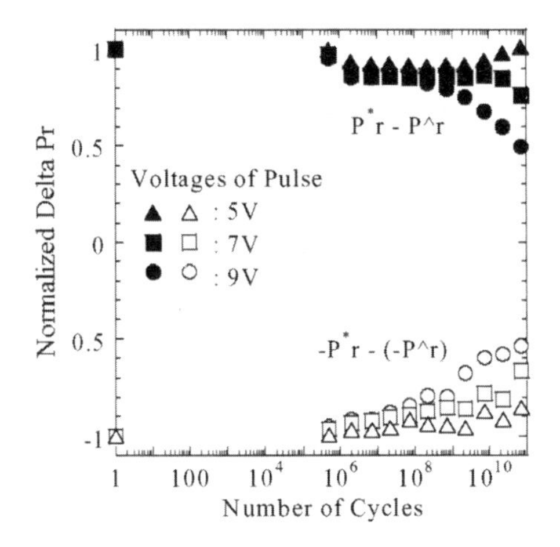

Fig. 9. The fatigue behavior of 650 °C-annealed $CaBi_4Ti_4O_{15}$ thin film [16, 17, 18]

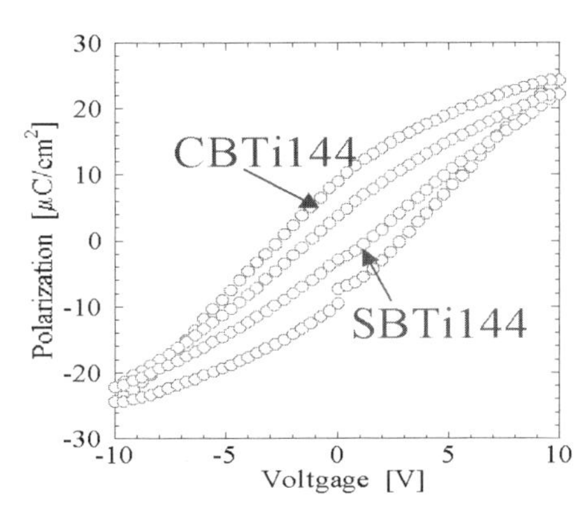

Fig. 10. Ferroelectric hysteresis loops of 650 °C-annealed $CaBi_4Ti_4O_{15}$ and $SrBi_4Ti_4O_{15}$ thin films

References

1. C. A. Paz de Araujo, J. D. Cuchiaro, L. D. McMillan, M. C. Scott, J. F. Scott: Nature **374**, 627 (1995)
2. K. Kato, C. Zheng, S. K. Dey, Y. Torii: Integr. Ferroelectr. **18**, 225 (1997)
3. K. Kato, J. M. Finder, S. K. Dey, Y. Torii: Integr. Ferroelectr. **18**, 237 (1997)
4. K. Kato, C. Zheng, J. M. Finder, S. K. Dey, Y. Torii: J. Am. Ceram. Soc. **81**, 1869 (1998)
5. K. Kato: Integr. Ferroelectr. **22**, 13 (1998)
6. K. Kato: Jpn. J. Appl. Phys. **37**, 5178 (1998)
7. K. Kato: Integr. Ferroelectr. **26**, 243 (1999)
8. K. Kato: Jpn. J. Appl. Phys. **38**, 5417 (1999)
9. K. Kato, K. Suzuki, K. Nishizawa, T. Miki: J. Appl. Phys. **88**, 3779 (2000)
10. K. Kato, K. Suzuki, K. Nishizawa, T. Miki: Jpn. J. Appl. Phys. **39**, 5501 (2000)
11. K. Kato, K. Suzuki, K. Nishizawa, T. Miki: J. Appl. Phys. **89**, 5088 (2001)
12. K. Kato: Mater. Res. Soc. Symp. Proc. **596**, 167 (2000)
13. K. Kato, K. Suzuki, K. Nishizawa, T. Miki: Mater. Res. Soc. Proc. **665**, CC12.12.1 (2001)

14. M. Toyoda, Y. Hamaji, K. Tomono, D. A. Payne: Jpn. J. Appl. Phys. **32**, 4158 (1993)
15. K. Kato, K. Suzuki, K. Nishizawa, T. Miki: Appl. Phys. Lett. **79**, 397 (2001)
16. K. Kato, K. Suzuki, K. Nishizawa, T. Miki: Appl. Phys. Lett. **78**, 1119 (2001)
17. K. Kato, K. Suzuki, K. Nishizawa, T. Miki: Integr. Ferroelectr. **36**, 321 (2001)
18. K. Kato, K. Suzuki, K. Nishizawa, T. Miki: Jpn. J. Appl. Phys. **40**, 5580 (2001)
19. K. Kato, K. Suzuki, D. Fu, K. Nishizawa, T. Miki: Jpn. J. Appl. Phys. **41**, 2110 (2002)
20. K. Kato, K. Suzuki, D. Fu, K. Nishizawa, T. Miki: J. Ceram. Soc. Jpn. **110**, 403 (2002)

Recent Development in the Preparation of Ferroelectric Thin Films by MOCVD

Hiroshi Funakubo

Department of Innovative and Engineered Materials, Interdisciplinary Graduate School of Science and Engineering, Tokyo Institute of Technology,
G1-405, 4259 Nagatsuta-cho, Midori-ku, Yokohama, 226–8502, Japan
funakubo@iem.titech.ac.jp

Abstract. Recent research by our group, concerned with the preparation of ferroelectric thin films by MOCVD, is summarized. MOCVD has been investigated as a most practical preparation method to realize a high-density FeRAM, because of the good step coverage and the uniform composition and film thickness over a large area. In addition, it has recently become possible to decrease the process temperature to obtain films retaining a large remanent polarization by MOCVD. In fact, $Pb(Zr,Ti)O_3$ (PZT) and $SrBi_2Ta_2O_9$ (SBT) films having good ferroelectricity were obtained at 415 and 570 °C, respectively. Moreover, new materials to overcome the problem of the present materials, such as PZT, SBT, and $(Bi,La)_4Ti_3O_{12}$ (BLT), were discovered by using MOCVD because the composition of the film was easily changed by the input source gas composition.

1 Introduction

Metalorganic chemical vapor deposition (MOCVD) is one of the most important preparation methods to realize the high-density ferroelectric random access memory (FeRAM) application [1]. Ferroelectric materials, such as $Pb(Zr,Ti)O_3$ (PZT) and $SrBi_2Ta_2O_9$ (SBT), contain elements with a high vapor pressure, such as Pb and Bi, so that the relatively high-pressure deposition in the MOCVD process has the advantage of diminishing the reevaporation of these elements. This realizes the highly compositional reproducibility of the film.

In general, there are two methods for obtaining the crystallized ferroelectric phase: the first is direct crystallization from the gas phase (the direct crystallization process) and the second is solid-phase transformation from other solid phases deposited at low temperature, such as the amorphous phase (the solid-phase process) [2]. The latter case is mainly used to get a high step coverage under reaction rate-limiting conditions [3]. However, these low-temperature conditions sometimes include imperfect decomposition of the starting source materials. This makes a large amount of the residue produced from the starting materials, such as the element carbon. On the other hand, the former process has been widely investigated in order to obtain epitaxially grown films to study the fundamental properties of ferroelectric

H. Ishiwara, M. Okuyama, Y. Arimoto (Eds.): Ferroelectric Random Access Memories,
Topics Appl. Phys. **93**, 95–103 (2004)
© Springer-Verlag Berlin Heidelberg 2004

thin films [4, 5, 6, 7, 8, 9, 10, 11, 12, 13, 14]. However, from the practical point of view, this process has hardly been employed because of the increase in the surface roughness and the risk of low step coverage under deposition in diffusion rate-limiting conditions except for the case of PZT film preparation.

In the present study, we have reported that the direct crystallization of the ferroelectric phase by MOCVD is useful in decreasing the process temperature to obtain a film that has good polarization hysteresis [4, 5, 6, 7, 8, 9, 10, 11, 12, 13, 14]. Moreover, our research into a new composition that will overcome the problems of the present ferroelectric materials is also mentioned.

2 Low-Temperature Deposition

The lowering of the deposition temperature is one of the most important key issues in realizing high-density FeRAM. This is because of the reduction of the interdiffusion between each layer in the devices. To obtain the crystalline ferroelectric phase, the solid-phase process was rate-limited by the solid-phase diffusion of species including the constituent elements, which is generally slower than the direct crystallization process from the gas phase. This means that direct crystallization from the gas phase is suitable to obtain the crystalline films at low temperature.

2.1 SBT Films

In SBT [15, 16, 17, 18, 19, 20, 21] it is widely known to be difficult to obtain sufficient ferroelectricity at a low process temperature. A process temperature above 650 °C is widely reported as a necessary temperature to get sufficient ferroelectricity by the process of solid-phase transformation. However, in that case, most of the crystalline SBT phase is obtained by transformation from the amorphous or fluorite phases.

We have succeeded in obtaining directly crystallized SBT film even at 570 °C by electron cyclotron resonance (ECR) plasma enhanced-MOCVD on (111)Ir/TiO_2/ SiO_2/Si substrates. The deposited film exhibited around (103) single orientation originated from the local epitaxial growth on (111)-oriented Ir grains, as shown in Fig. 1a. These films showed well-saturated P–E hysteresis loops without subsequent heat treatment, as shown in Fig. 1b,c. Twice the remanent polarization ($2P_r$) and the coercive field (E_c) of the resultant film were 16.1 μC/cm^2 and 83 kV/cm, respectively. This $2P_r$ value was 88% of the estimated value of the (103) orientation of the SBT films, both from the epitaxial films and from the detailed crystal structure analysis. Moreover, this film had a smooth surface with an average roughness (R_a) of 8.5 nm, which is reflected in the low leakage current density, of the order of 10^{-7} A/cm^2 up to 400 kV/cm. These results clearly show that direct crystallization of the ferroelectric phase is very effective in obtaining films with good ferroelectricity at a low process temperature.

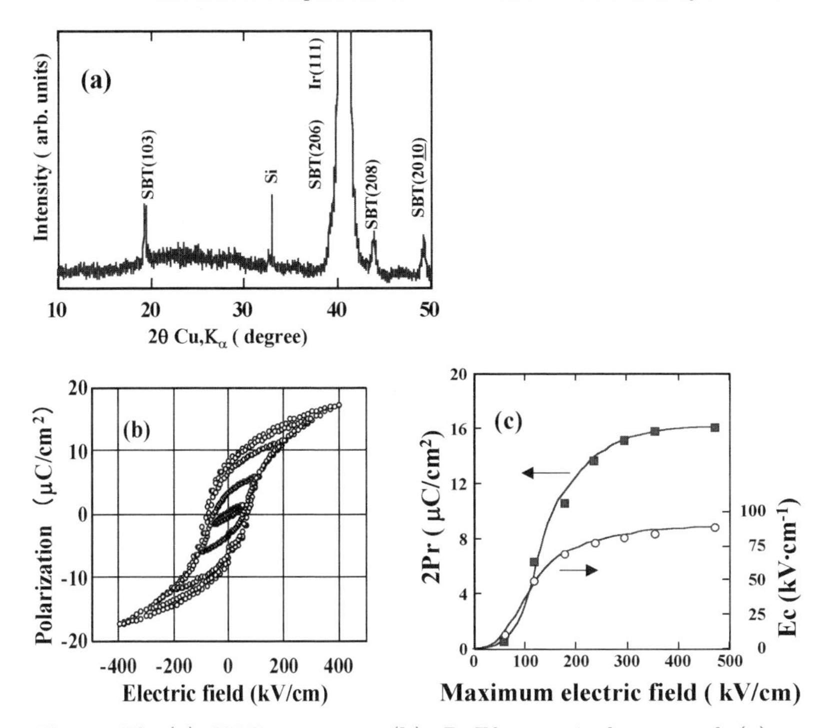

Fig. 1. The (**a**) XRD patterns, (**b**) P–E hysteresis loops, and (**c**) saturation properties of the P_r and E_c values of the films deposited at 570 °C on (111)Ir/TiO_2/SiO_2/Si substrate by ECR-MOCVD

2.2 PZT Films

MOCVD-PZT films [22,23,24] with large remanent polarization have been reported by some research groups [25,26]. In all of these reports, the PZT phase was crystallized directly from the gas phase. This is because the PZT film was easily crystallized at a relatively low temperature, such as 350 °C. We have developed epitaxial-grade polycrystalline PZT films even at 415 °C by source-gas-pulsed-introduced MOCVD (pulsed-MOCVD). The pulsed-MOCVD process was different from that of conventional MOCVD (continuous-MOCVD), as shown in Fig. 2, because the source gas was pulse-introduced by making interval times.

Polycrystalline PZT films with a Zr/(Zr + Ti) ratio of 0.35, prepared on (111)Pt/Ti/SiO_2/Si substrates, show highly (100)- and (001)-preferred orientations. Well-saturated ferroelectricity, with P_r and E_c values of 41.4 μC/cm^2 and 78.5 kV/cm, respectively, was obtained without subsequent heat treatment. This P_r value is almost the same as that of epitaxially grown films at 580 °C with the same composition and orientations, taking account of the volume fraction of the (100) and (001) orientations, as shown in Fig. 3. This result shows that the pulsed introduction of source gas is a very use-

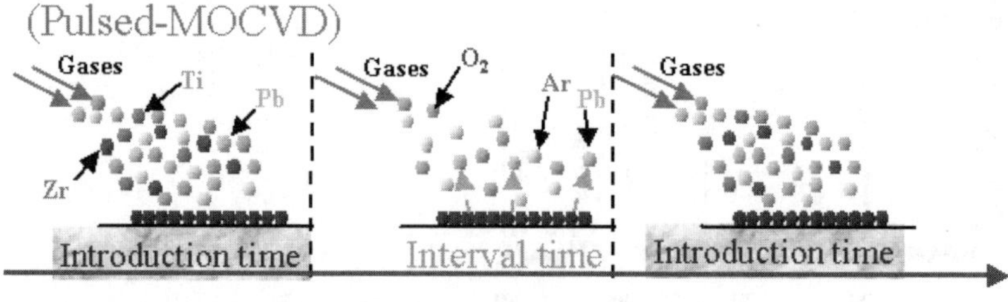

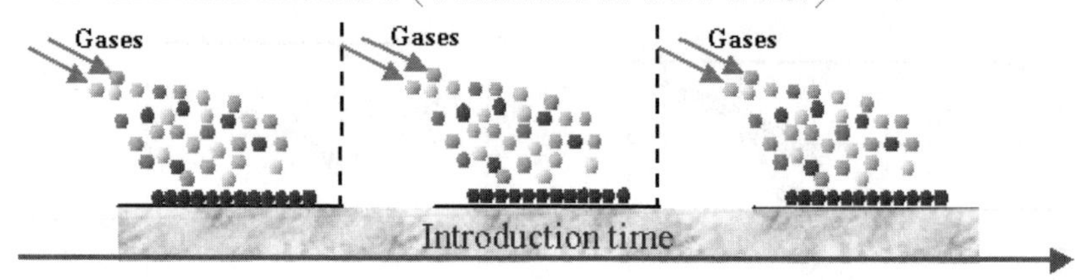

Fig. 2. Schematic diagrams of the deposition behavior in (**a**) pulsed-MOCVD and (**b**) continuous-MOCVD

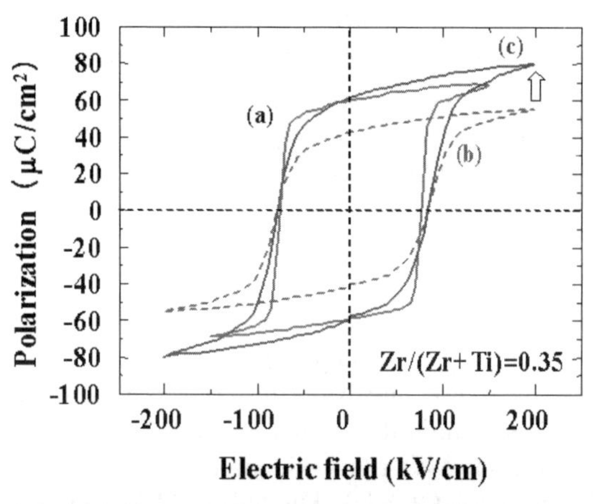

Fig. 3. *P–E* hysteresis loops of (**a**) epitaxial $Pb(Zr_{0.3}Ti_{0.7})O_3$ film deposited on $(100)SrRuO_3//(100)SrTiO_3$ substrate at 580 °C, (**b**) polycrystalline $Pb(Zr_{0.3}Ti_{0.7})O_3$ film deposited on $(111)Pt/Ti/SiO_2/Si$ substrate at 415 °C, together with (**c**) an estimated curve for polycrystalline films, assuming the same orientation with the epitaxially grown one

ful method for obtaining high-quality films at a low deposition temperature. This process is ascertained also to be effective for SBT films, to obtain good ferroelectricity at a low temperature.

3 Novel Materials Research

PZT and SBT films have been widely investigated for FeRAM applications, but still suffer from the problems of poor fatigue characteristics and a too small remanent polarization, respectively. Therefore, research into new materials suitable for FeRAM applications must be carried out to overcome these problems. In the search for new compositions, MOCVD is suitable because the film composition can be widely varied by adjusting the composition of the input source gas. Highly crystallized film with a controlled orientation is obtained by MOCVD.

3.1 Solid Solution of $SrBi_2(Ta_{0.7}Nb_{0.3})_2O_9 - Bi_3TaTiO_9$ [$(1-x)$SBT-xBTT]

An excess-Bi and Sr-deficient $Sr_{0.8}Bi_{2.2}(Ta_{0.7}Nb_{0.3})_2O_9$ ($Sr_{0.8}Bi_{2.2}TN$) thin film has been investigated for FeRAM applications because of its high fatigue resistance. In this composition, the excess Bi element is pointed out to be substituted for the Sr site together with a vacancy to maintain charge neutrality. This vacancy is mobile and may decrease the resistivity of the films. Therefore, we have attempted to search for a composition with a large remanent polarization but without a large number of vacancies, e.g. $Sr_{0.8}Bi_{2.2}(Ta_{0.7}Nb_{0.3})_2O_9$. *Irie* et al. [29] reported the Curie temperature (T_c) dependency of the spontaneous polarization (P_s) and the coercive field (E_c) along the a-axis of Bi-layer-structured ferroelectric materials (BLSFs). In this report, the P_s and E_c values increase almost linearly with a decrease of T_c, irrespective of the composition of the film. Therefore, we investigated materials with a T_c of about 450 °C, because that of $Sr_{0.8}Bi_{2.2}(Ta_{0.7}Nb_{0.3})_2O_9$ has been confirmed to be in the range from 400 to 450 °C.

$(1-x)SrBi_2(Ta_{0.7}Nb_{0.3})_2O_9 + xBi_3TiTaO_9$ ($x = 0$–0.5) solid-solution (SBTN + BTT) films with low defect contents were directly crystallized on (111)$Pt/Ti/SiO_2/Si$ substrates at 650 °C by MOCVD [27,28]. The deposited films showed a strong (103) orientation. The remanent polarization (P_r) of the directly crystallized SBTN ($x = 0$) was very small. However, the P_r value was increased to 7.1 $\mu C/cm^2$ by adding 30% of BTT ($x = 0.3$) and was almost equal to that of $Sr_{0.8}Bi_{2.2}(Ta_{0.7}Nb_{0.3})_2O_9$, which is widely studied for nonvolatile memory applications, as shown in Fig. 4. The leakage current density of the SBTN + BTT solid solution was of the order of 10^{-8} A/cm^2 for electric fields up to 200 kV/cm, due to its low defect contents characteristics, while that of $S_{0.8}B_{2.2}TN$ was above 10^{-6} A/cm^2 due to the existence of defects in the Sr sites, as shown in Fig. 5.

The solid-solution film showed a fatigue-free character. Therefore, this solid-solution film is a novel candidate ferroelectric material instead of SBT films, having a low leakage current density and good ferroelectricity.

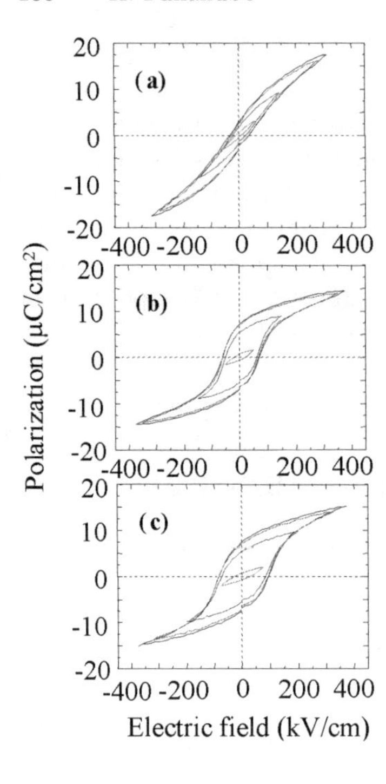

Fig. 4. *P–E* hysteresis loops of (**a**) $SrBi_2(Ta_{0.7}Nb_{0.3})_2O_9$, (**b**) $Sr_{0.8}Bi_{2.2}(Ta_{0.7}Nb_{0.3})_9$, and (**c**) $0.7SrBi_2(Ta_{0.7}Nb_{0.3})_2O_9 + 0.3Bi_3TaTiO_9$ thin films deposited at 650 °C on $(111)Pt/Ti/SiO_2/Si$ substrates

3.2 $(Bi,Ln)_4(Ti,V)_3O_{12}$

As an alternative to SBT, $(Bi,La)_4Ti_3O_{12}$ film [30, 31, 32] has been very attractive for its larger ferroelectricity than SBT and its good fatigue resistance. By substituting La for Bi, the essentially large ferroelectricity of BIT came into use [34]. However, the reports concerning BLT films with large ferroelectricity are limited to films prepared at a relatively high temperature, above 650 °C. On the other hand, a remarkable improvement of the ferroelectricity in BIT ceramics is obtained by adding a high-valence cation, V, to the B-site. This has been explained by a decrease in the defects responsible for domain pinning [35].

For the first time, we have prepared Ln (Ln = La and Nd) and V-co-substituted BIT (BLnTV) films on $(111)Pt/Ti/SiO_2/Si$ substrates at 600 °C by MOCVD. The films that were substituted for only the A-site by Ln, $(Bi,Ln)_4Ti_3O_{12}$, and for the B-site by V, $Bi_4(Ti,V)_3O_{12}$, showed no noticeable ferroelectricity; however, co-substituted films were shown to contribute to obtaining a large ferroelectricity. Furthermore, a superior ferroelectricity to $(Bi,La)_4(Ti,V)_3O_{12}$ was confirmed for $(Bi,Nd)_4(Ti,V)_3O_{12}$ films. Moreover, the (104) preferred orientation was changed to (110) and (111) orientations, with an advantage for large polarization when the substrate was changed to $(001)Ru/SiO_2/Si$ instead of $(111)Pt/Ti/SiO_2/Si$ and actually a larger ferroelectricity was obtained; the P_r and E_c values of the films were 17 μC/cm² and 145 kV/cm, respectively, as shown in Fig. 6.

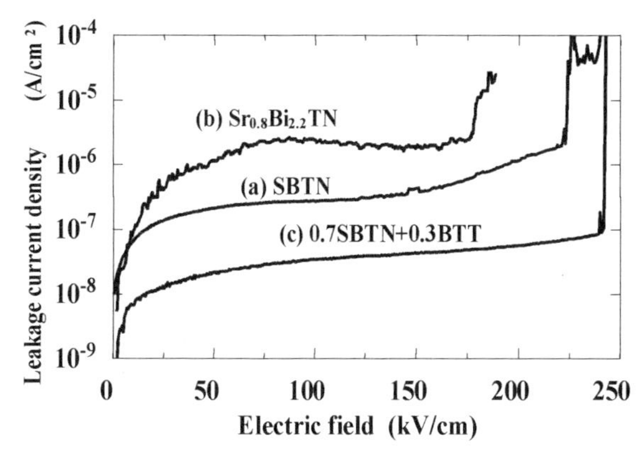

Fig. 5. Leakage current – electric field characteristics of the same films as shown in Fig. 4. (**a**) $SrBi_2(Ta_{0.7}Nb_{0.3})_2O_9$, (**b**) $Sr_{0.8}Bi_{2.2}(Ta_{0.7}Nb_{0.3})_9$, and (**c**) $0.7\,SrBi_2(Ta_{0.7}Nb_{0.3})_2O_9 + 0.3\,Bi_3TaTiO_9$ thin films

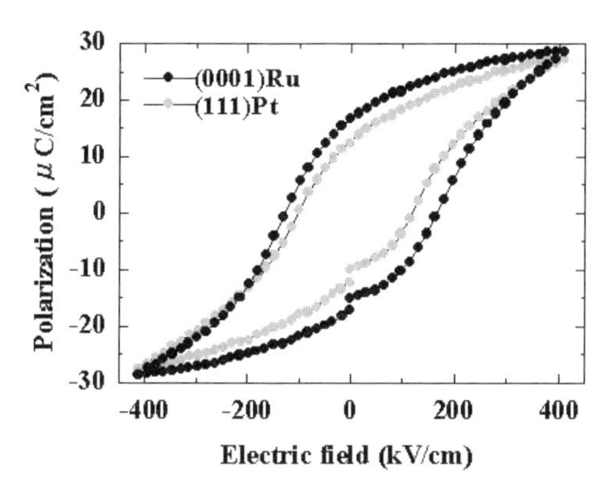

Fig. 6. *P–E* hysteresis loops of $(Bi,Nd)_4(Ti,V)_3O_{12}$ films deposited on $(001)Ru/SiO_2/Si$ and $(111)Pt/Ti/SiO_2/Si$ substrates at 600 °C

4 Summary

Our recent results for ferroelectric thin films prepared by MOCVD can be summarized as follows. The MOCVD process is useful to decrease the process temperature and to obtain film that retains a large remanent polarization. PZT and SBT films with good ferroelectricity were obtained at 415 °C and 570 °C, respectively. Moreover, novel materials to overcome the problem of the present materials, a SBT–BTT solid solution and $(Bi,Ln)_4(Ti,V)_3O_{12}$, were discovered by using the MOCVD process, because the composition of the film was easily changed by the input gas composition.

Acknowledgements

This work was partially performed under the auspices of the Grant-in-Aid or Scientific Research on Priority Area (B) "Control of Material Property of Ferroelectric Thin Films and Their Application to a Next-Generation Memory Device" and an R & D project in cooperation with Academic Institution (Next Generation Ferroelectric Memory) supported by the New Energy and Industrial Technology Development Organization (NEDO) and managed by FED (R & D Association for Future Electron Devices). I gratefully acknowledge the important contributions of M. Miyayama and Y. Noguch (University of Tokyo), M. Osada (JST), K. Saito and M. Mizuhira (Philips Japan), and K. Suzuki, Y. Nishi, and M. Fujimoto (Taiyo Yuden Co.), especially for the students in my laboratory.

References

1. P. Buskirk, J. Roeder, T. Baum, S. Bilodeau, M. Russell, S. Johnston, R. Carl, D. Dersrochers, B. Hendrix, F. Hintermaier: Integr. Ferroelectr. **21**, 273 (1998)
2. N. Nukaga, M. Mitsuya, T. Suzuki, Y. Nishi, M. Fujimoto, H. Funakubo: Jpn. J. Appl. Phys. **40**, 5595 (2001)
3. N. Nukaga, K. Ishikawa, H. Funakubo: Jpn. J. Appl. Phys. **38**, 5428 (1999)
4. K. Ishikawa, H. Funakubo: Appl. Phys. Lett. **75**, 1970 (1999)
5. T. Suzuki, Y. Nishi, M. Fujimoto, K. Ishikawa, H. Funakubo: Jpn. J. Appl. Phys. **38**, L1261 (1999)
6. T. Suzuki, Y. Nishi, M. Fujimoto, K. Ishikawa, H. Funakubo: Jpn. J. Appl. Phys. **38**, L1265 (1999)
7. K. Ishikawa, H. Funakubo, K. Saito, T. Suzuki, Y. Nishi, M. Fujimoto: J. Appl. Phys. **87**, 8018 (2000)
8. K. Ishikawa, A. Saiki, H. Funakubo: Jpn. J. Appl. Phys. **39**, 2102 (2000)
9. K. Saito, K. Ishikawa, A. Saiki, I. Yamaji, T. Akai, H. Funakubo: Integr. Ferroelectr. **33**, 59 (2001)
10. K. Ishikawa, T. Watanabe, H. Funakubo: Thin Solid Films **392**, 128 (2001)
11. T. Watanabe, A. Saiki, K. Saito, H. Funakubo: J. Appl. Phys. **89**, 3934 (2001)
12. T. Watanabe, H. Funakubo, K. Saito: J. Mater. Res. **16**, 303 (2001)
13. K. Nagashima, M. Aratani, H. Funakubo: J. Appl. Phys. **89**, 4517 (2001)
14. K. Ishikawa, T. Sakai, T. Watanabe, H. Funakubo: Key. Eng. Mater. **214–215**, 189 (2002)
15. M. Mitsuya, N. Nukaga, K. Ishikawa, H. Funakubo: Jpn. J. Appl. Phys. **39**, L822 (2000)
16. N. Nukaga, M. Mitsuya, H. Funakubo: Jpn. J. Appl. Phys. **39**, 5496 (2000)
17. N. Nukaga, M. Mitsuya, H. Funakubo: Technical Report of Institute of Electronics, Information and Communication Engineering, **SDM-237** (2001-3) 7 (2001)
18. M. Aratani, T. Oikawa, A. Ozeki, H. Funakubo: Jpn. J. Appl. Phys. **40**, L343 (2001)
19. M. Mitsuya, N. Nukaga, H. Funakubo: Jpn. J. Appl. Phys. **40**, 3337 (2001)
20. M. Mitsuya, N. Nukaga, T. Watanabe, H. Funakubo, K. Saito: Jpn. J. Appl. Phys. **40**, L758 (2001)

21. N. Nukaga, M. Mitsuya, H. Funakubo: Ferroelectr. **260**, 51 (2001)
22. M. Aratani, K. Nagashima, H. Funakubo: Jpn. J. Appl. Phys. **40**, 4126 (2001)
23. M. Aratani, T. Oikawa, T. Ozeki, H. Funakubo: Appl. Phys. Lett. **79**, 1000 (2001)
24. M. Aratani, K. Nagashima, H. Funakubo: Ferroelectr. **260**, 69 (2001)
25. H. Fujisawa, S. Nakashima, K. Kaibara, M. Shimizu, H. Niu: Jpn. J. Appl. Phys. **38**, 5392 (1999)
26. P. Buskirk, S. Bilodeau, J. Roeder, P. Kirlin: Jpn. J. Appl. Phys. **35**, 2520 (1996)
27. M. Mitsuya, N. Nukaga, T. Watanabe, H. Funakubo: Appl. Phys. Lett. **79**, 2067 (2001)
28. M. Mitsuya, M. Osada, H. Funakubo: J. Crystal Growth **237–239**, 473 (2002)
29. H. Irie, M. Miyayama, T. Kudo: Key. Eng. Mater. **181–182**, 27 (2000)
30. M. Osada, M. Kakihana, M. Mitsuya, T. Watanabe, H. Funakubo: Jpn. J. Appl. Phys. **40**, L891 (2001)
31. M. Osada, M. Tada, M. Kakihana, T. Watanabe, H. Funakubo: Jpn. J. Appl. Phys. **40**, 5527 (2001)
32. T. Watanabe, H. Funakubo, M. Osada, Y. Noguchi, M. Miyayama: Appl. Phys. Lett. **80**, 100 (2002)
33. T. Watanabe, K. Saito, H. Funakubo: J. Crystal Growth **235**, 389 (2002)
34. B. H. Park, B. Kang, S. Bu, T. Noh, L. Lee, W. Joe: Nature **401**, 682 (1999)
35. Y. Noguchi, M. Miyayama: Appl. Phys. Lett. **78**, 1903 (2001)

Materials Integration Strategies

Orlando Auciello[1], Anil M. Dhote[1], Bao T. Liu[2], Sanjeev Aggarwal[2], and Ramamoorthy Ramesh[2]

[1] Materials Science Division, Argonne National Laboratory,
Argonne, IL 60439, USA
auciello@msd.anl.gov

[2] Materials and Nuclear Engineering Department, University of Maryland,
College Park, MD 20742, USA

Abstract. In this chapter, materials integration strategies for the fabrication of high-density nonvolatile ferroelectric random access memories (FeRAMs), are discussed, in which unique combinations of in situ and ex situ analytical techniques capable of providing information about thin-film surface and interface processes at the atomic scale are used. These methods are also useful for establishing composition–microstructure–property relationships critical for the integration of ferroelectric capacitors with silicon microcircuits. We demonstrate that Ti–Al layers can be used as a material with a double diffusion barrier/bottom electrode functionality for integration of ferroelectric capacitors with complimentary metal oxide semiconductor (CMOS) devices for the fabrication of FeRAMs. We discuss here results from systematic studies designed to understand Ti–Al film growth and oxidation processes using sputter-deposition in conjunction with complementary in situ atomic layer-resolution mass spectroscopy of recoil ion (MSRI) and surface-sensitive X-ray photoelectron spectroscopy (XPS) and ex situ transmission electron microscopy and electrical characterization.

1 Introduction

Low-density nonvolatile ferroelectric random access memories (FeRAMs) have been introduced in the market in various devices, including "smart cards" [1] and microprocessors with embedded memories. The next major effort is focused on the development of high-density FeRAMs. In this case, conducting diffusion barrier layers will play an even more critical role, since the ferroelectric capacitors, involving the growth of oxide layers at relatively high temperatures, will be fabricated directly on top of complimentary metal oxide semiconductor (CMOS) transistors. There is an intense research and development effort to develop materials processing and integration strategies for the realization of high-density FeRAMs [1,2,3,4,5,6,7,8]. A stacked architecture is needed for the high-density memory application [9,10]. This requires the capacitor to be located directly on top of the drain of the CMOS transistor to minimize the total area of the memory cell [11].

Direct electrical contact between the drain of the transistor and the bottom electrode of the capacitor is established via a conducting polycrystalline-Si (poly-Si) plug. This geometry requires elimination of any inter-diffusion

H. Ishiwara, M. Okuyama, Y. Arimoto (Eds.): Ferroelectric Random Access Memories,
Topics Appl. Phys. **93**, 105–122 (2004)
© Springer-Verlag Berlin Heidelberg 2004

between the components of the transistor and capacitor, to prevent the degradation of the transistor electrical properties. The dramatically different materials chemistry between the ferroelectric oxides and Si makes necessary the introduction of a diffusion barrier layer. The fact that this barrier layer must be a good conductor and form an ohmic contact to Si further complicates this material integration problem.

Among a variety of metals, Pt was chosen as an electrode material primarily because of its refractive nature and resistance to oxidation (unlike, for example, Al). Studies on the integration of Pt into ferroelectric capacitors yielded devices with FeRAMs-compatible properties, such as large values of remanent polarization, film resistivity values higher than 10^{10} $\Omega\,$cm, and sufficient retention characteristics [12,13]. However, Pt does not prevent oxygen diffusion to the poly-Si plug [14]. Furthermore, these films suffered from some fundamental reliability limitations, such as fatigue and imprint. The problem of fatigue has, to a large extent, been solved through one of two approaches: (i) the use of conducting oxide electrodes with PZT as the ferroelectric layer [15, 16]; and (ii) the replacement of PZT with the layered ferroelectric material, strontium bismuth titanate (SBT) [17].

Despite success at addressing such fundamental reliability issues as those discussed above, among the various levels of process integration steps, the development of conducting barrier layers to directly contact the ferroelectric capacitor to the poly-Si plug of the pass-gate transistor in the memory cell is a key requirement. An obvious earlier choice for such a barrier layer was TiN, especially since it was already widely used in the semiconductor industry [18]. Unfortunately, TiN starts to oxidize at $\sim$ 200–300 °C [19], which is much lower than the optimum processing temperature for oxide ferroelectric materials. To overcome the shortcoming of the TiN layer in terms of temperature, Pt or Ir have been used as protective layers [20,21]. Another approach that has been adopted is that of doping TiN with Al to form (Ti,Al)N [7] or using silicides [22,23] or other complex structures [24]. The most common approaches currently being explored use a combination of at least two layers to create a composite barrier layer scheme. Taking the PZT system as a prototype example, one approach uses the combination of (Pt,Ir)/(Ti,Al)N as the composite barrier layer scheme. Many research groups, including ours, have been able to demonstrate the integration of PZT-based capacitors with silicon using the (Pt,Ir)/nitride approach.

Notwithstanding the potential for a successful process integration of ferroelectric layers on a poly-Si plug using this approach, there are several technological and possibly strategic issues related to the use of Pt and Ir that need to be considered as part of the barrier layer structure. From the technological point of view, the use of Pt or Ir in this stack poses some challenges. First, dry etching of Pt or Ir is still a very difficult problem, although there have been some breakthroughs in this regard. A dry etch process (such as a reactive ion beam etching process) is essential in order for the memory technology to be manufacturable with high yield. Since both Pt and Ir are relatively inert, the ability to form volatile reaction species during dry etching appears to be limited. Secondly, from the strategic point of view, the use of either Pt or Ir (both being precious metals) in such quantities may have an impact on the supply and demand economics of these metals (Hwang, personal communication). These issues prompted us to initiate some exploratory studies of alloys and compounds that would eliminate the use of these noble metals. Our preliminary studies have already shown some very striking results that are scientifically interesting and have strong technological potential.

Thin films of structurally amorphous metallic alloys, such as Cu–Zr and Ti–Si–N, have been widely used as diffusion barriers in semiconductor metallization, as ultrahard coatings, and as micromachining elements [25, 26]. These applications take advantage of one or more unique properties of these films, such as inertness to corrosive environments, electrical conduction or insulation, and the ability to remain X-ray amorphous up to processing temperatures. The reason for the formation of amorphous phases of binary metallic alloys appears to be the large differences in the atomic radii of the components, deep eutectics in the phase diagram, and a strong affinity between the components of the alloy [27]. In addition, the presence of a predominantly covalently bonded species such as Al favors the formation of these phases. Controlling extrinsic parameters such as introducing O or N impurities and extremely high cooling or deposition rates have also been demonstrated to favor the formation of amorphous phases [28]. Based on the above information, we decided to investigate Ti–Al (nominal composition 3 : 1) as a conducting ($300\,\mu\Omega\cdot\mathrm{cm}$) diffusion barrier deposited between poly-Si and the ferroelectric capacitor, and this is the topic of this chapter.

Because of its low resistivity ($\leq 300\,\mu\Omega\cdot\mathrm{cm}$), the Ti–Al layer can be used in a double functionality as diffusion barrier and bottom electrode, replacing conventional electrode-barrier structures such as Pt/TiN or Ir/TiN. The use of a material such as Ti–Al for double barrier/bottom electrode functionality may offer other advantages such as easier reactive ion etching than for Pt or Ir, and lower cost. High-density memory integration will require very thin films with smooth surfaces and sharp ferroelectric stack/Ti–Al/poly-Si interfaces. These interfaces play significant roles in defining device performance, especially because they will be exposed to various processing steps during the device fabrication.

The control and the microstructure of thin Ti–Al barrier layers depend strongly on the initial stages of film growth, which also affect interfaces such as those created in heterostructures such as La–Sr–Co–O (LSCO)/Ti–Al/poly-Si. We demonstrated that the LSCO/a[amorphous]-Ti–Al heterostructure exhibits a desirable ohmic behavior, while an LSCO/polycrystalline Ti–Al (c[crystalline]-Ti–Al) layer, where the c-Ti–Al layer has a columnar microstructure across the entire film thickness, exhibits a nonohmic behavior [29, 30]. The a- or c-microstructures are obtained via control of the sputtering process used to grow these layers [29, 30]. The nonohmic behavior of the LSCO/c-Ti–Al heterostructure described above was attributed to the diffusion of Al, through the grain boundaries of the columnar microstructure of the Ti–Al layer, to the LSCO/Ti–Al interface, and the formation of an Al_2O_3 interfacial layer, as shown by XPS analysis [29, 30].

Our own earlier work did not provide information to help us understand the fundamental processes underlying the interaction of oxygen atmospheres with the Ti–Al layer. Since this understanding is critical for optimization of the heterostructures described above and for device performance, we studied the growth and oxidation process of Ti–Al diffusion barriers and reported these results in a recent paper [31]. This chapter is therefore focused on a review of our recent work, directed at understanding the fundamental processes involved in the integration of PZT capacitors with silicon platforms, and on developing the appropriate materials integration strategies for the fabrication of FeRAMs based on PZT capacitors.

2 An Experimental Method for the Synthesis and Characterization of Ferroelectric Capacitor Layers

The Ti–Al layers were deposited on poly-Si/highly boron-doped (1×10^{19} atm/cm^3) single-crystal [100]Si substrates (or on undoped Si wafers for the four-probe measurements) at room temperature, using DC magnetron sputtering or ion beam sputter deposition, the latter in an integrated ion beam sputter-deposition (IBSD)/in situ mass spectroscopy of recoil ions (MSRI)/X-ray photoelectron spectroscopy (XPS) (implemented with Al X-rays provided by Al–K_α = 1486.6 eV line) system described elsewhere [32]. The Ti–Al layers synthesized by magnetron or ion beam sputter-deposition exhibit either polycrystalline or amorphous microstructures, depending on the deposition conditions. The polycrystalline structure is obtained when using high magnetron power (400 W or larger), while the amorphous structure is induced by low magnetron power (less than 400 W), in the case of magnetron sputter-deposition. The IBSD method generally results in amorphous a-Ti–Al layers.

Subsequent to the growth of the Ti–Al layers, $La_{0.5}Sr_{0.5}CoO_3$ (LSCO) electrode layers (70 nm thick) were grown by RF sputtering (300 W for 30 min using an Ar : O_2 (5 : 1) gas mixture) and annealed at 450 °C in an oxygen-flowing tube furnace. Ferroelectric layers, $PbZr_{0.4}Ti_{0.6}O_3$ (PZT) or $PbNb_{0.04}/Zr_{0.28}Ti_{0.68}O_3$ (PNZT), were grown using either a conventional sol-gel technique with lead (IV) acetate, titanium isopropoxide, and zirconium-*n*-butoxide precursor solutions dissolved in stoichiometric proportions in hexane (excess lead (7%) was incorporated in the lead precursor to compensate for lead loss during crystallization) or a new solution involving lead acetate trihydrate, titanium tetra-iso-propoxide, zirconium tetra-*n*-butoxide, acetylacetone, propylene glycol, and low-viscosity alcohol as the diluent. In both cases, the PZT solutions were spin-deposited onto the LSCO/Ti–Al in ~120 nm thick layers (in general, three layers were deposited), with intermediate curing steps at 250 °C for 5 min to dissociate and volatilize the organics. Once the growth of the total PZT layers was completed, they were annealed either at 650 °C for the conventional sol-gel method or at the record 450 °C for the new sol-gel process for about 1 h. The capacitor structures were completed by deposition of LSCO electrode layers (~70 nm thick) on top of the PZT layers, using the same protocol as described above for the bottom LSCO layer. Pt films (~150 nm thick) were deposited on top of the LSCO layers and lift-off and wet etching were performed to define Pt/LSCO/PZT or PNZT/LSCO/Ti–Al heterostructured capacitors with 50 μm diameter top electrodes.

X-ray diffraction (XRD), transmission electron microscopy (TEM), and electron energy loss spectroscopy (EELS) were used to study the structure, composition, and chemical changes in the layers and interfaces of the PZT capacitors. In addition, the integrated IBSD/MSRI/XPS system was used to study film growth and surface and Ti–Al/ferroelectric interface processes in a controlled environment. MSRI provides atomic-scale information of species on the Ti–Al surface of the growing films, via secondary ions ejected by impact of the probing ions (10 keV Ar^+) [32, 33, 34], while XPS yields information on the chemical environment at surfaces and interfaces. The ferroelectric properties were characterized using a RT6000 system.

3 The Magnetron-Based Synthesis and Characterization of Ti–Al Layers and LSCO/Ti–Al Heterostructures

Figure 1 shows the XRD spectra for Ti_3Al layers grown using low-power (a) and high-power magnetron sputter-deposition on poly-Si/Si. Figure 1a shows that the Ti_3Al layer deposited at low power is X-ray amorphous; all the peaks seen in the spectrum were unambiguously assigned to poly-Si and no XRD peaks could be assigned to the Ti_3Al composition. On the contrary, the XRD spectrum of Fig. 1b, corresponding to the Ti_3Al layer grown with high-power magnetron sputtering, shows a peak at ~ 39 °, which corresponds to (002)Ti_3Al. This demonstrates that the film deposited at high power is crystalline. In the rest of this chapter, the amorphous Ti_3Al film is referred to as a-Ti–Al and the crystalline Ti_3Al film is referred to as c-Ti–Al. Ferroelectric capacitors (LSCO/PNZT/LSCO) were fabricated on both a-Ti–Al and c-Ti–Al surfaces, with the intention of using them as a diffusion barrier layer. The crystallinity of the Ti–Al layer appeared to have an impact on the reactivity of the film. The heterostructure with c-Ti–Al had a milky color on the surface, as opposed to a shiny black color surface for the heterostructure with the a-Ti–Al film.

Transmission electron microscopy (TEM) studies were performed to investigate the difference between the two interfaces described above. Fig-

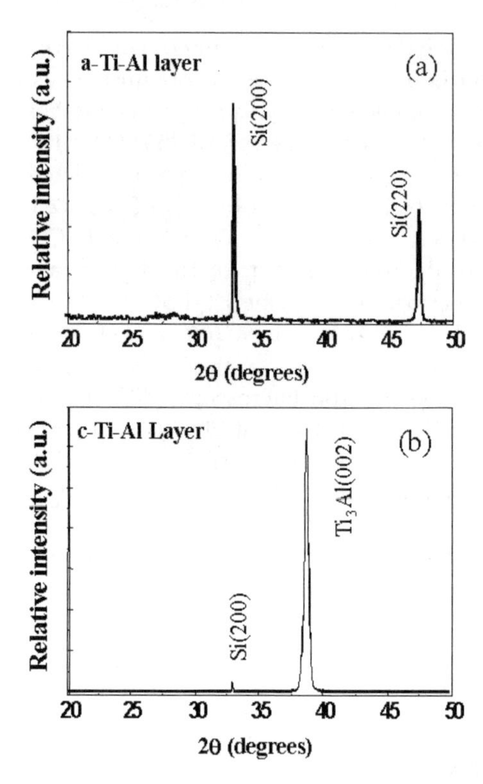

Fig. 1. XRD spectra for Ti_3Al grown with low-power (**a**) and high-power (**b**) magnetron sputter-deposition on poly-Si/Si

ure 2a is a cross-sectional TEM image of a PNZT/LSCO/a-Ti–Al/poly-Si/Si heterostructure. Notice the amorphous or nanocrystalline structure of the a-Ti–Al layer. In this figure, there is no reaction at the LSCO/a-Ti–Al interface. Further electron diffraction on the a-Ti–Al layer revealed that it is indeed amorphous. By contrast, the Ti–Al layer in Fig. 2b is clearly polycrystalline. The LSCO/Ti–Al interface (indicated by arrows) in Fig. 2a is abrupt, while there is a reaction zone at this interface for the sample cross-section shown in Fig. 2b. The difference between the microstructures of the two Ti–Al layers appears to be the only property correlated to the observed difference in oxidation/reaction properties. This observation is consistent with the electrical measurements and also the milky color of the LSCO film deposited on the c-Ti–Al surface, as opposed to the shiny black color of the LSCO film on the a-Ti–Al surface after deposition and annealing.

In order to determine the reaction layer at the LSCO/c-Ti–Al interface, we performed Rutherford backscattering spectrometry (RBS) and XPS. The signal for the Al from the Ti–Al film was too close to the Si signal, especially due to the thickness of the Ti–Al film, and could not be analyzed. The Ti signal suffered no interference from any other components in the heterostructure and was therefore analyzed to understand the nature of the reactions, if any. Figure 3a,b shows the RBS Ti spectra for LSCO/a-Ti–Al and LSCO/c-Ti–Al

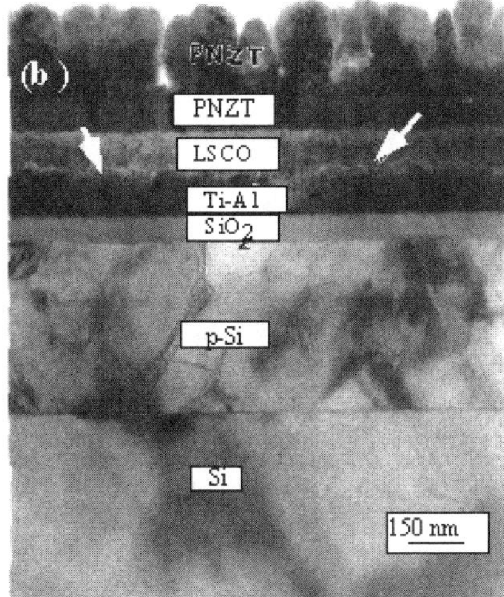

Fig. 2. A cross-sectional TEM image of an a-Ti–Al/LSCO interface (**a**) showing no reaction at the interface, and a c-Ti–Al/LSCO interface (**b**) showing significant reaction, as indicated by the *arrows*

interfaces, respectively, in both cases before and after LSCO deposition and subsequent annealing in oxygen at 650 °C for 1 h, to crystallize the LSCO layer. As clearly seen in Fig. 3a, there is almost no difference in the Ti peaks before and after LSCO deposition and subsequent annealing, which indicates that there is no macroscopic reaction between the a-Ti–Al layer and LSCO and no oxidation of the Ti–Al layer. The Ti peak for c-Ti–Al, as seen in Fig. 3b, broadens and the intensity drops after annealing, which indicates that Ti either oxidizes or reacts with the poly-Si or LSCO layers adjacent to it. The broadening toward the left indicates reaction with the poly-Si layer below it. This is consistent with the cross-section TEM image (Fig. 2b).

Ti–Al layers grown by the magnetron sputter-deposition method were also analyzed using XPS, which provides both a measure of the surface composition and the oxidation state of the species. The XPS measurements were made using Al–K_α X-rays in the integrated IBSD/MSRI/XPS system described in Sect. 2 [33]. We performed MSRI and XPS on both the amorphous and crystalline Ti–Al films as a function of temperature up to 650 °C. These results showed that Al diffuses to the surface to form Al_2O_3 in the case of c-Ti–Al structure, but not for the a-Ti–Al structure. This is again consistent with our electrical (presented later) and structural characterization of the LSCO/Ti–Al interface in the two cases.

In the experiments described above, the Ti–Al layers were grown in a magnetron sputtering system and then transferred to the XPS analysis system. In order to eliminate contamination effects due to the exposure of the Ti–Al sur-

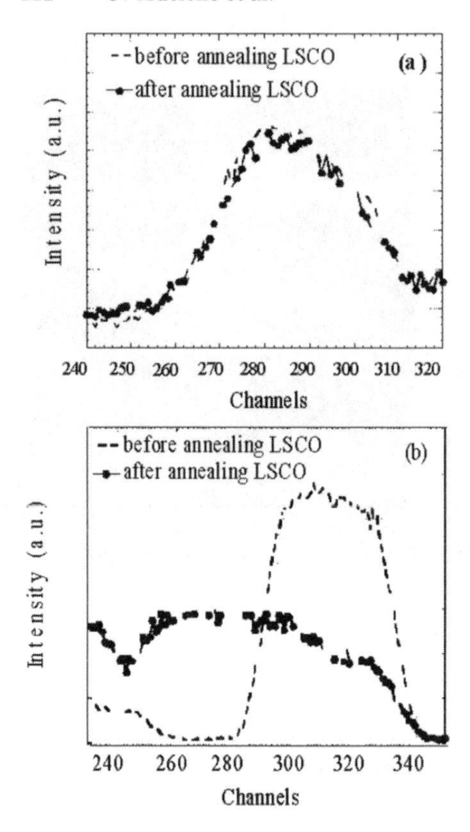

Fig. 3. Rutherford backscattering spectra around the Ti peak before and after LSCO deposition and annealing for an LSCO/a-Ti–Al (**a**) and an LSCO/c-Ti–Al (**b**) heterostructure

face to air, we deposited both the a-Ti–Al and LSCO films in situ by ion beam sputtering without exposing the surface to air. XPS spectra were obtained from the a-Ti–Al surface to confirm that Ti was in the metallic state prior to the LSCO deposition. The peak positions were calibrated with respect to the C 1s peak. Figure 4a is the XPS spectrum around the Ti [$2p_{1/2}$-459 eV, $2p_{3/2}$-453.8 eV] metallic peaks from the Ti–Al layer after room temperature deposition of a very thin ($\sim$ 2.6 nm) LSCO layer. The thickness of the LSCO layer is such that it completely covers the Ti–Al film, as determined using the monolayer-sensitive MSRI technique [33], but is still thin enough to permit the escape of photon-excited electrons to obtain the XPS spectrum from the Ti–Al film buried underneath the LSCO layer. The LSCO layer deposited at room temperature via ion beam sputtering is amorphous, as determined by X-ray diffraction, and does not appear to result in the oxidation of the Ti–Al layer, as demonstrated by the in situ XPS analysis shown in Fig. 4a. Figure 4b shows the XPS spectrum from the a-Ti–Al layer around the Ti [$2p_{1/2}$, $2p_{3/2}$] peaks after annealing the a-Ti–Al/LSCO heterostructure layer at 650 °C in 5×10^{-4} Torr O_2 for 5 min. The complementary TEM–XPS analysis of the a-Ti–Al/LSCO heterostructures discussed shows that the a-Ti–Al layer main-

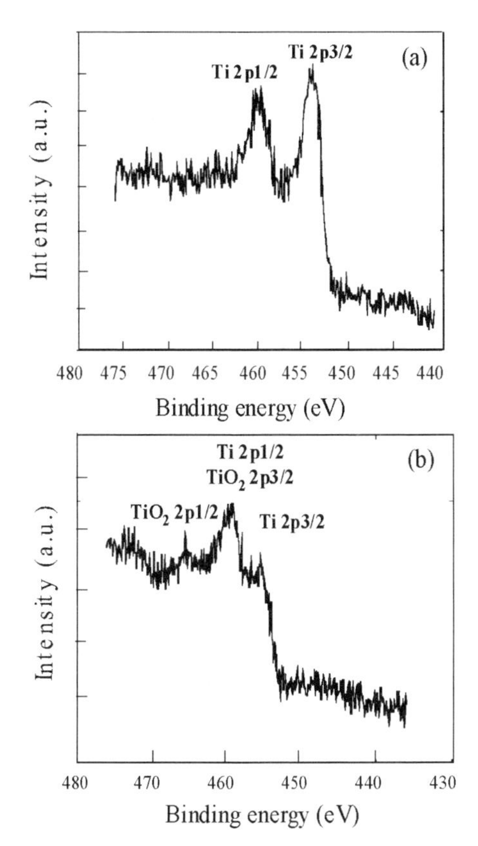

Fig. 4. XPS spectra of an LSCO/Ti–Al heterostructure about the Ti peaks after room-temperature LSCO deposition (**a**) and after annealing the LSCO/Ti–Al heterostructure (**b**)

tains a metallic character. The XPS spectrum reveals the presence of both TiO_2 [$2p_{3/2}$-458.5 eV] and Ti metallic [$2p_{1/2}$, $2p_{3/2}$] peaks (Figs. 4a,b). It remains to be determined if a thicker ($\sim$ 100–200 nm) LSCO layer may inhibit even the partial oxidation of the Ti–Al layer.

4 Studies of Ti–Al and LSCO Film Growth and Oxidation Processes

The studies discussed in this section were designed to investigate the a-Ti–Al film growth and oxidation processes. This was achieved using the integrated IBSD/MSRI/XPS system described in Sect. 2 and elsewhere [32]. The experiments discussed here involved in situ film growth and characterization to avoid surface exposure to air, since Ti is known to readily absorb oxygen. The work discussed in this section was focused on the investigation of the deposition environment and its effect on the microstructure of the Ti–Al layers, using ex situ analytical methods, and on studies of the oxidation process in Ti–Al films, using complementary in situ MSRI and XPS analysis during film growth.

4.1 Microstructure–Deposition Environment Relationships for Ti–Al Layers

Figure 5 displays XRD patterns of Ti–Al films deposited with different ambient pressures of oxygen, accurately controlled in the IBSD system via a metering oxygen valve. The intensity of the (002)-Ti_3Al peak at $\sim 39.5\,^{\circ}$ increases steadily with decreasing oxygen ambient pressure to a maximum intensity achieved for the lowest deposition pressure ($\sim 10^{-9}$ Torr), for which the smallest oxygen content is expected in the Ti–Al films. The (002) peak completely disappears for ambient pressures below $\sim 10^{-7}$ Torr, which is consistent with the data for magnetron sputter-deposited films [29].

Energy electron loss spectroscopy (EELS) reveals an important difference between the crystalline and amorphous Ti–Al layers; that is, the presence of dissolved oxygen in the latter. Figure 6 shows Ti and O edges from the EELS spectra of these two films. The Ti edge at about 460 eV has been normalized to make a comparison for the O edges. The O edge at 540 eV in Fig. 6 clearly reveals that the a-Ti–Al film has incorporated a higher concentration of oxygen (from the ambient) during deposition, resulting in the stabilization of the amorphous structure over the DO_{19} hexagonal structure of Ti_3Al. The systematic structural changes of the Ti–Al films are reflected in their electrical transport properties (see Sect. 5).

4.2 Studies of Ti–Al/LSCO Heterostructured Layer Growth and Oxidation Processes via Complementary in situ Analytical Techniques

An a-Ti–Al (100 nm) layer was deposited in situ via the IBSD method to prevent any oxidation of the Ti component of the a-Ti–Al layer prior to the free surface oxidation studies. The deposition of a-Ti–Al film on poly-Si/Si substrate was achieved at room temperature by IBSD (RF power = 43 W, beam energy = 600 eV, deposition rate = 0.03 nm/s). XPS analysis of the initial a-Ti–Al surface was performed to confirm that Ti was indeed in the metallic state. The peak positions were calibrated with respect to the C1s peak. MSRI spectra were recorded during the growth of the a-Ti–Al film. The work discussed here, involving the in situ MSRI and XPS characterization of Ti–Al layers [31], confirmed that in the case of a c-Ti–Al layer, Al diffuses to the surface upon heating in an oxygen atmosphere, resulting in the formation of an Al_2O_3 layer. This is consistent with structural and electrical characterization of the c-Ti–Al/LSCO interface discussed in Sects. 3 and 5, respectively, and elsewhere [11, 12]. The investigation by X-ray and cross-sectional TEM shows that Ti–Al films have amorphous microstructure

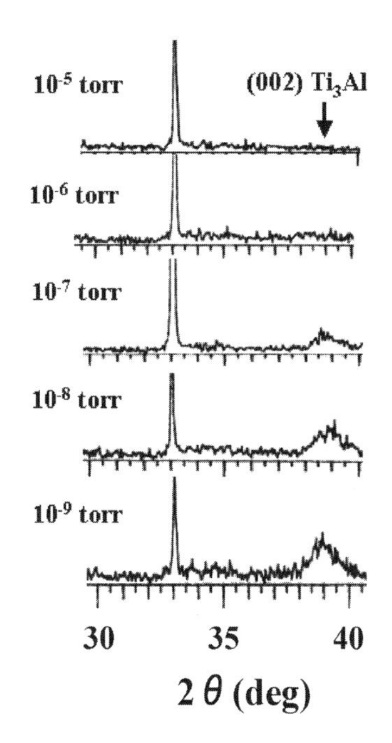

Fig. 5. XRD spectra for Ti–Al films deposited on poly-Si/Si wafers using the IBSD method under different background pressure; the Ti–Al films become increasingly crystalline as the deposition pressure is decreased below $\sim 10^{-6}$ Torr

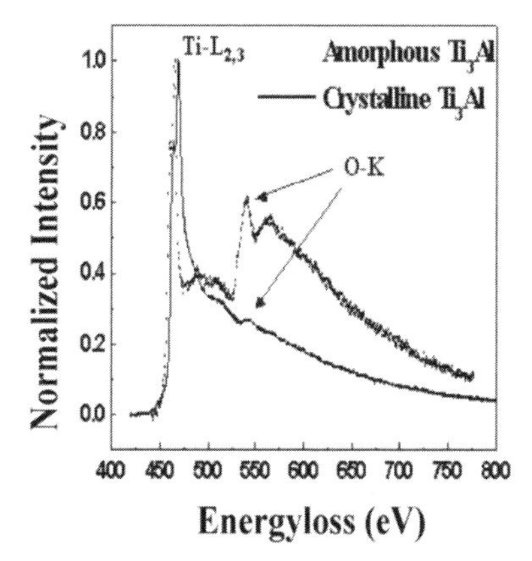

Fig. 6. EELS spectra for Ti–Al films with different crystallinity on poly-Si/Si, showing the role of dissolved oxygen in stabilizing the amorphous structure

and can be controlled by deposition parameters. The electrical characterization performed on an a-Ti–Al/poly-Si interface shows an ohmic behavior with a resistance of $\sim 45\ \Omega$, and the resistivity ($300\ \mu\Omega$ cm) value of the film matches well with the bulk samples (see Sect. 5).

Figure 7 shows the variations in MSRI signals for different species as a function of film thickness during the deposition of an a-Ti–Al thin film on poly-Si/Si substrate. A clear increase in the MSRI Ti^+ and Al^+ peak intensities is observed as a function of the a-Ti–Al film growth on the poly-Si/Si substrate. As the surface of the poly-Si/Si substrate is being covered

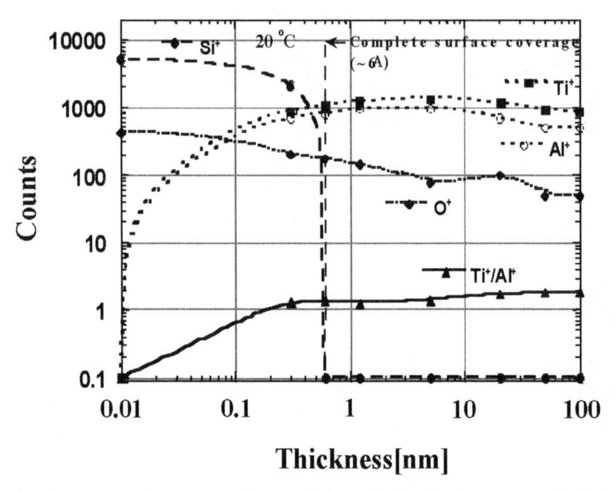

Fig. 7. Variations of MSRI signals for different species as a function of film thickness during the deposition of an a-Ti–Al thin film on a poly-Si surface

by Ti and Al species, a clear decrease in the Si^+ signal (from the poly-Si surface) is observed with the atomic-scale resolution characteristic of the MSRI method. As the a-Ti–Al layer thickness reaches 0.6 nm (measured with a quartz crystal resonator balance side-by-side with the substrate), the MSRI Si^+ signal disappears, which indicates a complete surface coverage of the poly-Si/Si substrate. Further growth of the a-Ti–Al layer results in the stabilization of the MSRI Ti^+/Al^+ signals corresponding to the formation of a continuous film. Following the in situ growth of the a-Ti–Al layer, oxygen was admitted in the system to study the surface oxidation process.

Figure 8 shows the variation of MSRI signals for individual atomic species evolving from the a-Ti–Al surface and the Ti^+/Al^+ peak ratio, as a function of the oxidation temperature. As seen in Fig. 8, there is almost no variation in the intensities of the Ti^+, Al^+, O^+, and Ti^+/Al^+ ratio signals up to 600 °C. However, there is a substantial increase of the MSRI Ti^+ and Ti^+/Al^+ ratio signals as the substrate temperature changes from 600 °C to 700 °C. The large increase in the Ti^+/Al^+ ratio and Ti^+ signals is due to the fact that metallic Ti species segregate to the surface of the a-Ti–Al layer and are oxidized to TiO_2. The segregated Ti layer is only one to two monolayers thick, as indicated by the sharp nature of the Ti peak. The increase in the secondary ion signal (e.g., Ti^+) upon surface oxidation is due to a lower rate of electron transfer from the oxidized surface to the departing ions (a neutralization process) – hence the larger survival probability of the latter (a well-documented phenomenon in surface analysis based on secondary ions detection [13]). The oxidation process on the surface of the a-Ti–Al layer was confirmed by in situ XPS analysis, which also provided insights into the chemical attachment of oxygen to the individual atomic species.

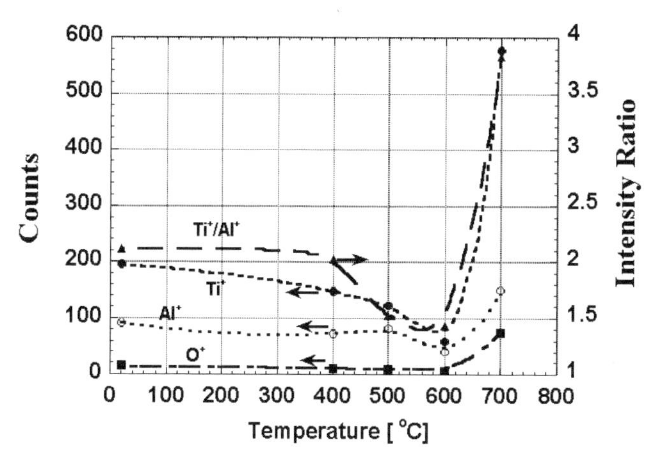

Fig. 8. Variation of MSRI signals for different species, and the Ti^+/Al^+ intensity ratio as a function of the oxidation temperature for an a-Ti–Al barrier layer on poly-Si

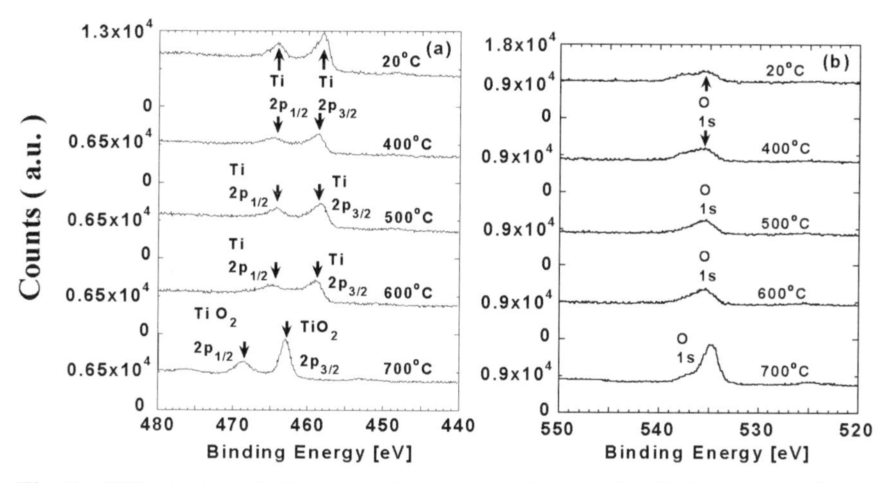

Fig. 9. XPS spectra of a-Ti–Al surface around the metallic Ti [$2p_{1/2}$, $2p_{3/2}$] and TiO_2[$2p_{1/2}$, $2p_{3/2}$] peaks (**a**); and the O[1 s] peak (**b**) as a function of the oxidation temperature for an a-Ti–Al barrier layer on poly-Si

Figure 9 shows XPS spectra of the a-Ti–Al surface around the Ti [$2p_{1/2}$, $2p_{3/2}$] and TiO_2 [$2p_{1/2}$, $2p_{3/2}$] peaks before (spectrum at 20 °C) and after free surface oxidation [$P(O_2) = 5 \times 10^{-4}$ Torr] at 400, 500, 600, and 700 °C. Each XPS spectrum shown here represents one scan in a given energy window with a pass energy of 25 eV, a scan speed of 500 ms/step, and a step width of 0.1 eV. The XPS peaks of Fig. 9a correspond to metallic Ti for the surface of an a-Ti–Al layer exposed to 10^{-3} Torr of oxygen in the temperature range of 400–600 °C, while the peaks observed after increasing the film temperature from 600 °C to 700 °C are clearly correlated with the presence of TiO_2 formation on the a-Ti–Al surface. In addition, the evolution of the XPS oxygen peak (Fig. 9b) shows a one-to-one correlation with the evolution of the Ti and TiO_2 peaks in Fig. 9a. The combined MSRI/XPS data indicate that the a-Ti–Al layer is stable up to 600 °C in oxygen, and that upon transition from 600 to 700 °C there is a segregation of Ti to the surface of the alloy, concurrently with a chemical attachment of oxygen to the segregated Ti species to form TiO_2. This is consistent with the MSRI data in Fig. 8 (increase in Ti^+ and Ti^+/Al^+) that indicate oxidation of Ti.

In conclusion, in situ sputter-deposition and MSRI analysis of growing a-Ti–Al layers on poly-Si/Si substrates revealed that the substrate is completely covered with a pinhole-free layer of only 0.6 nm of film thickness. Concurrent MSRI/XPS analysis of the a-Ti–Al surface revealed that this layer is resistant to oxidation up to at least 600 °C. Upon a film temperature increase from 600 to 700 °C in an oxygen environment, oxygen is chemically attached to a segregated Ti layer on the surface of the Ti–Al alloy.

5 Electrical Characterization of Ferroelectric Capacitors with a Ti–Al Diffusion Barrier/Electrode Layer

The systematic structural changes of the Ti–Al films are reflected in their electrical transport properties. Four-point I–V data from these samples (Fig. 10) show that films grown at oxygen ambient pressures below 10^{-6} Torr are metallic, with the resistivity progressively increasing from 30 μΩ cm at 10^{-9} Torr to 185 μΩ cm at 10^{-6} Torr; for films grown at 10^{-5} Torr, the resistivity increases dramatically to about 7 μΩ cm.

Ferroelectric LSCO/PZT [35]/LSCO capacitors were produced on the various Ti–Al layers described above, using two different approaches. In the case of the crystalline Ti–Al barrier layer, cross-section TEM [29] and in situ XPS [29] studies revealed a reaction between the Ti–Al and the LSCO layer (the bottom electrode of the PZT capacitor), especially when processing the capacitor heterostructure at 650 °C. These capacitors yielded poor ferroelectric properties. On the other hand, capacitors fabricated on a-Ti–Al barriers exhibited excellent ferroelectric properties (i.e., relatively high polarization, a low coercive field, negligible polarization fatigue and imprint, and long polarization retention).

Figure 11 displays typical polarization hysteresis loops, measured at 5 V, for LSCO/PZT/LSCO/a-Ti–Al ferroelectric capacitors produced by two sol-gel processes [14, 15]. The inset at the top left of Fig. 11 is the remanent polarization ($\Delta P = P^* - P$) as a function of the applied voltage, while the inset at the bottom right is the pulse width dependence of ΔP at 5 V for the two types of capacitor. The ferroelectric films typically exhibit resistivities in the range of 1–3×10^{10} Ω cm at 5 V, with ΔP values $\sim$ 18–20 μC/cm^2 at 5 V and a 1 μs pulse width. An interesting (and technologically important)

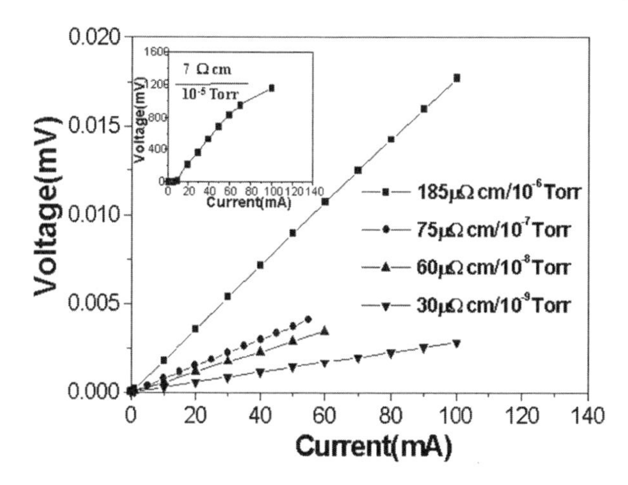

Fig. 10. Four-probe, current–voltage (I–V) data from the Ti–Al barrier layers deposited at various oxygen pressures in the IBSD process. Note that the films are metallic for deposition pressures below 10^{-6} Torr

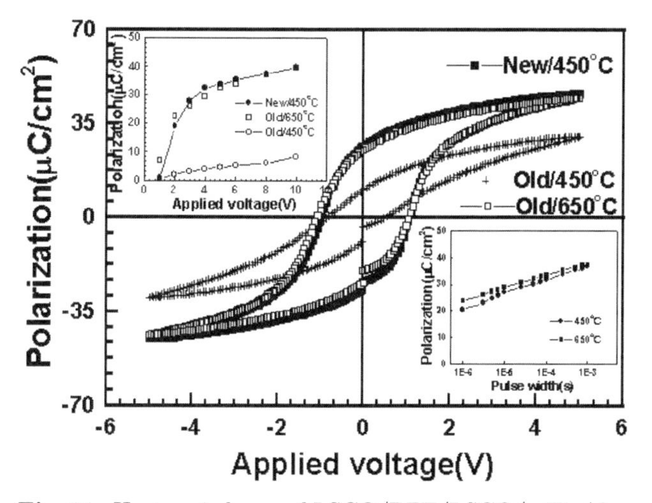

Fig. 11. Hysteresis loops of LSCO/PZT/LSCO/a-Ti–Al capacitors produced with different PZT processing temperatures, measured at 5 V. The *insets* show the voltage dependence of pulsed remanent polarization as well as its pulse width dependence

point is that the conventional sol-gel precursor does not yield high-quality ferroelectric capacitors when processed at 450 °C. By contrast, the new sol-gel precursor yields capacitors processed at 450 °C with essentially the same properties as those obtained with the conventional process at 650 °C. Finally, the superior resistance of these capacitors to polarization fatigue is demonstrated in Fig. 12, which also shows the fatigue behavior for PZT capacitors

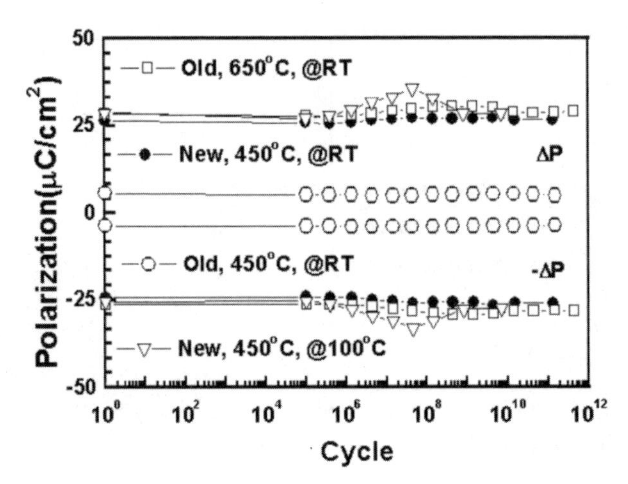

Fig. 12. Fatigue characteristics of test capacitors annealed at 450 °C as a function of polarization switching cycles; also shown is the data for PZT capacitors produced using the conventional sol-gel PZT method with annealing of the PZT layer at 450 °C or 650 °C

produced with the old PZT sol-gel precursor and annealed at 450 °C and 650 °C.

The studies reported here, in conjunction with prior electrical characterization of the LSCO/PNZT/LSCO/a-Ti–Al capacitors integrated on poly-Si/Si substrates [10, 11], indicate that the a-Ti–Al layer developed by our group can be used for the fabrication of nonvolatile FeRAMs based on PZT ferroelectric layers. It is also possible that the a-Ti–Al layer could be used for integration of $Ba_xSr_{1-x}TiO_3$ (BST) and $SrBi_2Ta_2O_9$ (SBT) based capacitors with Si substrates, and this issue will be explored in forthcoming work.

6 Summary

In summary, we have presented a systematic study of the dependence of the crystallinity of Ti–Al barrier films on the sputtering power density, base pressure, and dissolved oxygen of the films. We have shown that a high sputtering power in magnetron sputtering methods, in conjunction with low gas pressure during deposition, leads to the formation of crystalline Ti_3Al layers. This can be understood as a consequence of the interplay between the high kinetic energy of the inert gas species used for sputtering at high power and the low background pressure that enables energetic inert ions to arrive at the growing film and induces mobility of the depositing Ti and Al species, leading to crystallization at low temperature. On the other hand, the growth at low power and high pressure leads to low mobility of depositing species and dissolved oxygen in the Ti–Al film, stabilizing the amorphous phase.

The rather high solubility of oxygen in the Ti–Al system is well known [36]. The fact that this dissolved oxygen can stabilize an amorphous structure is very valuable and is in general agreement with the work of *Gasser* et al. [37]. We find that up to a film growth pressure of about 10^{-6} Torr, the amorphous Ti–Al is highly metallic, while below a pressure of about 10^{-7} Torr it begins to crystallize. This provides the window of processing conditions that will lead to the desirable amorphous microstructure in the barrier layer. Finally, PZT-based test capacitors fabricated on the a-Ti–Al barriers at 450 °C exhibit ferroelectric properties compatible with FeRAMs. We believe that this novel, inexpensive approach is now ready for insertion into an advanced IC process development program that will test the various ferroelectric process integration issues, such as wafer-level scaling, the development of etching processes, ferroelectric performance, and reliability.

Acknowledgements

This work was supported by the US Department of Energy, BES-Materials Sciences, under Contract W-13-109-ENG-38; by the NSF/ONR, under Contract N00014-89-J1178; and by the University of Maryland NSF-MRSEC, under Grant #DMR 00-80008.

References

1. O. Auciello, J. F. Scott, R. Ramesh: Phys. Today (July 1998) p. 22
2. C. A. Paz de Araujo, O. Auciello, R. Ramesh (Eds.): *Science and Technology of Integrated Ferroelectrics: Selected Papers of the Past Eleven Years of the International Symposium on Integrated Ferroelectrics Proceedings* (Gordon & Breach, Switzerland), "*Ferroelectricity and Related Phenomena*", **11** (2000)
3. S. K. Dey: *Ferroelectric Thin Films: Synthesis and Basic Properties*, C. A. Paz de Araujo, J. F. Scott, G. W. Taylor (Eds.) (Gordon & Breach, Switzerland 1997) p. 329
4. S. B. Krupanidhi, H. Hu, G. R. Fox: *Ferroelectric Thin Films: Synthesis and Basic Properties*, C. A. Paz de Araujo, J. F. Scott, G. W. Taylor (Eds.) (Gordon & Breach, Switzerland 1997) p. 427
5. P. McIntyre, S. R. Summerfelt: Appl. Phys. Lett. **70**, 711 (1997)
6. K. L. Saenger, A. Grill, D. E. Kotecki: J. Appl. Phys. **83**, 802 (1998)
7. D.-F. Lii: J. Mater. Sci. **33**, 2137 (1998)
8. D. McIntyre, J. E. Greene, G. Hakansson, J. E. Sundgren, W. D. Munz: J. Appl. Phys. **67**, 1542 (1990)
9. A. M. Dhote, S. Madhukar, D. Young, T. Venkatesan, R. Ramesh, C. M. Cotell, J. Benedetto: J. Mater. Res. **12**, 1589 (1997)
10. A. M. Dhote, S. Madhukar, W. Wei, T. Venkatesan, R. Ramesh, C. M. Cottel: Appl. Phys. Lett. **68**, 1350 (1996)
11. A. M. Dhote, R. Ramesh: US Patent No. 5798903
12. J. M. Benedetto, R. A. Moore, F. B. McLean: J. Appl. Phys. **75**, 460 (1994)
13. J. J. Lee, C. L. Thio, S. B. Desu: Phys. Status Solidi (a) **151**, 171 (1995)

14. J. O. Olowolafe, R. E. Jones, Jr., A. C. Campbell, R. I. Hegde, C. J. Mogab, R. B. Gregory: J. Appl. Phys. **73**, 1764 (1993)
15. H. N. Al-Shareef, A. I. Kingon: *Ferroelectric Thin Films: Synthesis and Basic Properties*, C. A. Paz de Araujo, J. F. Scott, G. W. Taylor (Eds.) (Gordon & Breach, Switzerland 1997) p. 193
16. S. Aggarwal, B. Yang, R. Ramesh: *Thin Film Ferroelectric Materials and Devices*, R. Ramesh (Ed.) (Kluwer, Dordrecht 1998) p. 221
17. C. A. Paz de Araujo, J. D. Cuchlaro, L. D. McMillan, M. C. Scott, J. F. Scott: Nature **374**, 627 (1995)
18. M. Wittmer, J. Noser, H. Melchoir: J. Appl. Phys. **52**, 6659 (1981)
19. H. G. Tompkins: J. Appl. Phys. **71**, 980 (1992)
20. P. McIntyre, S. R. Summerfelt: J. Appl. Phys. **82**, 4577 (1997)
21. K. L. Saenger, A. Grill, D. E. Kotecki: J. Appl. Phys. **83**, 802 (1998)
22. K. L. Saenger, A. Grill, C. Cabral, Jr.: J. Mater. Res. **13**, 462 (1998)
23. S. Madhukar, S. Aggarwal, A. M. Dhote, R. Ramesh, S. B. Samavedam, S. Choopun, R. P. Sharma: J. Mater. Res. **14**, 940 (1999)
24. H. D. Bhatt, S. B. Desu, D. P. Vijay, Y. S. Hwang, X. Zhang, M. Nagata, A. Grill: Appl. Phys. Lett. **71**, 719 (1997)
25. S. Veprek, S. Reiprich, S. H. Li: Appl. Phys. Lett. **66**, 2640 (1995)
26. X. Sun, J. S. Reid, E. Kolawa, M.-A. Nicolet: J. Appl. Phys. **81**, 664 (1997)
27. A. L. Greer: *Intermetallic Compounds*, vol. 1, J. H. Westbrook, R. L. Fleischer (Eds.) (Wiley, New York 1995) p. 371
28. S. M. Gasser, E. Kolawa, M.-A. Nicolet: J. Appl. Phys. **86**, 1974 (1999)
29. S. Aggarwal, B. Nagaraj, H. Li, R. P. Sharma, L. Salamanca-Riba, R. Ramesh, A. M. Dhote, A. R. Krauss, O. Auciello: Acta Mater. **48**, 3387 (2000)
30. S. Aggarwal, A. M. Dhote, H. Li, S. Ankem, R. Ramesh: Appl. Phys. Lett. **74**, 230 (1999)
31. A. M. Dhote, O. Auciello, D. M. Gruen, R. Ramesh: Appl. Phys. Lett. **79**, 800 (2001)
32. O. Auciello, A. R. Krauss, J. Im, J. A. Schultz: Ann. Rev. Mater. Sci. **28**, 375 (1998)
33. A. R. Krauss, O. Auciello, A. M. Dhote, J. Im, E. A. Irene, Y. Gao, A. H. Mueller, S. Aggarwal, R. Ramesh: Integr. Ferroelectr. **32**, 121 (2001)
34. S. Aggarwal, A. M. Dhote, H. Li, S. Ankem, R. Ramesh: Appl. Phys. Lett. **74**, 230 (1999)
35. N. Soyama, K. Maki, S. Mori, K. Ogi: Jpn. J. Appl. Phys. **39**, 5434 (2000)
36. M. X. Zhang, K. C. Hsieh, J. DeKock, Y. A. Chang: Scr. Metal. Mater. **27**, 1361 (1992)
37. S. M. Gasser, E. Kolawa, M.-A. Nicolet: J. Appl. Phys. **86**, 1974 (1999)

Characterization by Scanning Nonlinear Dielectric Microscopy

Yasuo Cho

Research Institute of Electrical Communication, Tohoku University,
2-1-1 Katahira, Aoba-ku, Sendai 980-8577, Japan
cho@riec.tohoku.ac.jp

Abstract. A sub-nanometer resolution scanning nonlinear dielectric microscopy (SNDM) was developed for the observation of ferroelectric polarization. We also demonstrate that the resolution of the SNDM is higher than that of conventional piezo-response imaging. Next, we report a new SNDM technique detecting the higher nonlinear dielectric constants ε_{3333} and ε_{33333}. It is expected that higher-order nonlinear dielectric imaging will provide higher lateral and depth resolution. Finally, a new type of scanning nonlinear dielectric microscope probe, called the ε_{311}-type probe, and a system to measure the ferroelectric polarization component parallel to the surface are developed.

1 Introduction

Recently, ferroelectric materials, especially in thin-film form, have attracted the attention of many researchers. Their large dielectric constants make them suitable as dielectric layers of microcapacitors in microelectronics. They are also investigated for applications in nonvolatile memory, using the switchable dielectric polarization of ferroelectric materials. To characterize such ferroelectric materials, a high-resolution tool is required to observe the microscopic distribution of remanent (or spontaneous) polarization of ferroelectric materials.

With this background, we have proposed and developed a new purely electrical method for imaging the state of the polarizations in ferroelectric and piezoelectric materials and their crystal anisotropy. It involves the measurement of point-to-point variations of the nonlinear dielectric constant of a specimen and is termed "scanning nonlinear dielectric microscopy" (SNDM) [1, 2, 3, 4, 5, 6, 7]. This is the first successful purely electrical method for observing the ferroelectric polarization distribution without the influence of the shielding effect from free charges. To date, the resolution of this microscope has been improved down to the sub-nanometer order.

Here, we describe the theory for detecting polarization and the technique for the nonlinear dielectric response, and report the results of the imaging of ferroelectric domains in single crystals and thin films using SNDM. Especially in a measurement on the PZT thin film, it was confirmed that the resolution was of sub-nanometer order. We also describe the theoretical resolution of

H. Ishiwara, M. Okuyama, Y. Arimoto (Eds.): Ferroelectric Random Access Memories,
Topics Appl. Phys. **93**, 123–136 (2004)
© Springer-Verlag Berlin Heidelberg 2004

SNDM. Moreover, we demonstrate that the resolution of SNDM is higher than that of a conventional piezo-response imaging using the scanning force microscopy (SFM) technique [8, 9].

Next, we report a new SNDM technique. In the above conventional SNDM technique, we measure the lowest-order nonlinear dielectric constant ε_{333}, which is a third-rank tensor. To improve the performance and resolution of SNDM, we have modified the technique such that higher nonlinear dielectric constants ε_{3333} (fourth-rank tensor), and ε_{33333} (fifth-rank tensor) are detected. It is expected that higher-order nonlinear dielectric imaging will provide higher lateral and depth resolution. We have confirmed this improvement over conventional SNDM imaging experimentally, and used the technique to observe the growth of a surficial paraelectric layer on periodically poled $LiNbO_3$ [10, 11, 12].

In addition to this technique, a new type of scanning nonlinear dielectric microscope probe, called the ε_{311}-type probe, and a system to measure the ferroelectric polarization component parallel to the surface have been developed. This is achieved by measuring the ferroelectric material's nonlinear dielectric constant ε_{311} instead of ε_{333}, which is measured in conventional SNDM. Experimental results show that the probe can satisfactorily detect the direction of polarization parallel to the surface [13].

2 The Principles and Theory of SNDM

First, we will briefly describe the theory for detecting polarization. Precise descriptions of the principle of the microscope have been reported elsewhere [3, 4]. We also report the results of the imaging of ferroelectric domains in single crystals and in thin films using SNDM. Especially in the PZT thin-film measurement, we have succeeded in obtaining a domain image with a sub-nanometer resolution.

2.1 Nonlinear Dielectric Imaging with Sub-Nanometer Resolution

Figure 1 shows the system setup of SNDM using the LC lumped constant resonator probe [4]. In the figure, $C_\mathrm{S}(t)$ denotes the capacitance of the specimen under the center conductor (the tip) of the probe. $C_\mathrm{S}(t)$ is a function of time because of the nonlinear dielectric response under an applied alternating electric field, $E_{\mathrm{P}3}(= E_\mathrm{P} \cos \omega_\mathrm{P} t, f_\mathrm{P} = 5\text{–}100\,\mathrm{kHz})$. The ratio of the alternating variation of capacitance $\Delta C_\mathrm{S}(t)$ to the static value of capacitance C_S0 without time dependence is given as [3]

$$\frac{\Delta C_\mathrm{S}(t)}{C_\mathrm{S0}} = \frac{\varepsilon_{333}}{\varepsilon_{33}} E_\mathrm{P} \cos \omega_\mathrm{P} t + \frac{\varepsilon_{3333}}{4\varepsilon_{33}} E_\mathrm{P}^2 \cos 2\omega_\mathrm{P} t, \tag{1}$$

where ε_{33} is a linear dielectric constant and ε_{333} and ε_{3333} are nonlinear dielectric constants. The even rank tensor, including the linear dielectric constant

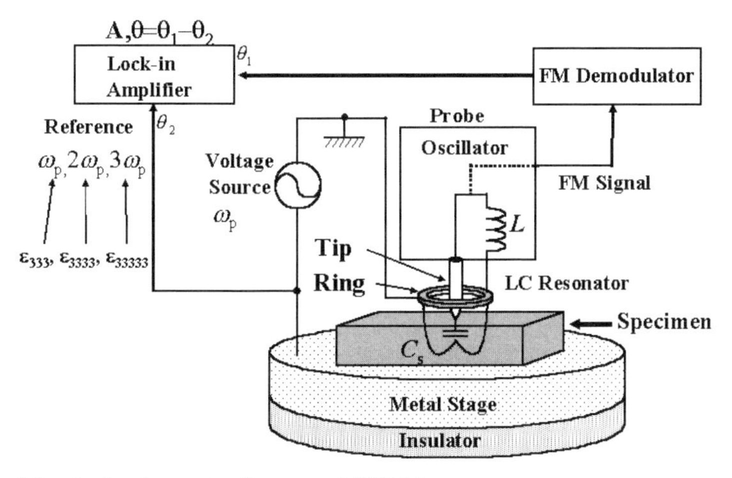

Fig. 1. A schematic diagram of SNDM

ε_{33}, does not change with 180° rotation of the polarization. On the other hand, the lowest order of the nonlinear dielectric constant ε_{333} is a third-rank tensor, similar to the piezoelectric constant, so that there is no ε_{333} in a material with a center of symmetry, and the sign of ε_{333} changes in accordance with the inversion of the spontaneous polarization.

This LC resonator is connected to the oscillator tuned to the resonance frequency of the resonator. The above-mentioned electrical parts (i.e., the needle, ring, inductance, and oscillator) are assembled into a small probe for SNDM. The oscillating frequency of the probe (or oscillator) (around 1.3 GHz) is modulated by the change of capacitance $\Delta C_{\mathrm{S}}(t)$ due to the nonlinear dielectric response under the applied electric field. As a result, the probe (oscillator) produces a frequency-modulated (FM) signal. By detecting this FM signal using the FM demodulator and lock-in amplifier, we obtain a voltage signal proportional to the capacitance variation. Each signal corresponding to ε_{333} and ε_{3333} was obtained by setting the reference signal of the lock-in amplifier at the frequency ω_{P} of the applied electric field and at the doubled frequency $2\omega_{\mathrm{P}}$, respectively. Thus we can detect the nonlinear dielectric constant just under the needle and we can obtain the fine resolution determined by the diameter of the pointed end of the tip and the linear dielectric constant of specimens. The capacitance variation caused by the nonlinear dielectric response is quite small ($\Delta C_{\mathrm{S}}(t)/C_{\mathrm{S0}}$ is in the range from 10^{-3} to 10^{-8}). Therefore, the sensitivity of the SNDM probe must be very high. The measured value of the sensitivity of the above-mentioned lumped constant probe is 10^{-22} F.

For this study, the needle of the lumped constant resonator probe was fabricated using electrolytic polishing of a tungsten wire or a metal-coated conductive cantilever. The radius of curvature of the chip was between 1 μm and 25 nm. To check the performance of the new SNDM, we first measured the

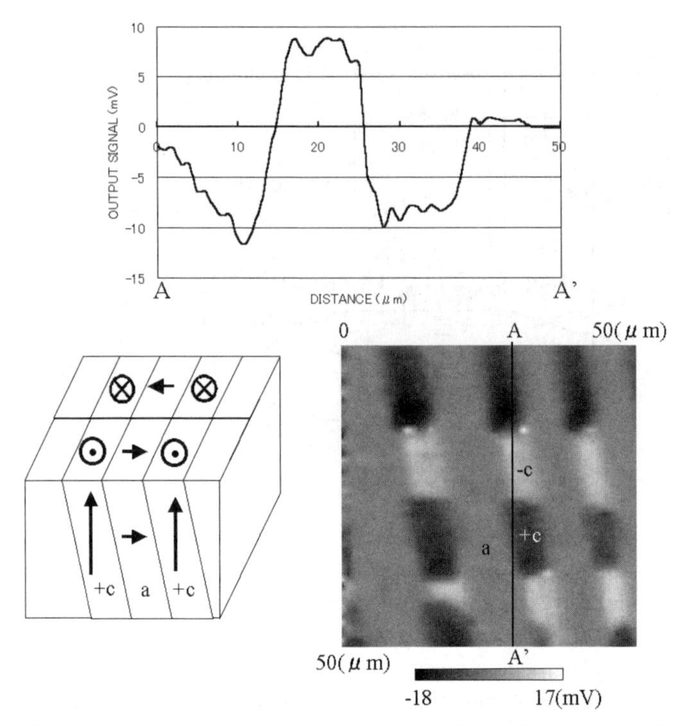

Fig. 2. A two-dimensional image of the 90° a–c domain in a $BaTiO_3$ single crystal and the cross-sectional (one-dimensional) image along the line A–A′

macroscopic domains in a multi-domain $BaTiO_3$ single crystal. Figure 2 shows the two-dimensional image of the so-called 90° a–c domain which is obtained by a coarse scanning over a large area. The sign of the nonlinear dielectric constant ε_{333} of the $+c$ domain is negative, whereas it is positive in the $-c$ domain. Moreover, the magnitude of $\varepsilon_{111} = \varepsilon_{222}$ is zero in the a domain, because $BaTiO_3$ belongs to the tetragonal system at room temperature. Thus, we can easily distinguish the type of the domains.

To demonstrate that this microscopy is also useful for the domain measurement of thin ferroelectric films, we measured a PZT thin film. Figure 3 shows the SNDM (a) and AFM (b) images taken from the same location of the PZT thin film, deposited on a $SrTiO_3$ (STO) substrate using metal organic chemical vapor deposition. From the figure, it is apparent that the film is polycrystalline (see Fig. 3b) and that each grain in the film is composed of several domains (see Fig. 3a). From X-ray diffraction analysis, this PZT film belongs to the tetragonal phase and the diffraction peaks, corresponding to both the c-axis and the a-axis, were observed. Moreover, in Fig. 3a, the observed signals were partially of zero amplitude, and partially positive. Thus, the images show that we succeeded in observing 90° a–c domain distributions in a single grain of the film.

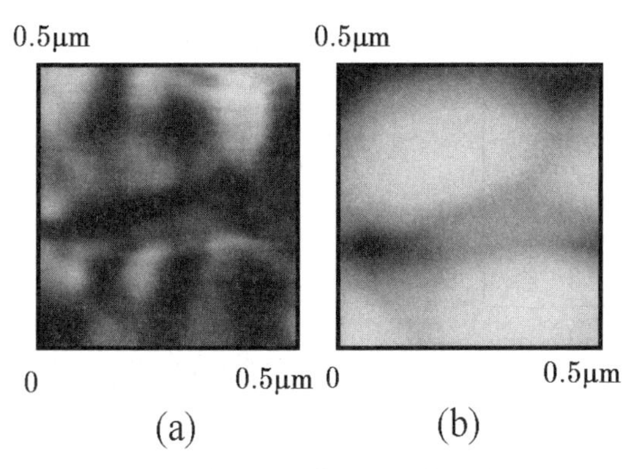

Fig. 3. Images of a PZT film on a $SrTiO_3$ substrate: (**a**) domain patterns by SNDM; (**b**) surface morphology by AFM

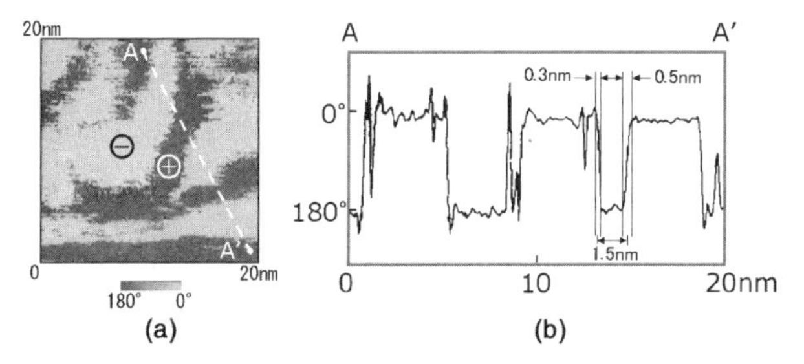

Fig. 4. The nanoscale ferroelectric domain on the PZT thin film. (**a**) The domain image of the nanoscale 180° *c*–*c* domain; (**b**) the cross-sectional (one-dimensional) image of the phase signal along the line A–A′

These images of the film were taken from a relatively large area. Therefore, we also tried to observe very small domains in the same PZT film on STO substrate. The results are shown in Fig. 4a and b. The bright and dark areas correspond to negative polarization and positive polarization, respectively. This shows that we can successfully observe a nanoscale 180° *c*–*c* domain structure. Figure 4b shows a cross-sectional image taken along line A–A′ in Fig. 4a. As shown in this figure, we measured the *c*–*c* domain with a width of 1.5 nm. Moreover, we found that the resolution of the microscope is less than 0.5 nm.

To clarify the reason why such a high resolution can be easily obtained, even if a relatively thick needle is used for the probe, we show the calculated results of the one-dimensional image of the 180° *c*–*c* domain boundary lying at $y = 0$ (we chose y direction as the scanning direction) [14, 15]. Figure 5 shows the calculated results, where Y_0 is the tip position normalized with respect to the tip radius a. The resolution of the SNDM image is heavily dependent on the dielectric constant of the specimen. For example, for the

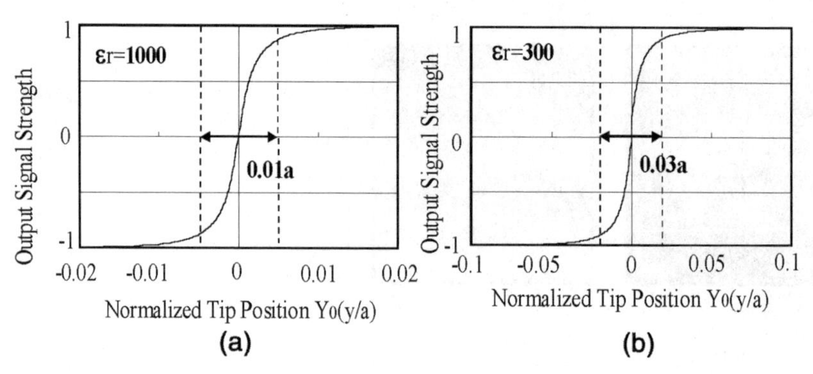

Fig. 5. Theoretical images of the 180° c–c domain boundary

case of $\varepsilon_{33}/\varepsilon_0 = 1000$ and $a = 10\,\mathrm{nm}$, an atomic-scale image will be able to be captured by SNDM.

2.2 A Comparison between SNDM Imaging and Piezo-Response Imaging

Another frequently reported high-resolution tool for observing ferroelectric domains is piezoelectric response imaging using SFM [8, 9]. From the viewpoint of resolution for ferroelectric domains, SNDM will surpass piezo-response imaging, because SNDM measures the nonlinear response of a dielectric material, which is proportional to the square of the electric field, whereas the piezoelectric response is linearly proportional to the electric field. The concentration of the distribution of the square of the electric field in the specimen underneath the tip is much higher than that of the linear electric field. Thus, SNDM can resolve smaller domains than those measured by the piezo-imaging technique. To prove this fact experimentally, we also performed simultaneous measurements at the same location on the above-mentioned PZT film sample by using AFM (topography)-, SNDM-, and piezo-imaging, as shown in Fig. 6. These images were captured under just the same condition, using the same metal-coated cantilever. From the images, we can prove that SNDM can resolve greater detail than conventional piezo-response imaging by using the SFM technique.

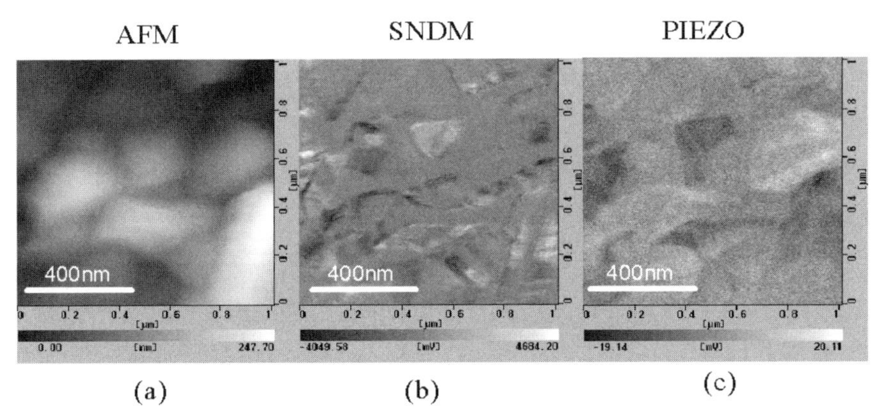

Fig. 6. Simultaneously captured images of a PZT film. (**a**) Topography by AFM; (**b**) domain patterns by SNDM; (**c**) domain patterns by SFM (piezo-imaging)

3 Higher-Order Nonlinear Dielectric Microscopy

A higher-order nonlinear dielectric microscopy technique with higher lateral and depth resolution than conventional nonlinear dielectric imaging is investigated. The technique is demonstrated to be very useful for observing surface layers of the order of unit cell thickness on ferroelectric materials.

3.1 The Theory of Higher-Order Nonlinear Dielectric Microscopy

Equation (2) is a polynomial expansion of the electric displacement D_3 as a function of electric field E_3

$$D_3 = P_{s3} + \varepsilon_{33}E_3 + \frac{1}{2}\varepsilon_{333}E_3^2 + \frac{1}{6}\varepsilon_{3333}E_3^3 + \frac{1}{24}\varepsilon_{33333}E_3^4 + \cdots . \quad (2)$$

Here, ε_{33}, ε_{333}, ε_{3333}, and ε_{33333} correspond to linear and nonlinear dielectric constants and are tensors of second, third, fourth, and fifth rank, respectively. Even-ranked tensors including the linear dielectric constant ε_{33} do not change with polarization inversion, whereas the sign of the odd-ranked tensors reverses. Therefore, information regarding polarization can be elucidated by measuring odd-ranked nonlinear dielectric constants such as ε_{333} and ε_{33333}.

Considering the effect up to E^4, the ratio of the alternating variation of capacitance ΔC_S underneath the tip to the static capacitance C_S0 is given by

$$\frac{\Delta C_\mathrm{S}(t)}{C_\mathrm{S0}} \approx \frac{\varepsilon_{333}}{\varepsilon_{33}}E_\mathrm{P}\cos\omega_\mathrm{P}t + \frac{1}{4}\frac{\varepsilon_{3333}}{\varepsilon_{33}}E_\mathrm{P}^2\cos 2\omega_\mathrm{P}t + \frac{1}{24}\frac{\varepsilon_{33333}}{\varepsilon_{33}}E_\mathrm{P}^3\cos 3\omega_\mathrm{P}t + \cdots . \quad (3)$$

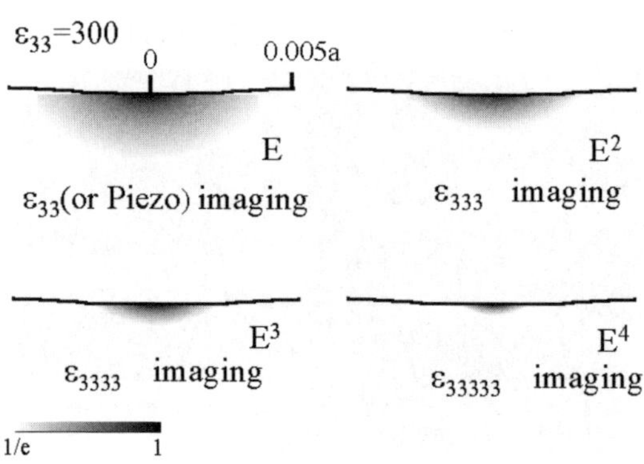

Fig. 7. The distribution of the E, E^2, E^3 and E^4, fields under the needle tip. a denotes the tip radius

This equation shows that the alternating capacitance of different frequencies corresponds to each order of the nonlinear dielectric constant. Signals corresponding to ε_{333}, ε_{3333}, and ε_{33333} were obtained by setting the reference signal of the lock-in amplifier in Fig. 1 to the frequencies ω_P, $2\omega_P$, and $3\omega_P$ of the applied electric field, respectively.

Next, we consider the resolution of SNDM. From (2), the resolution of SNDM is found to be a function of the electric field E. We note that the electric field under the tip is more highly concentrated with the increase of ε_{33} [16], and the distributions of the E^2, E^3, and E^4 fields underneath the tip become much more concentrated in accordance with their power than that of the E field, as shown in Fig. 7. From this figure, we find that higher-order nonlinear dielectric imaging has higher resolution than lower-order nonlinear dielectric imaging.

3.2 Experimental Details of Higher-Order Nonlinear Dielectric Microscopy

We have experimentally confirmed that ε_{33333} imaging has a higher lateral resolution than ε_{333} imaging using an electroconductive cantilever as a tip with a radius of 25 nm. Figure 8a,b shows ε_{333} and ε_{33333} images of the two-dimensional distribution of lead zirconate titanate (PZT) thin film. The two images can be correlated, and it is clear that the ε_{33333} image resolves greater detail than the ε_{333} image due to the higher lateral and depth resolution.

Next, we investigated the surface layer of periodically poled $LiNbO_3$ (PPLN) by ε_{333}, ε_{3333}, and ε_{33333} imaging. Figure 9a shows ε_{333}, ε_{3333}, and ε_{33333} signals of virgin unpolished PPLN. In this figure, only ε_{333} imaging detects the c–c domain boundary, while ε_{33333} imaging does not. The ε_{3333} signal shows weak peaks at domain boundaries. This is because ε_{3333} and ε_{33333} imaging are affected by the surface paraelectric layer. To prove the existence of a surface paraelectric layer, we polished and measured the PPLN,

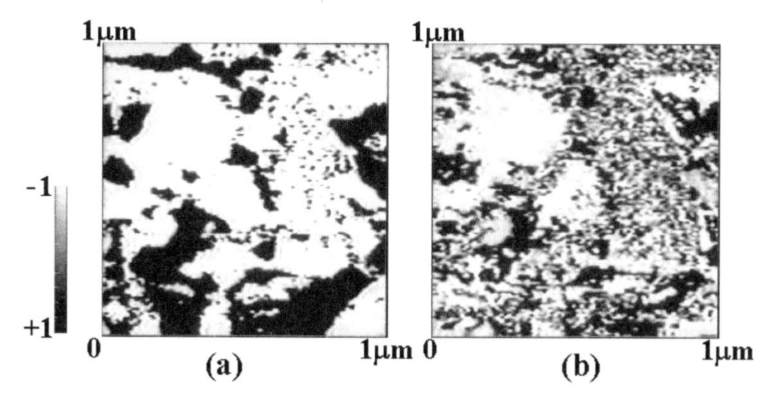

Fig. 8. (**a**) ε_{333} and (**b**) ε_{33333} images of PZT thin film

images of which are shown in Fig. 9b. In this figure, it is clear that ε_{33333} imaging can detect the *c*–*c* domain boundary after removal of the paraelectric layer. Moreover, ε_{3333} imaging can also detect periodic signals, in contrast to our expectations. The nonlinear dielectric signals of a positive area of PPLN are stronger than those of a negative area immediately after polishing, possibly because the negative area is more easily damaged than the positive area and has already been covered by a very thin surface paraelectric layer with weak nonlinearity, even immediately after polishing.

One hour after polishing, we conducted the ε_{333}, ε_{3333}, and ε_{33333} imaging again, and the results are shown in Fig. 9c. In this figure, the ε_{33333} signal disappears and the ε_{3333} signal becomes flat again, whereas ε_{333} imaging clearly detects the *c*–*c* domain boundary (Fig. 9a). This implies that the entire surface area of the PPLN is again covered by the surface paraelectric layer. From theoretical calculations, on the $LiNbO_3$ substrate, the ε_{33333} and ε_{3333} imaging processes have sensitivities down to 0.75 nm depth and 1.25 nm, respectively, whereas ε_{333} imaging has a sensitivity down to 2.75 nm depth when a tip of 25 nm radius is used. Thus, we conclude that the thickness of this surface paraelectric layer ranges between 0.75 nm and 2.75 nm. From these results, we confirm that the negative area of $LiNbO_3$ can be more easily damaged than the positive area.

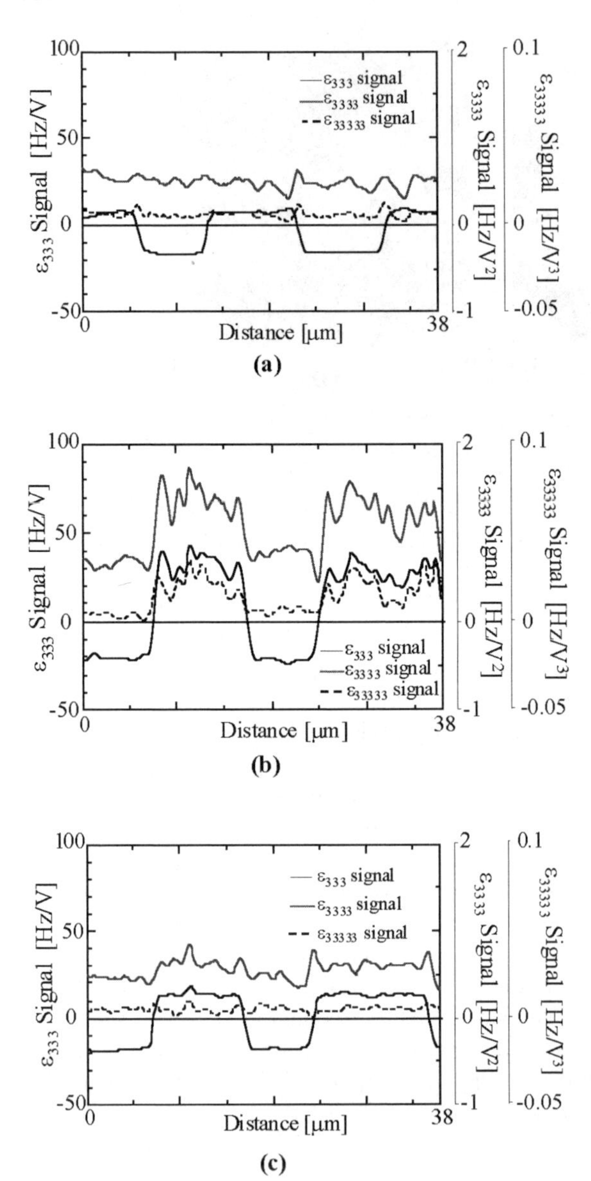

Fig. 9. (**a**) ε_{333}, ε_{3333}, and ε_{33333} images of virgin PPLN, (**b**) immediately after polishing, and (**c**) 1 h after polishing

4 Three-Dimensional Measurement Techniques

A new type of scanning nonlinear dielectric microscopy (SNDM) probe, called the ε_{311}-type probe, and a system to measure the ferroelectric polarization component parallel to the surface using SNDM have been developed [17]. This is achieved by measuring a nonlinear dielectric constant of the ferroelectric material, ε_{311}, instead of ε_{333}, which is measured in conventional SNDM. Experimental results show that the probe can detect the polarization direction parallel to the surface with high spatial resolution. Moreover, we propose an advanced measurement technique using a rotating electric field. This technique can be applied to measure three-dimensional polarization vectors.

4.1 The Principle and the Measurement System

Figure 10 shows parallel plate models of nonlinear dielectric constant measurements. Since precise descriptions of the ε_{333} measurement have been mentioned above, we explain only the ε_{311} measurement. We consider the situation in which a relatively large electric field $\bar{E_3}$ with amplitude E_{P} and angular frequency ω_{P} is applied to the capacitance C_{S}, producing a change of the capacitance that results from the nonlinear dielectric response. We detect the capacitance variation $\Delta\, C_{\mathrm{S}}$, which is perpendicular to the polarization direction (the z-axis) by a high-frequency electric field with small amplitude along the x-axis ($\tilde{E_1}$), as shown in Fig. 10b. (In the ε_{333} measurement, we detect $\Delta\, C_{\mathrm{S}}$ along the direction of spontaneous polarization $P_{\mathrm{S}3}$.) That is, in the ε_{311} measurement, $\bar{E}$ is perpendicular to $\tilde{E}$. We call this kind of measurement, which uses the crossed electric field, an "ε_{311}-type" measurement. In this case, the final formula is given by

$$\frac{\Delta\, C_{\mathrm{S}}(t)}{C_{\mathrm{S}0}} = \frac{\varepsilon_{311}}{\varepsilon_{11}} E_{\mathrm{P}} \cos\omega_{\mathrm{P}} t + \frac{\varepsilon_{3311}}{4\varepsilon_{11}} E_{\mathrm{P}}^2 \cos 2\omega_{\mathrm{P}} t\,, \tag{4}$$

where ε_{11} is the linear dielectric constant, and ε_{311} and ε_{3311} are nonlinear dielectric constants. From this equation, by detecting the component of capacitance variation with the angular frequency of the applied electric field ω_{P}, we can detect the nonlinear dielectric constant ε_{311}. According to this principle, we have developed a ε_{311}-type probe for measuring the polarization direction parallel to the surface. Figure 11 shows a schematic diagram of the measurement system. We put four electrodes around the probe tip to supply the electric filed $\bar{E}$, which causes the nonlinear effect. The electrodes A and B supply $\bar{E_3}$, which is along the z-axis, and electrodes C and D supply $\bar{E_2}$, which is along the y-axis. The applield voltage was so adjusted as to satisfy the condition that $\bar{E_2}$ and $\bar{E_3}$ just under the tip become parallel to the surface without becoming concentrated at the tip, as shown in Fig. 11b. (In Fig. 11b, the components related to y-direction are omitted for simplification.) On the other hand, the electric field $\tilde{E}$ for measuring the capacitance variation is

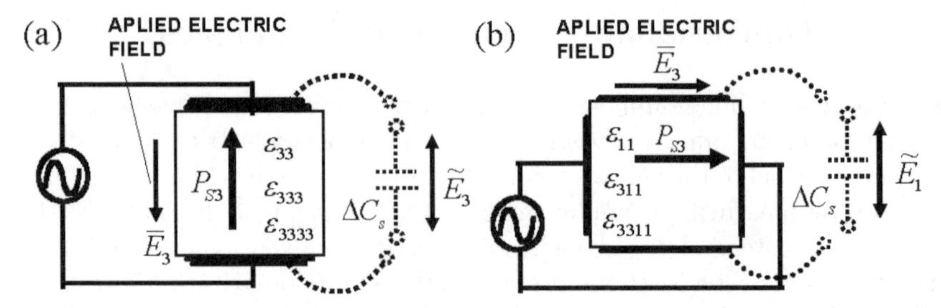

Fig. 10. Capacitance variation with an alternating electric field: (**a**) ε_{333} measurement; (**b**) ε_{311} measurement

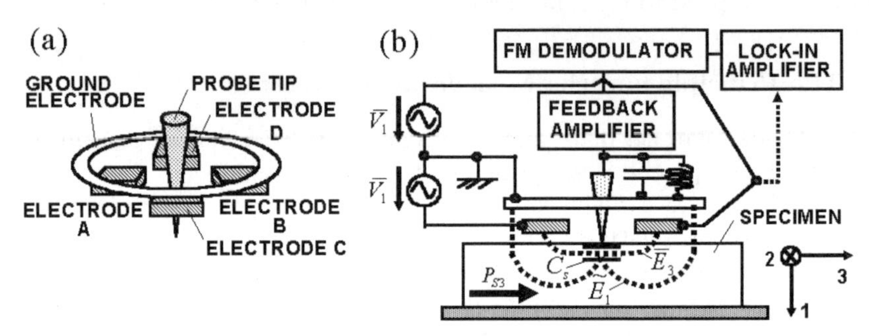

Fig. 11. The schematic configuration of (**a**) the new ε_{311} probe and (**b**) the measurement system

concentrated at the probe tip, as for the conventional measurement. It is sufficient to consider only the x-component of $\tilde{E}$, because we have confirmed that most of $\tilde{E}$ underneath the tip is perpendicular to the surface.

Moreover, we can obtain any electric field vector $\bar{E}$ with an arbitrary rotation angle by combining the amplitudes of $\bar{E}_2$ and $\bar{E}_3$. Therefore, we need not rotate the specimen to detect a lateral polarization with an arbitrary direction.

4.2 Experimental Results

Figure 12 shows the measurement results for PZT thin film as changing the direction of the applied electric field $\bar{E}$. In Fig. 12a, when $\bar{E}$ is parallel to the polarization direction, the pattern corresponding to the polarization is observed, while no pattern is observed in Fig. 12b because, in this case, $\bar{E}$ is perpendicular to the polarization direction. Figure 12c,d shows the cases in which $\bar{E}$ is applied along the intermediate direction. The pattern can be observed from both Fig. 12c and Fig. 12d, because the vector $\bar{E}$ can be divided by the component along the polarization direction. However, the opposite contrast was obtained, because the signs of the component along the polarization direction are opposite.

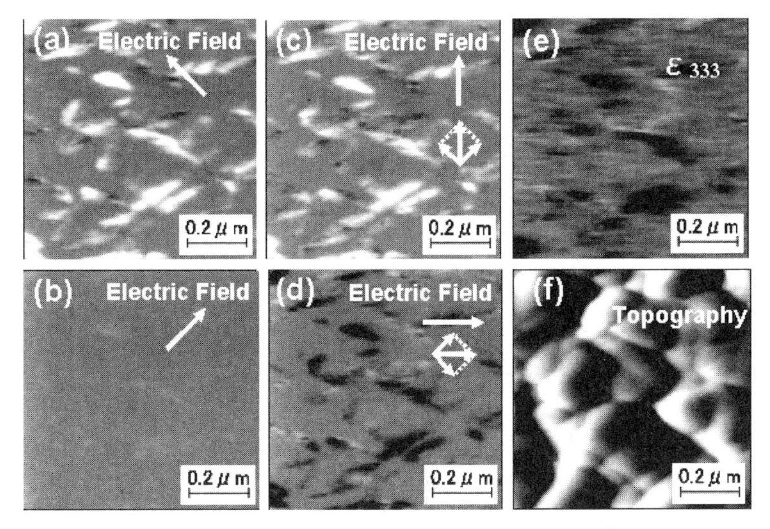

Fig. 12. Images of PZT thin film: (**a**)–(**d**) ε_{311} images, (**e**) a ε_{333} image, and (**f**) the topography

Moreover, the new probe can measure both ε_{311} and ε_{333} independently. Figure 12e shows an ε_{333} image, which corresponds to the perpendicular component of the polarization. From the same position in Fig. 12a, signals are observed. This means that this polarization has both parallel and perpendicular components; that is, the polarization tilts away from the surface. Figure 12f is the topography, which is also measured simultaneously. From these results, we have confirmed that the new probe and system can be applied to three-dimensional polarization measurements.

References

1. Y. Cho, A. Kirihara, T. Saeki: Denshi Joho Tsushin Gakkai Ronbunshi **J78-C-1**, 593 (1995) [in Japanese]
2. Y. Cho, A. Kirihara, T. Saeki: Electronics and Communication in Japan, Part 2, Scripta Technica **79**, 68 (1996)
3. Y. Cho, A. Kirihara, T. Saeki: Rev. Sci. Instrum. **67**, 2297 (1996)
4. Y. Cho, S. Atsumi, K. Nakamura: Jpn. J. Appl. Phys. **36**, 3152 (1997)
5. Y. Cho, S. Kazuta, K. Matsuura: Appl. Phys. Lett. **72**, 2833 (1999)
6. H. Odagawa, Y. Cho: Surf. Sci. **463**, L621 (2000)
7. H. Odagawa, Y. Cho: Jpn. J. Appl. Phys. **39** 5719 (2000)
8. A. Gruverman, O. Auciello, R. Ramesh, H. Tokumoto: Nanotechnol. **8**, A38 (1997)
9. L. M. Eng, H.-J. Gü, G. A. Schneider, U. Köpke, J. Muñoz Saldaña: Appl. Phys. Lett. **74**, 233 (1999)
10. Y. Cho, K. Ohara, A. Koike, H. Odagawa: Jpn. J. Appl. Phys. **40**, 3544 (2001)
11. K. Ohara, Y. Cho: Jpn. J. Appl. Phys **40**, 5833 (2001)

12. Y. Cho, K. Ohara: Appl. Phys. Lett. **79**, 3842 (2001)
13. H. Odagawa, Y. Cho: Ferroelectr. **268**, 149 (2002)
14. Y. Cho, K. Ohara, S. Kazuta, H. Odagawa: J. Eur. Ceram. Soc. **21**, 2135 (2001)
15. H. Odagawa, Y. Cho: Jpn. J. Appl. Phys. **39**, 5719 (2000)
16. K. Matsuura, Y. Cho, H. Odagawa: Jpn. J. Appl. Phys. **40**, 3534 (2001)
17. H. Odagawa, Y. Cho: Appl. Phys. Lett. **80**, 2159 (2002)

Part III

The Fabrication Process and Circuit Design

The Current Status of FeRAM

Glen R. Fox, Richard Bailey, William B. Kraus, Fan Chu, Shan Sun, and Tom Davenport

Ramtron International Corporation,
1850 Ramtron Drive, Colorado Springs, CO 80921, USA
tom.davenport@ramtron.com

Abstract. Ferroelectric nonvolatile memories (FeRAM) have been mass-produced since 1992 and densities up to 256 kb are currently available in a range of products. Both memory density and the market for FeRAM are increasing at an exponential rate due to the demands for nonvolatility, high read/write endurance, fast access speed, and low-power operation. Current applications include smart cards, data collection and storage (e.g., power meters), configuration storage, and buffers. FeRAM cell designs utilize a $PbZr_xTi_{1-x}O_3$ based bistable ferroelectric capacitor structure that is integrated with a transistor (1T-1C) or a complementary capacitor and two transistors (2T-2C). The small memory cell sizes under development will not only enable high-density stand-alone memories but also extend the application of FeRAM, which is already used in embedded and system-on-chip devices. A review of ferroelectric materials performance in current memory products will be presented. Recent development has led to capacitor performance with an endurance beyond 10^{12} read/write cycles and operation at 1.8 V. A roadmap for future FeRAM development will be presented.

1 Introduction

Over the past 15 years, the nonvolatile ferroelectric random access memory (FeRAM) has evolved from a concept on paper to a memory product that is used in a variety of consumer products and industrial applications. FeRAM is a vital component of applications including smart cards, power meters, printers, and video games. The advantage of using FeRAM instead of other types of nonvolatile memory, such as EEPROM and battery-backed SRAM, varies according to the application. FeRAM exhibits a unique combination of performance features, including low power consumption, high read/write endurance, a fast read/write access time, and long-term retention. Not only is FeRAM of interest as a high-performance stand-alone memory, but it also provides a low-cost embedded memory solution that can be scaled to high densities.

FeRAM was first demonstrated in 1988 using $PbZr_xTi_{1-x}O_3$ (PZT) as the ferroelectric material for data storage (Fig. 1). The FeRAM cell consisted of two transistors and two capacitors, commonly referred to as a 2T-2C cell. The ferroelectric capacitors were added to the underlying CMOS by applying backend ferroelectric, electrode, metallization, and inter-level dielectric

H. Ishiwara, M. Okuyama, Y. Arimoto (Eds.): Ferroelectric Random Access Memories,
Topics Appl. Phys. **93**, 139–148 (2004)
© Springer-Verlag Berlin Heidelberg 2004

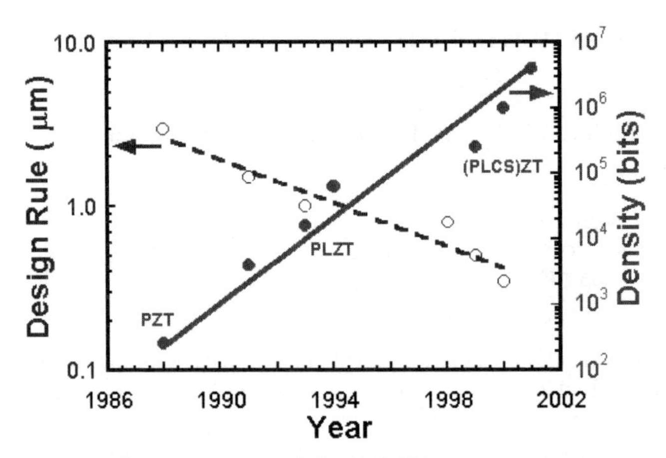

Fig. 1. The progression of the FeRAM memory density and design rule in manufacturing

processes. With a cell factor of approximately 110 and a CMOS feature size of 3 μm, these initial FeRAM devices had a very large cell size of nearly 1000 μm^2.

FeRAM mass production commenced in 1992 [1]. Soon after production was started, the cell size was significantly reduced by decreasing the minimum feature size to 1.0 μm. At almost the same time, La doping of the PZT was introduced in order to improve the read/write cycling endurance and data retention. Access to new process lines has led to a current feature size of 0.35 μm, and steady advances in PZT processing and dopant additions have allowed production of FeRAM products with a density of 256 kb, 10 yr retention (−45 to 85 °C), and a read/write endurance as high as 10^{16} cycles. Memories that have been fabricated and tested but are still under development have achieved densities as high as 4 Mb.

A new era is emerging for FeRAM development and product manufacturing. FeRAM is now being manufactured in state-of-the-art facilities. In addition, advances in materials, cell design, and architecture are allowing for rapid decreases in cell size, which will enable the production of low-cost high-density memories. With a well-organized development effort, FeRAM-based products have the potential to capture the majority of the stand-alone and embedded nonvolatile memory market, as well as to penetrate the SRAM and DRAM markets. The rest of this chapter presents a roadmap that outlines the requirements for high-density FeRAM and describes the status of ferroelectric materials development with respect to this roadmap.

2 The High-Density FeRAM Roadmap

In order to achieve cost-effective production of high-density FeRAM, both the cell factor and the feature size must be reduced. Most commercial FeRAM devices that are currently available utilize a 2T-2C cell with a 0.5 μm minimum CMOS feature size. In these devices, the ferroelectric capacitors are formed on top of the field oxide such that the capacitors are next to the accompanying transistors and each component of the cell occupies a uniquely defined area. This architecture will be referred to as the capacitor-over-field-oxide (COFO) architecture.

The large, 110, cell factor for the 2T-2C devices can be reduced by changing both the cell design and the architecture. The introduction of a 1T-1C-type cell decreases the cell factor to approximately 63; therefore, the cell area is nearly half that for a 2T-2C cell. Because a 2T-2C cell is self-referencing, the change to a 1T-1C cell places stricter requirements on the ferroelectric capacitor performance with respect to switchable polarization, retention, and cycling endurance. An additional decrease in the cell factor is achieved by building the ferroelectric capacitor on top of a plug that contacts an underlying access control transistors, as shown in Fig. 2. Cell factors as small as 4 have been proposed for 1T-1C designs with the capacitor-over-plug (COP) architecture [2]. The COP architecture also introduces new demands on the ferroelectric capacitor materials, since the capacitor is grown on a conducting plug material (as opposed to field oxide) that must be protected from oxidation to insure low contact resistance between the capacitor and plug.

Cell factors, architectures, and cell sizes for 1T-1C-based FeRAM are given in Table 1 for existing processes based on 0.5 and 0.35 μm CMOS feature sizes. Targets for future 1T-1C FeRAM generations with smaller feature sizes are also shown in Table 1. A range of cell factors and sizes is given for each generation, because these values depend on several factors, including the ferroelectric capacitor stability, the reference scheme for detecting the 0 and 1 states of the cell, and the minimum ferroelectric switched charge detectable by the CMOS sensing circuitry. Although FeRAM has a potentially low processing cost, with a requirement for only two or three additional mask steps (making it particularly attractive for embedded applications) the current cell factor of 63 must continue to be reduced in advanced generations in order to maintain cost competitiveness with other memory technologies.

In addition to cell size and architecture, Table 1 also lists the targets for memory density and performance specifications for feature sizes down to 0.1 μm. The reduction of the feature size is accompanied by scaling of not only the cell factor but also the operating voltage, the access time, and the read/write endurance. FeRAM memory for all feature sizes must maintain data for a period of 10 yr over the specified operating temperature (typically −45 to 85 °C) and it is desirable to keep the active current below 15 mA. Meeting all performance requirements, while scaling cell size and density for

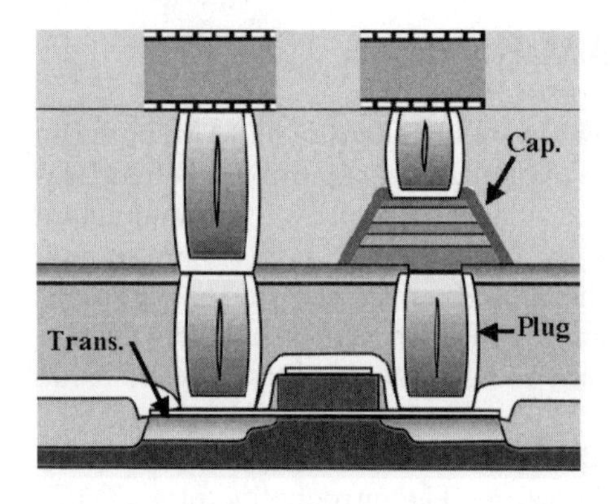

Fig. 2. A diagram of the FeRAM capacitor-over-plug (COP) architecture

Table 1. The roadmap for 1T-1C FeRAM memory

Generation (μm)	0.5	0.35	0.25	0.18	0.13	0.1
Density (Mb)	1	16	64	256	512	2000
Architecture	COFO	COFO/ COP	COP	COP	COP	COP/ 3DCOP
Cell factor	63	63–30	30–15	20–8	16–8	12–6
Cell size (μm^2)	15.8	8.0–3.7	1.9–0.9	0.65–0.26	0.27–0.14	0.12–0.06
Operating voltage (V)	5/3.3	3.3	2.5	1.8	1.5	1.2
Data retention (yr)	10	10	10	10	10	10
Read/write endurance (cycles)	$10^{10}/10^{16}$	10^{16}	10^{16}	10^{16}	10^{16}	10^{17}
Access time (ns)	70	50	30	20	15	10
Active current (mA)	< 15	< 15	< 15	< 15	< 15	< 15

each generation, requires increased design efficiency and control of ferroelectric capacitor performance.

Table 2 provides a summary of the ferroelectric capacitor requirements at each feature size. For the 0.5 and 0.35 μm generations, which employ the COFO architecture, sufficient ferroelectric switchable charge can be generated from a planar capacitor that occupies less than 20% of the total cell size. The capacitor dimensions and performance metrics for these two generations are typical of those found for capacitors in FeRAM products. The capacitor switchable charge is determined by the ferroelectric switchable polarization (Q_{SW}) multiplied by the area of the capacitor. Since mechanisms such as relaxation, imprint, and fatigue can reduce the magnitude of Q_{SW}, the capacitor must be sized such that sufficient switched charge is available for data state detection during the specified memory lifetime; for example 10 yr. One metric that can be used for capacitor sizing is the extrapolated Q_{SW} remaining after a 10 yr 150 °C bake [Q_{SW} (10 yr) at 150 °C]. This metric gives a conservative measure of the switchable polarization remaining after 10 yr retention of a data state. For the 0.5 and 0.35 μm generations, a Q_{SW} of approximately 5 μC/cm^2 is required to generate the switched charge necessary for data state detection. Since Q_{SW} (10 yr) at 150 °C is 20 μC/cm^2, a significant switched charge margin remains even if other Q_{SW} degradation mechanisms occur. The capacitors in these FeRAM products have purposely been oversized to allow a larger than required margin for switched charge in order to increase product reliability.

Table 2. Ferroelectric capacitor properties for current generations and target requirements for 1T-1C FeRAM

Generation (μm)	0.5	0.35	0.25	0.18	0.13	0.1
Capacitor area (μm^2)	3.0	1.5–1.0	1.0–0.5	0.33–0.13	0.14–0.07	0.06–0.03
PZT thickness (nm)	200	180	140	100	80	65
$V_{90\%}$ (V)	4.0/2.8	2.2	1.8	1.5	1.2	0.9
Q_{SW} (μC/cm^2)	30	30	30	35	40	45
Q_{SW} (10 yr) at 150 °C (μC/cm^2)	20	20	20	20	35	40
Switching endurance (cycles)	$10^{10}/10^{14}$	10^{14}	10^{15}	10^{15}	10^{15}	10^{16}
Switching time (ns)	< 10	< 10	< 10	< 10	< 5	< 5
Energy/bit (pJ)	2.3/1.4	0.68	0.19	0.04	0.02	0.02

Because the capacitor is stacked on top of the transistor for the COP architecture, the relative cell area occupied by the capacitor increases to approximately 50%. Starting with the 0.35 μm COP generation, it can be seen that the capacitor size scales proportionally to the cell size. It is desirable to maintain the largest possible capacitor area, since this increases the ferroelectric switchable charge that can be used to detect the difference between a 0 and 1 data state. Although larger capacitor areas provide a larger margin for Q_{SW} degradation over the lifetime of a memory, it is not desirable to achieve this margin at the expense of an increased cell size. For this reason, it is required that Q_{SW} increases for generations with feature sizes smaller than 0.25 μm in order to maintain sufficient switched charge for data state retention. In addition, less degradation of Q_{SW} is allowed for advanced generations, as can be seen from the reduced difference between Q_{SW} and Q_{SW} (10 yr) at 150 °C. For the 0.1 μm generation, meeting the Q_{SW} scaling requirement with a planar PZT-based capacitor becomes difficult. A Q_{SW} of at least 45 μC/cm^2 is required even with a cell factor of 12. Smaller cell factors will require even larger Q_{SW} values. In order to meet the switched charge requirement for an aggressive cell factor of 6 at the 0.1 μm generation, three-dimensional capacitor structures are expected to be required.

Since a high number of switching cycles can degrade the magnitude of Q_{SW}, Table 2 lists the switching endurance; that is, the number of cycles – required without appreciable loss of Q_{SW}. For the 0.5 and 0.35 μm generations, some reduction of Q_{SW} with cycling is allowable due to the margin afforded by the oversized capacitors. Modeling of reliability test data has confirmed that 3.3 V FeRAM products can achieve 10^{16} read/write cycle endurance [3]. But as the feature size is decreased, less margin in Q_{SW} is available and nearly fatigue free films are required for the endurance specification. A capacitor endurance specification of at least 10^{16} cycles is required for use in a parallel FeRAM to insure a 10 yr lifetime. Taking into account differences in bit access within a FeRAM memory and the continuous cycling in a capacitor test allows a capacitor with an extrapolated fatigue-free 10^{15} cycles to achieve 10^{16} read/write cycles in a memory. As the memory access time decreases, a corresponding increase in read/write endurance is required. At the 0.1 μm generation, an endurance of 10^{17} read/write cycles is required for the memory, which indicates that the capacitor within a bit must have an endurance of at least 10^{16} cycles.

A PZT thickness reduction is introduced with each generation for two reasons. One reason is to maintain a reasonable capacitor height to width aspect ratio to minimize difficulties with a single-step capacitor stack etch process. But the primary reason for scaling the PZT thickness is to meet the operating voltage requirements for each generation. In order to insure proper data retention and data state detection, PZT must switch at voltages below the standard operating voltage of the FeRAM product. Saturation of ferroelectric switching as a function of the applied voltage is quantified by determining the voltage at which 90% of the maximum Q_{SW} is achieved ($V_{90\%}$). Because of the operating voltage reduction with each generation (Table 1), $V_{90\%}$ must be scaled accordingly (Table 2). Typically, $V_{90\%}$ should be 20–30% below the operating voltage to insure performance over a range of temperature and stress conditions. Since ferroelectric switching is determined by the magnitude of the electric field applied to the ferroelectric, the most direct method for reducing $V_{90\%}$ is to reduce the PZT thickness. This approach is expected to be viable even for the 0.1 μm generation, where it is estimated that PZT films as thick as 65 nm can be used to achieve a $V_{90\%}$ of 0.9 V. Dopants can also be introduced to lower $V_{90\%}$, but strong effects on processing and other ferroelectric properties, such as Q_{SW}, can complicate this approach.

By combining the Q_{SW} and operating voltage requirements for each technology node, the energy per bit; that is, the energy required to change data states within a single capacitor, can be calculated. This energy is extremely low even at the 0.5 μm generation and it decreases with advanced generations. The energy required to switch the data state within a ferroelectric capacitor is in fact negligible in comparison with the energy required to drive the supporting circuitry; therefore, the active current for FeRAM memory is primarily determined by the circuit design.

3 The Current Status of PZT Capacitor Materials

Development of the capacitor stack for the 0.5 and 0.35 μm generations is for the most part complete. The capacitor properties for both the 0.5 and 0.35 μm generations and the mass-produced memories at the 0.5 μm generation (memories for the 0.35 μm generation have not yet been qualified) meet the specifications listed in Tables 1 and 2. For both generations, the capacitor stack consists of a Pt bottom electrode, a PZT ferroelectric, and an IrO_X top electrode. The Pt has a {111} texture, which seeds the growth of {111}-textured PZT. It has been confirmed that the PZT and Pt have corresponding rocking curve angular widths for films that give the most efficient ferroelectric switching performance.

Because FeRAM materials development is currently driven by access to production facilities, the next generation of FeRAM products could be introduced with 0.25, 0.18, or 0.13 μm minimum feature sizes. For this reason, basic materials studies are focused on thickness scaling to meet the reduced operating voltages for these generations. These studies also determine whether sufficient Q_{SW} and retention can be achieved in films with reduced thickness.

Figure 3 shows the change in $V_{90\%}$ and Q_{SW} as a function of the PZT film thickness. The PZT has a Zr:Ti ratio of 40:60 and also contains dopants of La, Ca, and Sr. The PZT sputter deposition and anneal processes were established at a film thickness of 180 nm and then thinner films were produced with the same process by reducing the deposition time. For all film thicknesses, greater than 90% (by volume) {111} texture was maintained. It is observed that the $V_{90\%}$ decreases linearly with thickness until 80 nm, where it departs from linear behavior. With $V_{90\%} = 1.8$ V, the 100 nm thick film could be used for the 0.25 μm generation. Process reoptimization for this thickness is likely to provide the 1.5 V $V_{90\%}$ needed for the 0.18 μm generation. The Q_{SW} for the 100 nm thick film is below the value required for both the 0.25 and 0.18 μm generations. In addition, the Q_{SW} (10 yr) at 150 °C for both the same-state and the opposite-state retention tests is insufficient even though the absolute rate of Q_{SW} degradation with bake time remains constant with thickness scaling. As with the $V_{90\%}$ metric, it is expected that careful attention to

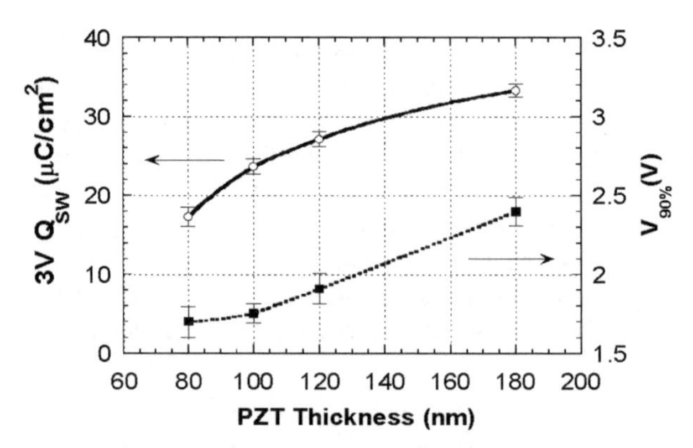

Fig. 3. The dependence of Q_{SW} and $V_{90\%}$ on the PZT film thickness

Pb concentration gradients and anneal thermal budgets can bring the Q_{SW} and Q_{SW} (10 yr) at 150 °C values in line with requirements for the 0.25 and 0.18 μm generations. In all cases, the thickness-scaled films exhibit excellent cycling endurance when compared at equal fields. The films exhibit no fatigue when tested beyond 10^{11} cycles and voltage-accelerated testing indicates that 10^{13} cycles could be achieved without fatigue. Because of testing limitations, new fatigue models and testing methodology must be established to confirm fatigue-free behavior beyond 10^{13} cycles. In general, these reduced-thickness films, which were deposited in a mass-production sputtering tool, provide capacitor performance that is close to that which is needed for the 0.25 and 0.18 μm FeRAM memory generations even without full process optimization.

In order to achieve capacitor properties that meet the requirements for the 0.13 μm and 0.10 μm generations, changes in PZT composition will likely be required. Low-voltage (1.5 V) switching has been demonstrated for capacitor sizes as small as 0.12 μm^2 with undoped 20:80 PZT grown by MOCVD. A Q_{SW} of 40 $\mu C/cm^2$ was achieved, indicating that process optimization for these films may lead to capacitor specifications that match the 0.13 and 0.1 μm generation requirements [4].

In addition to meeting the capacitor performance specifications for generations with a 0.18 μm and smaller feature size, it is believed that control of the ferroelectric PZT crystallographic texture will become a critical factor. Texture becomes important in two ways. For planar capacitors with areas below 0.5 μm^2, the capacitor size is of the same order of magnitude as the grain size of the PZT. Randomly oriented films are expected to show a wide capacitor-to-capacitor Q_{SW} distribution due to the random orientation of the PZT grains and the corresponding angular distribution of the ferroelectric polarization. The distribution will be broadest in the case in which no 90 ° domain switching occurs, which is the commonly observed case for films with thicknesses below 300 nm. For 3D capacitor structures, better capacitor-to-capacitor uniformity may be achieved with random PZT, since

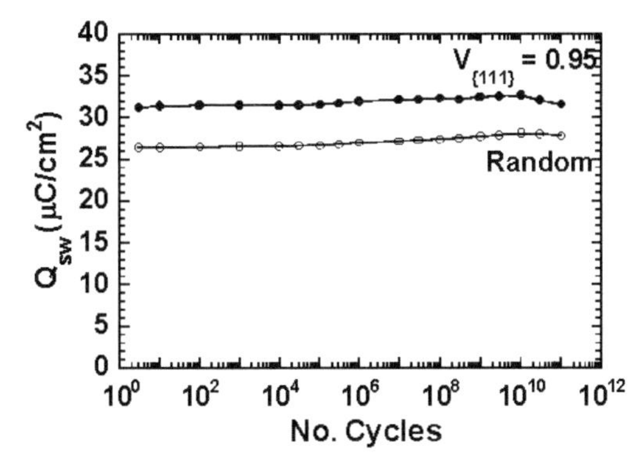

Fig. 4. A comparison of the voltage-accelerated cycling endurance of {111}-textured and randomly orientated 180 nm thick PZT films cycled at 5 V and 1 MHz

it is expected to be difficult to uniformly control the texture over the surface of a 3D structure, but this will depend on the physical area of the capacitor and the domain switching mechanisms of the film. It has been demonstrated that both random and {111}-textured PZT films can be produced with similar fatigue-free performance up to more than 10^{11} switching cycles, as shown in Fig. 4. If the voltage acceleration is taken into account, these films will continue to show fatigue-free behavior to greater than 10^{12} cycles under the 3 V operating voltage for which the films were designed. Although it may be less difficult to control the growth of random PZT films, this relaxation of processing constraints is gained at the expense of ferroelectric switching performance. Randomly oriented films exhibit a reduced Q_{SW} and a higher $V_{90\%}$ than films of the same thickness with {111} texture. The decreased switching performance of the random film is believed to result primarily from the absence of 90° domain switching. Both the {111}-textured and the randomly oriented PZT materials show potential for meeting the requirements of the 0.1 μm generation, but the trade-off between ease of manufacturing, ferroelectric anisotropy, and size effects will need to be clearly understood.

4 Summary

The current status and a roadmap for 1T-1C FeRAM memory were presented for generations having feature sizes between 0.5 and 0.1 μm. The corresponding roadmap for ferroelectric capacitor requirements was described in relation to the memory roadmap. The ferroelectric capacitor requirements for the 0.5 and 0.35 μm generations have been achieved with {111}-textured PZT containing La, Ca, and Sr dopants. With process modifications, similar PZT films are expected to meet the requirements for the 0.25 and 0.18 μm generations. At the 0.13 and 0.1 μm generations, a PZT composition modification will be required and careful consideration of the trade-off between ease of manufacturing, crystallographic texture, and cell size will be needed.

References

1. T. Davenport, S. Mitra: Integr. Ferroelectr. **31**, 213 (2000)
2. D. Takashima, S. Shuto, I. Kunishima, H. Takenaka, Y. Oowaki, S. Tanaka: IEEE J. Solid-State Circuits **34**, 1557 (1999)
3. T. D. Hadnagy, D. Dalton, B. Thomas, S. Sun, R. Bailey: Integr. Ferroelectr. **37**, 215 (2001)
4. T. S. Moise, S. R. Summerfelt, G. Xing, L. Colombo, T. Sakoda, S. R. Gilbert, A. Loke, S. Ma, R. Kavari, L. A. Wills, T. Hsu, J. Amano, S. T. Johnston, D. J. Vestyck, M. W. Russell, S. M. Bilodeau: Tech. Dig. Intern. Electron Devices Meeting (1999) p. 940

Operation Principle and Circuit Design Issues

Ali Sheikholeslami

Department of Electrical and Computer Engineering,
University of Toronto, Toronto, Canada
ali@eecg.utoronto.ca

Abstract. This chapter provides an introduction to the circuit design aspects of ferroelectric random access memories (FeRAM). A FeRAM stores binary data in an array of FeRAM cells, each consisting of either two transistors and two capacitors (2T-2C cell) or one transistor and one capacitor (1T-1C cell). The 2T-2C cell stores both the binary data and its complement on the two capacitors. Compared to the 1T-1C cell, the 2T-2C cell provides data storage that is robust to process variation at the price of using twice the silicon area. The 1T-1C cell stores the binary data only, and not its complement, to save silicon area at the price of more complex data sensing. Two well-known approaches for data sensing are step sensing and pulse sensing. There are a variety of circuit architectures for FeRAMs. This chapter presents a few well-known architectures and points to the literature for the most recently proposed FeRAM architectures.

1 Introduction

Ferroelectric random access memories (FeRAM) belong to a family of semiconductor memories that are known as nonvolatile memories. A nonvolatile memory stores digital information in an array of memory cells and retains this information indefinitely without electrical power. Two other types of nonvolatile memories are electrically-erasable-and-programmable read-only-memories (EEPROM) and Flash memories. Table 1 compares a FeRAM, an EEPROM, and a Flash memory in terms of density, read access-time, write-access time, and the energy consumed in a 32 bit read/write operation. Enjoying a mature process technology, EEPROMs and Flash memories [1, 2] are superior to ferroelectric memories in terms of density. Also, they require less power compared to ferroelectric memories for read operations, a factor that will keep them popular in applications that demand numerous memory reads but only occasional memory writes. An example of such applications is an identity card, where an identity code is programmed into the memory once but read many times afterwards.

Ferroelectric memories, on the other hand, are superior to EEPROMs and Flash memories in terms of write-access time and overall power consumption, hence target applications where a nonvolatile memory is required with such features. Two examples of such applications are contactless smart cards and digital cameras. Contactless smart cards require nonvolatile memories with low power consumption as they use only electromagnetic coupling to power

H. Ishiwara, M. Okuyama, Y. Arimoto (Eds.): Ferroelectric Random Access Memories,
Topics Appl. Phys. **93**, 149–163 (2004)
© Springer-Verlag Berlin Heidelberg 2004

Table 1. Nonvolatile memories and their typical features

Nonvolatile memory	EEPROM	Flash memory	FeRAM
Area per cell (norm.)	2	1	5
Read access-time	50 ns	50 ns	100 ns
Write (prog.) access-time	10 μs	1 μs	100 ns
Energy* per 32 b write	1 μJ	2 μJ	1 nJ
Energy* per 32 b read	150 pJ	150 pJ	1 nJ

* Data are given for the basic memory cell and may be different when designed into an integrated circuit

up the electronic chips on the card. Digital cameras require both low power consumption and fast-frequent writes in order to store and re-store an entire image into the memory in less than 0.1 s. Another advantage of ferroelectric memories over EEPROMs and Flash memories is that they can be easily embedded as part of a larger integrated circuit, to provide system-on-a-chip solutions for various applications [3, 4].

Despite all the potential advantages of ferroelectric memories, at the moment commercially available ferroelectric memories have a density less than 256 kb. This is far below the density level of a commercially available 64 Mb Flash memory. In order to reduce this density gap, a number of technical challenges must be addressed, including integration challenges and design challenges.

Integration challenges relate to reducing the memory cell area while minimizing unwanted material interactions among the silicon substrate, the ferroelectric material, and the electrodes. These interactions are known to degrade both the ferroelectric material properties and the MOS transistor characteristics (e.g., the transistor threshold voltage) [5]. Design challenges include developing circuit techniques and architectures around memory cells using only one transistor and one capacitor, known as the 1T-1C cell. Another design challenge is to develop circuit techniques to circumvent various process shortcomings, and, more importantly, to develop a ferroelectric capacitor model for circuit simulations to evaluate the performance of proposed circuit techniques.

A general block diagram for a ferroelectric memory architecture is shown in Fig. 1. This is in fact a generic architecture for all semiconductor memories such as SRAM, DRAM, EEPROM, and Flash memories. The only distinction made here for ferroelectric memories is the presence of a plateline (PL) in the architecture. In order to read a datum with a specific address from the memory, the address bits are presented to the x-decoder, where a single wordline (WL) and plateline (PL) pair is selected among $2r$ pairs (also known as rows) in the cell array. The selected wordline and plateline receive proper signals to access all the cells in that row, whereas the unselected rows

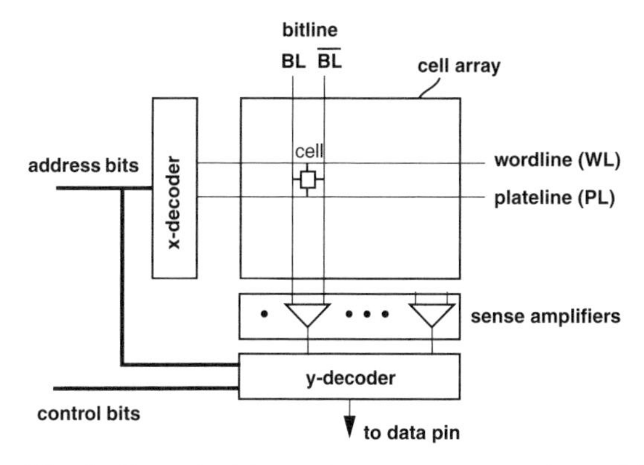

Fig. 1. Ferroelectric random access memories architecture

receive no signal and hence remain inactive. All of the cells along a selected row are activated and their data contents are deposited onto their respective bitlines, which run vertically through the array. The storage cells store one bit of information; and hence the term "bitlines" for the wires that carry these bits.

The signal available on the bitlines is directed to a parallel set of sense amplifiers at the bottom of the memory array. A sense amplifier acts as a buffer and a decision device, that accurately estimates the original content or state of the selected memory cell. The column address is used to select a subset of sense amplifier outputs (possibly as few as one) from a set of $2c$ columns and to route this to an output buffer that drives the result out of the memory system. A write operation is identical to a read operation, except for the data being provided from outside the system to drive the selected bitlines.

In this chapter, we focus on specifics of ferroelectric memories, such as ferroelectric capacitors, ferroelectric memory cells, their read and write operations, and sensing schemes. First, we review the characteristics of ferroelectric capacitors from a circuit point of view. A more detailed review of the ferroelectric capacitor characteristics along with their circuit modeling issues can be found in [6]. Then, we introduce a basic ferroelectric memory cell, consisting of a single ferroelectric capacitor and a single transistor, along with its read and write operations, and sensing. Finally, we present conventional ferroelectric memory architectures and some of their terminologies. We refer the readers to [7] for a more comprehensive survey of sensing schemes and FeRAM architectures.

2 The Ferroelectric Capacitor

A ferroelectric capacitor is physically distinguished from a regular capacitor by substituting the dielectric with a ferroelectric material [8]. In a regular di-

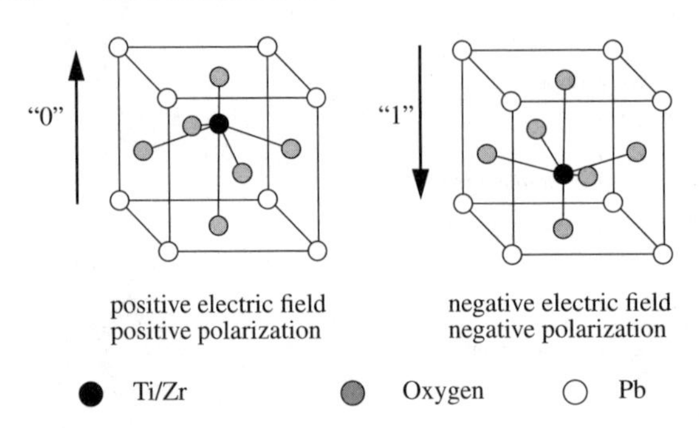

Fig. 2. Two stable states in PZT: the orientation of the spontaneous polarization is reversed by applying a proper electric field

electric, upon the application of an electric field, positive and negative charges will be displaced from their original positions – a concept that is characterized by polarization. This polarization, or displacement, will vanish, however, when the electric field returns back to zero. In a ferroelectric material, on the other hand, there is a spontaneous polarization – a displacement that is inherent to the crystal structure of the material and does not disappear in the absence of an electric field. In addition, the direction of this polarization can be reversed or reoriented by applying an appropriate electric field.

One of the well-known ferroelectric materials that has been used for over a decade is lead zirconate titanate (PZT). Figure 2 illustrates a unit cell of this material. The central atom in this unit cell is either titanium (Ti) or zirconium (Zi), depending on the contribution of each atom to the material's chemical formula, $Pb(Zr_X Ti_{1-X})O_3$. An appropriate electric field is capable of displacing the central atom from its previous stable position and, thereby, changing the polarization state of the unit cell. Although the polarization of each individual unit cell is tiny, the net polarization of several domains – each consisting of a number of aligned unit cells – can be large enough for detection using standard sense amplifier designs. The gross effect of polarization is a nonzero charge per unit area of the ferroelectric capacitor that exists at 0 V and does not disappear over time. The polarization charge, which is simply referred to as the capacitor charge in this chapter, responds to the voltage across the capacitor in the same way as the magnetic flux of a ferromagnetic core responds to the current through the core. In this sense, ferroelectric capacitors are duals of ferromagnetic cores, and justify their names. Ferroelectric capacitors sit on top of a conventional CMOS process and provide a typical polarization charge of $150\,\mathrm{fC}/(\mu\mathrm{m})^2$ in a typical area of $2(\mu\mathrm{m})^2$. This nonvolatile polarization charge level is about 20 times larger than the volatile polarization level on gate oxide (SiO_2), in a typical CMOS process, when subjected to a full power supply voltage.

There are two common methods of characterizing a ferroelectric capacitor behavior. One is to monitor the polarization charge on the ferroelectric

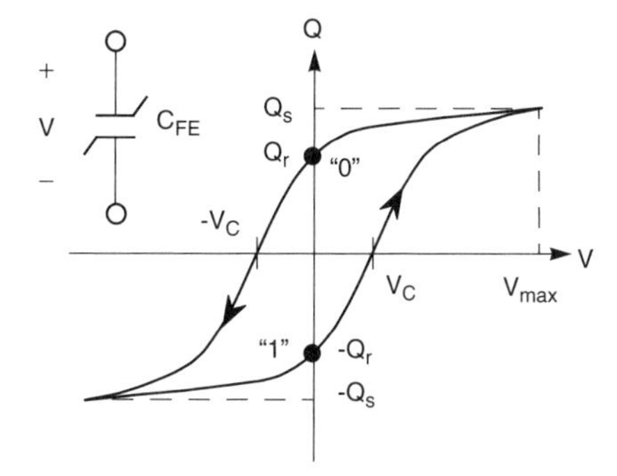

Fig. 3. The hysteresis loop characteristic of a ferroelectric capacitor. Three important parameters of the loop are: remanent charge (Q_r), saturation charge (Q_s), and coercive voltage (V_C)

capacitor while a low-frequency sinusoidal waveform is applied to a series combination of the ferroelectric capacitor and a large linear capacitor, a circuit setup known as the Sawyer–Tower circuit. The other method is to monitor the transient current of the capacitor while a sequence of voltage pulses is applied to a series combination of the ferroelectric capacitor and a linear resistor. We describe both characterization methods in the following sections.

2.1 The Hysteresis Loop Characteristic

A hysteresis loop for a ferroelectric capacitor, as shown in Fig. 3, displays the total charge on the capacitor as a function of the applied voltage. When the voltage across the capacitor is 0 V, the capacitor assumes one of the two stable states: "0" or "1". The total charge stored on the capacitor is Q_r, for a "0", or $-Q_r$, for a "1". Both Q_r and $-Q_r$ are bound charges that cannot be released for sensing until a voltage pulse is applied to the ferroelectric capacitor. A "0" can be switched to a "1" by applying a negative voltage pulse across the capacitor. By doing so, the total charge on the capacitor is reduced by $2Q_r$, a change of charge that can be sensed by the sense circuitry as explained later in this chapter. Similarly, a "1" can be switched back to a "0" by applying a positive voltage pulse across the capacitor, hence restoring the capacitor charge to $+Q_r$.

Three parameters are often extracted from a hysteresis loop: the remanent charge, Q_r, the saturation charge, Q_s, and the coercive voltage, V_C. The remanent charge and the saturation charge are self-explanatory from Fig. 3. The coercive voltage is defined as the voltage at which the net charge on the capacitor is zero. For a hysteresis loop with sharp transitions at V_C, this voltage can be interpreted as the approximate switching voltage for all the ferroelectric domains. For a hysteresis loop with typically smooth transitions at V_C, there is no fixed switching voltage but a range of voltage during which domains switch gradually. This is referred to as the partial switching of domains in a ferroelectric capacitor. Partial switching has a voltage distribution that can be quantified by plotting the capacitor charge increment per unit applied voltage versus applied voltage or, equivalently, the voltage derivative

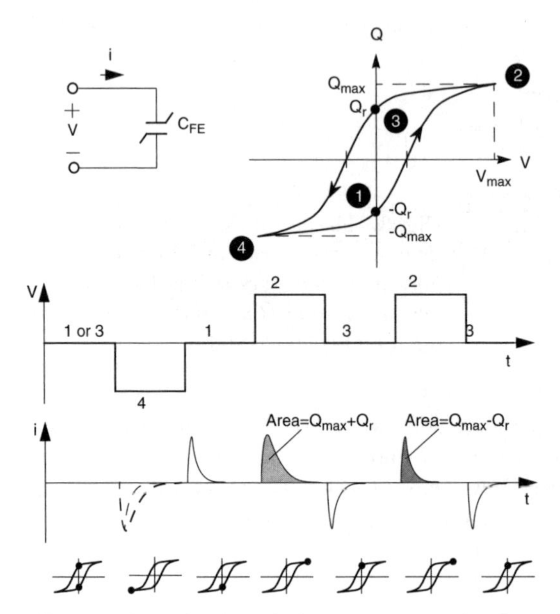

Fig. 4. The pulse-based characteristics of a ferroelectric capacitor. The transient current of the first positive pulse consists of both switching and nonswitching currents. The transient current of the second positive pulse consists of only the nonswitching current

of the capacitor charge in a hysteresis loop versus the applied voltage [9]. However, such a distribution cannot be obtained using a single hysteresis loop datum. Therefore, it is recommended [9] to use a set of hysteresis loops, instead of just one, to obtain this distribution for characterizing ferroelectric capacitors.

There is another consideration in characterizing a ferroelectric capacitor using its hysteresis loop: the sinusoidal waveform applied to the capacitor is different from the pulse-type voltage waveform applied to a ferroelectric capacitor in a ferroelectric memory. In the next section, we explore the behavior of the ferroelectric capacitor in response to pulse-type voltage waveforms.

2.2 Pulse-Based Characteristics

Let us apply a sequence of voltage pulses, as shown in Fig. 4, to a series combination of a ferroelectric capacitor and a small resistor, and monitor the transient current of the ferroelectric capacitor via the voltage across the series resistor. The pulse sequence consists of a negative pulse followed by two positive pulses. Each pulse forces the ferroelectric capacitor to move along a path on a hysteresis loop that is used here as a state diagram. On the falling edge of the first negative pulse, the ferroelectric capacitor goes to State 4. This is independent of the capacitor's initial state (State 1 or 3). On the rising edge of this pulse, the capacitor goes to State 1. This first negative pulse helps to establish the initial state of the capacitor before applying the next two positive pulses. In a ferroelectric memory, this pulse can be considered as a signal that writes a binary digit “1” into the cell.

On the rising edge of the first positive voltage pulse, the capacitor acquires a charge increment of $Q_{\mathrm{max}}+Q_{\mathrm{r}}$ and, as a result, its state changes to State 2. On the falling edge of the same pulse, the capacitor will lose $Q_{\mathrm{max}}-Q_{\mathrm{r}}$ as it goes to State 3. The transient current corresponding to the falling edge, therefore, will have less area underneath it than the one corresponding to the rising edge of the first pulse. The net effect of the first positive pulse is a charge increment of $2Q_{\mathrm{r}}$. In a ferroelectric memory, this pulse can be considered as a signal which reads a binary digit “1” into the bitline.

The second positive pulse causes a state transition from 3 to 2 and back to 3 again, a total charge increment of 0. In a ferroelectric memory, this corresponds to reading a binary digit “0” into the bitline.

3 The Ferroelectric Memory Cell: Read and Write Operations

Figure 5 shows a simplified schematic of a fully differential 2T-2C cell (two-transistor, two-capacitor) memory cell and that of a 1T-1C cell. The 1T-1C cell occupies half the semiconductor area required by the 2T-2C cell. Therefore, a memory architecture based on the 1T-1C cell achieves twice the density of an architecture based on the 2T-2C cell, albeit at the price of a more complex sensing scheme [7]. In the rest of this section, we concentrate on the read/write operations of the 1T-1C cell.

The 1T-1C cell consists of a single ferroelectric capacitor that is connected to a plateline (PL) at one end and, via an access transistor, to a bitline (BL) at the other end. The cell is accessed by raising the wordline (WL) and, hence, turning ON the access transistor. The access is one of two types: a write access or a read access.

The timing diagram for a write operation is shown in Fig. 6a. To write a “1” into the memory cell, the BL is raised to V_{DD}, then the WL is raised to $V_{\mathrm{DD}}+V_{\mathrm{T}}$ (known as boosted V_{DD} [10]), where V_{T} is the threshold voltage of the access transistor. This allows a full V_{DD} to appear across the ferroelectric capacitor ($-V_{\mathrm{DD}}$ according to the voltage convention adopted in Fig. 5). At this time, the state of the ferroelectric capacitor is independent of the initial state of the ferroelectric capacitor, as shown in Fig. 6b. Next, the plateline (PL) is pulsed; that is, pulled up to V_{DD} and subsequently pulled back down to ground. Note that the WL stays activated until the PL is pulled down completely and the BL is driven back to zero. The final state of the capacitor is a negative charge state S1 (defined as digital “1” in this chapter). Finally, deactivating the WL leaves this state undisturbed until the next access.

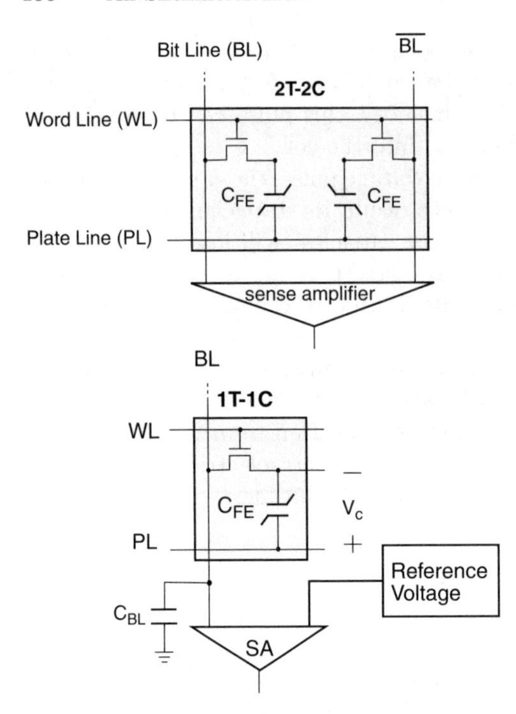

Fig. 5. The ferroelectric 2T-2C memory cell and the 1T-1C memory cell. C_{BL} represents the total parasitic capacitance of the bitline

To write a "0" into the cell, the BL is driven to 0 V prior to activating the WL. The rest of operation is similar to that of writing a "1", as shown in Fig. 6.

The timing diagram for a read access is shown in Fig. 7. A read access begins by precharging the BL to 0 V, followed by activating the WL (Δt_0). This establishes a capacitor divider consisting of C_{FE} and C_{BL} between the PL and the ground. During Δt_1, the PL is raised to V_{DD}. This voltage is divided between C_{FE} and C_{BL}, the parasitic capacitance of the bitline, according to their relative capacitance. Depending on the data stored, the capacitance of the ferroelectric capacitor can be approximated by C_0 or C_1, as shown in Fig. 8. Therefore, the voltage developed on the bitline (V_x) can be one of the two values, V_0 or V_1:

$$V_x = \begin{cases} V_0 = \dfrac{C_0}{C_0 + C_{\mathrm{BL}}} V_{\mathrm{DD}} & \text{if the stored data is a } 0\,, \\ V_1 = \dfrac{C_1}{C_1 + C_{\mathrm{BL}}} V_{\mathrm{DD}} & \text{if the stored data is a } 1\,. \end{cases} \tag{1}$$

At this point, the sense amplifier is activated to drive the BL to full V_{DD} if the voltage developed on the BL is V_1, or to 0 V if the voltage on the BL is V_0. The WL is kept activated until the sensed voltage on the BL restores the original data back into the memory cell and the BL is precharged back to 0 V.

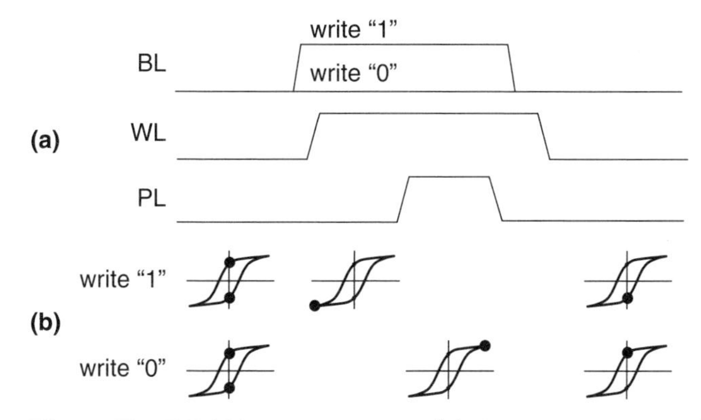

Fig. 6. The FeRAM write operation: (**a**) the timing diagram; (**b**) the state sequence. The initial state does not affect the subsequent states of the capacitor

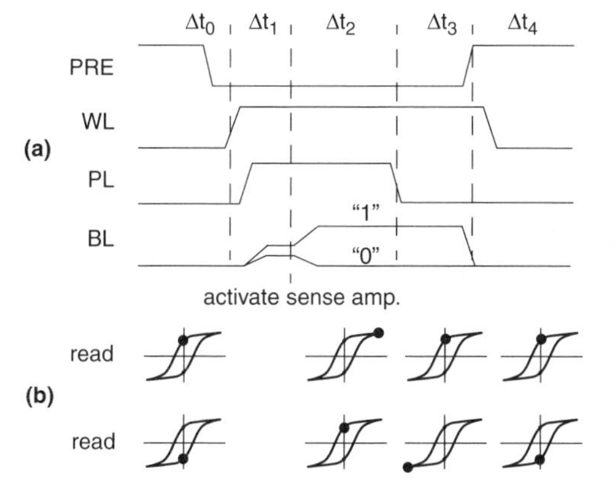

Fig. 7. The FeRAM read operation: (**a**) the timing diagram; (**b**) the state sequence for an initially-0 and an initially-1 memory cell capacitor as it undergoes the read operation

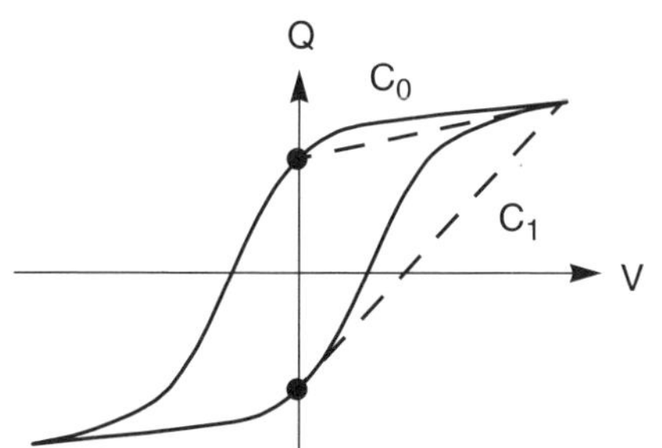

Fig. 8. A simplified model of a hysteresis loop. Linear capacitor C_0 models the nonswitching branch of the hysteresis loop, whereas linear capacitor C_1 approximates its switching branch

4 Sensing Schemes

The read access as presented above is known in the literature as the step-sensing approach, since a step voltage (the rising edge of a pulse) is applied to the PL prior to sensing. An alternative is the pulse-sensing approach, in which a full pulse is applied to the PL prior to activating the sense amplifiers. The charge transferred to the BL in a pulse-sensing scheme is either 0, for a stored "0", or $2Q_r$ for a stored "1". Equivalently, the voltage developed on the BL is either 0 V, for a stored "0", or $V_1 - V_0$ (refer to (1)) for a stored "1".

In both step- and pulse-sensing schemes, the voltage difference between a stored "1" and a stored "0" on the BL is equal to $V_1 - V_0$. Their common-mode voltage, however, is equal to $(V_1 + V_0)/2$ in the step-sensing approach, as compared to $(V_1 - V_0)/2$ in the pulse-sensing approach. Therefore, the step-sensing approach provides a higher common-mode voltage on the BL that simplifies the sense amplifier design when a bias voltage is required. Another advantage of the step-sensing approach is that it provides a faster read-access, as the sensing does not wait for the PL to be pulled low.

Both the step-sensing and the pulse-sensing approaches restore "1", but only the step-sensing approach fully reinforces a "0". To substantiate this point, note that during a read operation in a step-sensing approach, a ferroelectric capacitor storing a "0" experiences a voltage sequence [11] of 0 V, $V_{DD} - V_0$, V_{DD}, and 0 V (a full-V_{DD} excursion). In a pulse-sensing approach, the corresponding voltage sequence is 0 V, $V_{DD} - V_0$, and 0 V (a $V_{DD} - V_0$ excursion). Neither of the two voltage sequences upsets the original data (i.e., "0") of the ferroelectric capacitor. However, the latter provides a weak reinforcement of "0" by applying a voltage less than V_{DD} across the capacitor. This seems to cause the capacitor's long-term retention performance to deteriorate [12]. To remedy this situation, a second pulse must be applied to the PL to fully restore the "0" into the capacitor. This implies that the cycle time for the pulse-sensing approach can be twice as large as that for the step-sensing approach.

The pulse-sensing approach applies to both the leading edge and the trailing edge of the voltage pulse to the ferroelectric capacitor prior to sensing. The trailing edge eliminates the nonswitching part of the polarization that was introduced on the BL by the rising edge and, therefore, bypasses the effect of the nonswitching part of the polarization and its process variations altogether. This seems to be the only advantage of the pulse-sensing approach over the step-sensing approach.

So far, we have assumed that the sense amplifier can discriminate between a "0" and "1" voltage signal on the BL. This is only possible if a reference voltage, midway between a "0" and a "1" signal, is provided for the sense amplifier. We refer the readers to [7] for a survey of reference generation schemes.

5 Ferroelectric Memory Architectures

Ferroelectric memories have borrowed many circuit techniques from dynamic random access memories (DRAMs), due to similarities of their cells and DRAM's mature architecture [13]. A folded-bitline architecture, for example, that was first introduced to replace an older open-bitline architecture in DRAM [14,15] is now well adopted in FeRAM [16]. The bitlines are folded to lie on the same side of a sense amplifier, as shown in Fig. 9, instead of lying open on opposite sides of the sense amplifier, to reduce chances of any bitline mismatch that could occur due to process variations.

On the other hand, the requirement for having an extra line – that is, the plateline – in an FeRAM has called for original circuit architectures that were not required in a DRAM. There are various memory architectures that have been developed for FeRAM. We only discuss three well-known architectures in this chapter, and we refer the readers to the original papers [20,21,22,23, 24,25,26,27,28] for the most recently proposed FeRAM architectures.

5.1 The Wordline-Parallel Plateline (WL//PL)

Figure 10a shows a simplified block diagram of a WL//PL architecture. As its name suggests, the PL is run parallel to the WL in this architecture. When a WL and PL pair is activated, an entire row that shares the same WL and PL is accessed at once. It is impossible, in this architecture, to access a single cell without accessing an entire row. In fact, this is common in almost every RAM, since the adjacent cells in a row store the adjacent bits of a byte, which are accessed simultaneously. Sometimes, the PL in this architecture is shared between two adjacent rows to reduce the array area by eliminating a metal

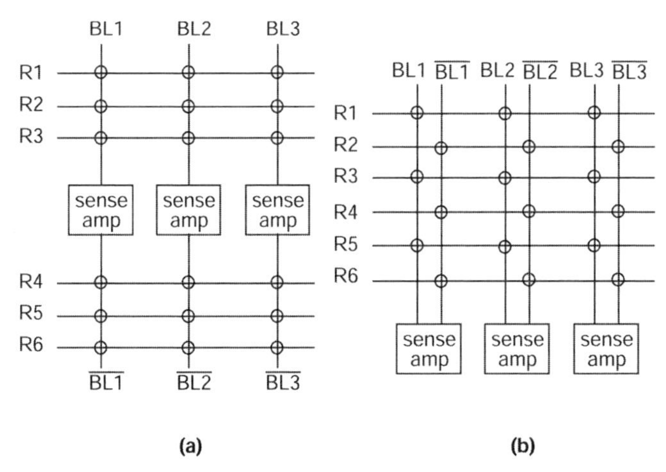

Fig. 9. Ferroelectric memory architectures: (**a**) the open-bitline architecture, and (**b**) the folded-bitline architecture

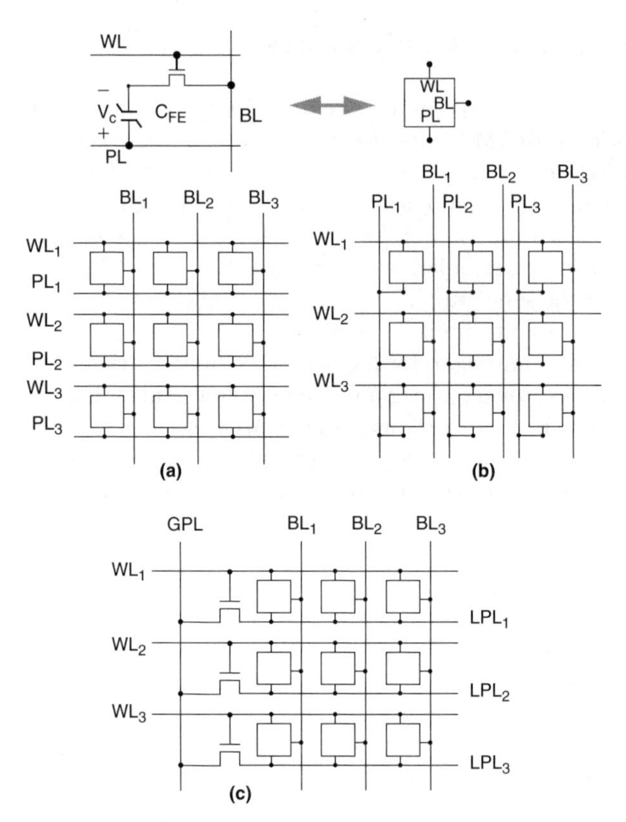

Fig. 10. A block diagram of a ferroelectric memory with (**a**) WL//PL architecture, (**b**) BL//PL architecture, and (**c**) segmented-PL architecture

line [20]. In this case, the unaccessed cells connected to an activated PL can be disturbed. This is due to the voltage that develops across the ferroelectric capacitors of the nonselected cells with the active PL. Ideally, one expects this voltage to be zero, because the storage nodes of the cells should be floating. However, the parasitic capacitance of a storage node forms a capacitor divider with the ferroelectric capacitor itself and produces a nonzero voltage across the ferroelectric capacitor. For a stored "0" datum, the disturb voltage is in the direction that reinforces the "0". For a stored "1" datum, however, the disturb voltage is in the direction of changing the state of the memory cell. If this voltage is small enough (much less than the coercive voltage of the ferroelectric capacitor), it can be ignored. Otherwise, a data "1" can be flipped by a sequence of small voltage disturbances.

5.2 The Bitline-Parallel Plateline (BL//PL)

Figure 10b shows an array architecture in which the PL is run parallel to the BL [17]; hence the name BL//PL for the architecture. Unlike the previous architecture, only a single memory cell can be selected by simultaneous activation of a WL and a PL. This is the memory cell that is located at the intersection of the WL and the PL. It is possible to select more than one memory cell in a row by activating their corresponding platelines.

This architecture absorbs the function of a y-decoder in the selection of the platelines. In fact, the activation of the sense amplifiers is controlled by the same signal as the PL. Therefore, only one sense amplifier is activated if only one memory cell needs to be accessed. This reduces the power consumption significantly. On the other hand, if an entire row needs to be accessed, then all of the platelines are selected simultaneously, hence increasing the dynamic power consumption due to charging and discharging of the platelines.

The main disadvantage of this architecture is that activating a PL could disturb all of the cells in the corresponding column [17]. This is very similar to the situation discussed for the WL//PL architecture with PL shared between two adjacent rows.

5.3 The Segmented Plateline (Segmented PL)

A WL//PL architecture is power-consuming and relatively slow, because the PL is activated in its full length to access all of the cells in the row at once. Also, a BL//PL architecture could be power-consuming if multiple platelines were activated to access multiple memory cells in the selected row. For larger arrays, the PL can be segmented into local plate lines (LPL) that run parallel to the WL and controlled by a global plate line (GPL) that runs parallel to the BL [18, 19]. As shown in Fig. 10c, a GPL is ANDed with the WL to generate the signal for the LPL. Since the LPL is only connected to a few memory cells (eight in this example), it can respond much faster than a PL in the WL//PL architecture. Also, since the GPL is gated by the WL, there is no disturbance to the nonselected cells in the column as it was in the BL//PL architecture.

Among the three architectures discussed so far, the segmented-PL architecture seems to be the most feasible architecture for a large-density ferroelectric memory. A compromise between speed and power consumption can be made by choosing the number of LPLs per GPL.

6 Summary

Ferroelectric capacitors are the basic storage elements for ferroelectric memories. We presented an overview of ferroelectric capacitor behavior in response to a low-frequency sinusoidal waveform and a high-frequency pulse sequence. Also, we described the read and write operations for a basic ferroelectric memory cell, as well as sensing methods. In the last section, we briefly discussed conventional architectures for ferroelectric memories.

Acknowledgements

Generous financial support from Fujitsu Labs, Japan, and the Natural Sciences and Engineering Council of Canada (NSERC) are acknowledged.

References

1. B. Ricco, G. Torelli, M. Lanzoni, A. Manstretta, H. Maes, D. Montanari, A. Modelli: Proc. IEEE **86**, 2399 (1998)
2. P. Pavan, R. Bez, P. Olivi, E. Zanoni: Proc. IEEE **85**, 1248 (1997)
3. K. Amanuma, T. Tatsumi, Y. Maejima, S. Takahashi, H. Hada, H. Okizaki, T. Kunio: Tech. Dig. Intern. Electron Devices Meeting, (1998) p. 363
4. J. Yamada, T. Miwa, H. Koike, H. Toyoshima, K. Amanuma, S. Kobayashi, T. Tatsumi, Y. Maejima, H. Hada, H. Mori, S. Takahashi, H. Takeuchi, T. Kunio: Tech. Dig. Intern. Solid-State Circuits Conf. (2000) p. 270
5. R.E. Jones Jr.: IEEE Custom Integrated Circuits Conf. (1998) p. 431
6. A. Sheikholeslami, P.G. Gulak: IEEE Trans. Ultrason. Ferroelectr. Freq. Control **44**, 917 (1997)
7. A. Sheikholeslami, P.G. Gulak: Proc. IEEE **88**, 667 (2000)
8. J.C. Burfoot, G.W. Taylor: *Polar Dielectrics, Their Applications* (University of California Press, Berkeley 1979)
9. A. Sheikholeslami, P.G. Gulak, H. Takauchi, H. Tamura, H. Yoshioka, T. Tamura: IEEE Trans. Ultrason. Ferroelectr. Freq. Control **44**, 784 (2000)
10. W. Kraus, L. Lehman, D. Wilson, T. Yamazaki, C. Ohno, E. Nagai, H. Yamazaki, H. Suzuki: Tech. Dig. Symp. VLSI Circuits (1998) p. 242
11. N. Abt: Mater. Res. Soc. Symp. Proc. **200**, 303 (1990)
12. Y. Chung, M. Choi, S. Oh, B. Jeon, K. Suh: Tech. Dig. Symp. VLSI Circuits (1999)
13. W. Kinney: Integr. Ferroelectr. **4**, 131 (1994)
14. R.F. Harland: US Patent No. 4045783 (Aug. 30, 1977)
15. R.C. Foss: Tech. Dig. Intern. Solid-State Circuits Conf. (1979) p. 140
16. W.L. Larson: US Patent No. 5371699 (Dec. 6, 1994)
17. R. Womak, D. Tolsch: Tech. Dig. Intern. Solid-State Circuits Conf. (1989) p. 242
18. T. Sumi, N. Moriwaki, G. Nakane, T. Nakakuma, Y. Judai, Y. Uemoto, Y. Nagano, S. Hayashi, M. Azuma, E. Fujii, S. Katsu, T. Otsuki, L. McMillan, C.A. Paz de Araujo, G. Kano: Tech. Dig. Intern. Solid-State Circuits Conf. (1994) p. 268
19. R.E. Jones Jr.: US Patent No. 5373463 (Dec. 13, 1994)
20. M. Aoki, H. Takauchi, H. Tamura: Tech. Dig. Intern. Electron Devices Meeting (1997) p. 942
21. D. Takashima, I. Kunishima: IEEE J. Solid-State Circuits **33**, 787 (1998)
22. D. Takashima, S. Shuto, I. Kunishima, H. Takenaka, Y. Oowaki, S. Tanaka: Tech. Dig. Intern. Solid-State Circuits Conf. (1999) p. 102
23. N. Tanabe, S. Kobayashi, T. Miwa, K. Amanuma, H. Mori, N. Inoue, T. Takeuchi, S. Saitoh, Y. Hayashi, J. Yamada, H. Koike, H. Hada, T. Kunio: Tech. Dig. Symp. VLSI Circuits (1998) p. 124
24. N. Tanabe, S. Kobayashi, H. Hada, T. Kunio: Tech. Dig. Intern. Electron Devices Meeting (1997) p. 863

25. S. Kawashima, T. Endo, A. Yamamoto, K. Nakabayashi, M. Nakazawa, K. Morita, M. Aoki: Tech. Dig. Symp. VLSI Circuits (2001) p. 127
26. H. B. Kang, D. M. Kim, K. Y. Oh, J. S. Roh, J. J. Kim, J. H. Ahn, H. G. Lee, D. C. Kim, W. Jo, H. M. Lee, S. M. Cho, H. J. Nam, J. W. Lee, C. S. Kim: Tech. Dig. Intern. Solid-State Circuits Conf. (1999) p. 108
27. H. Koike, T. Otsuki, T. Kimura, M. Fukuma, Y. Hayashi, Y. Maejima, K. Amanuma, M. Tanabe, T. Matsuki, S. Saito, T. Takeuchi, S. Kobayashi, T. Kunio, T. Hase, Y. Miyasaka, N. Shohota, M. Takada: Tech. Dig. Intern. Solid-State Circuits Conf. (1996) p. 368
28. H. Hirano, T. Honda, N. Moriwaki, T. Nakakuma, A. Inoue, G. Nakane, S. Chaya, T. Sumi: IEEE J. Solid-State Circuits **32**, 649 (1997)

High-Density Integration

Kinam Kim

Semiconductor R & D Center, Semiconductor Business, Samsung Electronics Co.,
San #24, Nongseo-Ri, Kiheung-Eup, Yongin-City, Kyungki-Do, 449-711, Korea
kn_kim@samsung.com

Abstract. Ferroelectric random access memory (FeRAM) has been pursued as a promising new memory due to its ideal memory properties, such as fast random access in read/write mode and nonvolatility with unlimited usage. However, FeRAM has only achieved limited success in low-density applications because of its large cell size (large cell size factor). Recently, mega-bit density scaled FeRAMs have been developed with many key process and new integration technologies. Among those technologies, 1T-1C COB (capacitor over bit line) cell structure is the most important, because it can greatly reduce the cell size of FeRAM. With these technologies, 0.25 μm 32 Mb FeRAM with the 15 F^2 cell size factor of a 1T-1C COB cell has been successfully developed. In order to further reduce the cell size factor to its theoretical limits of 6 F^2 or 8 F^2, it is highly desirable to develop three-dimensional ferroelectric capacitor technology, where CVD technology is essential for fabricating ferroelectrics as well as electrodes.

1 Introduction

Ideal memory should have properties such as random access, fast read and fast write operations, unlimited usage with nonvolatility, and a small cell size. None of the existing memories, such as DRAM, SRAM, and Flash memory, can satisfy these requirements for an ideal memory. For instance, DRAM and SRAM do not possess nonvolatility, while Flash memory satisfies neither fast write operation nor unlimited usage. On the way to pursuing the ideal memory, FeRAM, which potentially satisfies all of the requirements for an ideal memory, has been introduced in the past decade [1].

In spite of the excellent properties of the ferroelectric memory, only low-density FeRAM products have been commercialized compared to its counterparts among memory devices, such as DRAM, SRAM, and Flash memory. The low-density FeRAM products have resulted from the large cell size of its 2T2C cell architecture. Cell size has been one of the most important figures of merit to measure the competitiveness of memory products, regardless of the memory type. The cell size of the memory is typically evaluated by the cell size factor for a given technology generation. The cell size factor is determined using the ratio of the unit cell area to the square of the minimum feature size F. The minimum feature size F in memory devices is typically given by the half pitch of the minimum pitch of the bit line or word line. Therefore, the cell size factor denotes how much area is consumed in storing

H. Ishiwara, M. Okuyama, Y. Arimoto (Eds.): Ferroelectric Random Access Memories,
Topics Appl. Phys. **93**, 165–176 (2004)
© Springer-Verlag Berlin Heidelberg 2004

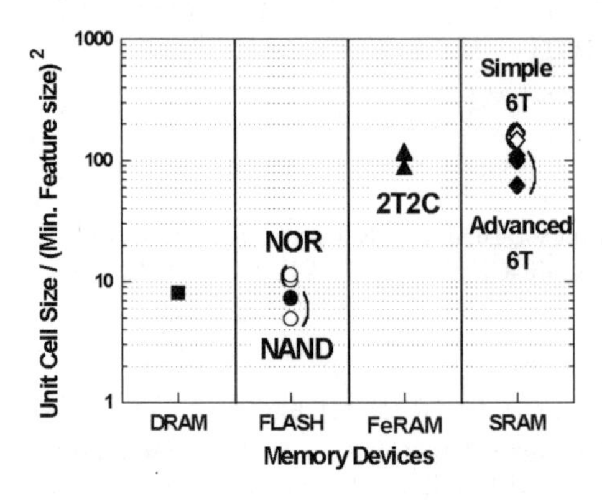

Fig. 1. The cell size factor for various memories (DRAM, SRAM, Flash, and FeRAM)

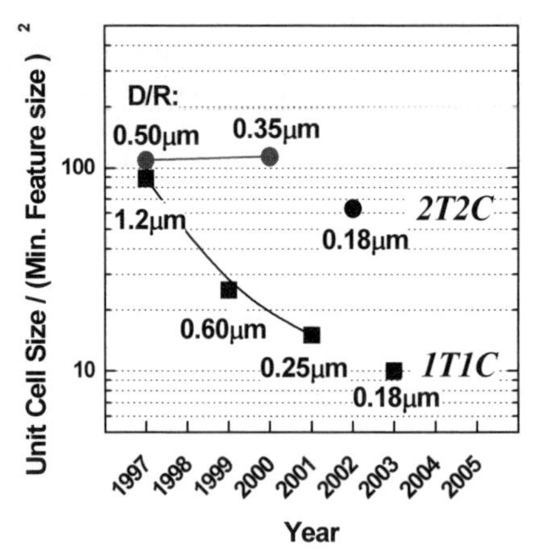

Fig. 2. The evolution of the cell size factor for 2T2C and 1T-1C FeRAMs

one bit of data for the given technology. Figure 1 shows the cell size factors for various types of commercially available memory products. As indicated in Fig. 1, the cell size factor of DRAM approaches its theoretical limit of $8\,F^2$ in folded bit line architecture [2]. NAND Flash has the smallest cell size factor due to its one-transistor (1T) cell architecture. SRAM has the biggest cell size factor, ranging from 120 to 60, and depending on the configurations of a six-transistor (6T) cell. The cell size factor of the FeRAM with 2T2C cell architecture ranges from 100 to 60. Therefore, current FeRAM is less competitive than DRAM and SRAM in cost-effectiveness.

In order to improve the cost-effectiveness of FeRAM so that it is close to that of DRAM and/or Flash memory, it is necessary to reduce the cell size factor of FeRAM close to that of DRAM and/or Flash memory by using a 1T-1C cell architecture [3]. Since the first 1T-1C FeRAM was demonstrated, much effort has been put into developing a high-density 1T-1C FeRAM whose cell size factor can compete with that of DRAM. Figure 2 shows the trend of the cell size factor of the 1T-1C FeRAM. Since the 1T-1C 64kb FeRAM with a CUB (capacitor under bit line) cell architecture was successfully developed [4], the cell size factor of the 1T-1C FeRAM has been greatly reduced, from 80 F^2 for the 64kb FeRAM to 25 F^2 for the 4 Mb FeRAM, where the COB (capacitor over bit line) cell architecture was first developed [5]. The appreciable reduction of the cell size factor is mainly due to the fact that the ferroelectric capacitor is placed directly on top of the array transistor, thereby eliminating the unwanted area that must be insured in CUB cell architecture. Recently, the cell size factor of the 1T-1C FeRAM has been further decreased to 15 F^2 for the 32 Mb FeRAM [6], where many novel technologies have been developed. Table 1 summarizes the key technology features used for advancing the 1T-1C FeRAM from 80 F^2 64 kb FeRAM to 15 F^2 32 Mb FeRAM. The fabricated cell structures corresponding to each cell size factor are also illustrated in Fig. 3.

Table 1. A summary of the key technologies used for advancing the 1T-1C FeRAM

Technology features	80 F^2 CUB 1T-1C	25 F^2 COB 1T-1C	15 F^2 COB 1T-1C
Density	64 kb	4 Mb	32 Mb
Feature size (μm)	1.2	0.6/0.4	0.25
Plug material	–	Poly	W
F-cap size	17.4 F^2	14.2 F^2	7 F^2
Effective F-cap size	6.3 F^2	4.5 F^2	4 F^2
F-cap structure	Ir/IrO_2/PZT/ Pt/IrO_2/TiO_2	Ir/IrO_2/PZT/ Pt/IrO_2/Ir/Ti	Ir/IrO_2/PZT/Pt/ IrO_2/Ir/TiAlN
F-cap height (μm)	0.75	0.75	0.4
EBL (encapsulated hydrogen barrier layer)	TiO_2	Al_2O_3	Al_2O_3
P/L	Separate (1PL/cell)	Separate (1PL/cell)	Common (1PL/2cell)
Metallization	Double metal	Triple metal	Triple metal
Operational voltage (V)	5.0	3.3	3.3

Innovative cell architecture and integration technology, together with advanced process technology, will further decrease the cell size factor in future. It is eventually expected to reach to its ultimate cell size factor limit of either 6 F^2 in open bit line cell architecture or 8 F^2 in folded bit line cell architecture [7]. In this chapter, the key technologies for reducing the cell size factor of the 1T-1C FeRAM down to 15 F^2 will be discussed. The key technologies are summarized as highly reliable ferroelectric capacitor technology, low damage and vertical shape reactive ion etching (RIE) technology for capacitor, plug technology for COB cell structure, encapsulated hydrogen barrier layer technology, and novel integration technology. Finally, emerging technologies will also be suggested for the sub-10 F^2 cell size factor in the future.

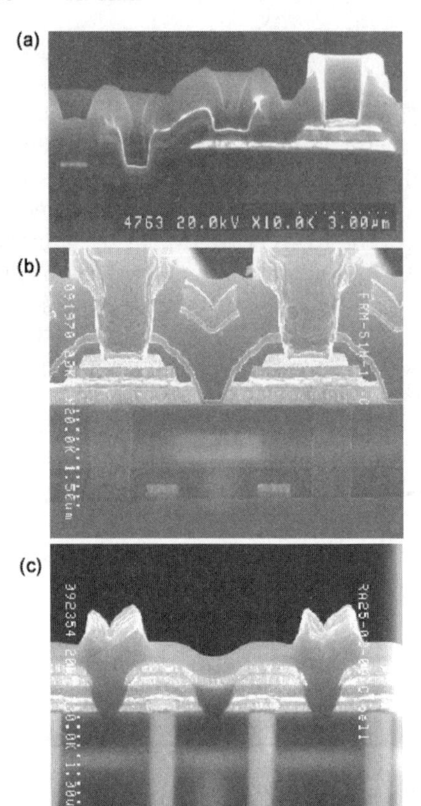

Fig. 3. Advances in the 1T-1C FeRAM from 80 F^2 to 15 F^2 cell size factor. (**a**) A cross-sectional SEM image of an 80 F^2 cell of a 64 kb 1T-1C CUB FeRAM with 1.2 µm technology. (**b**) A cross-sectional SEM image of a 25 F^2 cell of a 4 Mb 1T-1C COB FeRAM with 0.60 µm technology. (**c**) A cross-sectional SEM image of a 15 F^2 cell of 32 Mb 1T-1C COB FeRAM with a 0.25 µm technology

2 Key Technologies for High-Density FeRAM with a Small Cell Size Factor

2.1 Reliable Ferroelectric Capacitor Technology

A reliable ferroelectric capacitor is essential for FeRAM technology regardless of its cell architecture (2T2C or 1T-1C) or its cell structure (CUB or COB). It is well known that a high-quality ferroelectric film and appropriate bottom/top electrodes are the most crucial factors in achieving a reliable ferroelectric capacitor. The high-quality ferroelectric film must have high directionality of the crystal orientation without any second phase, such as a pyrochlore phase. In order to fabricate highly directional ferroelectric film, it is very important to optimize the crystallization process as well as the material parameters of the ferroelectric film, such as the composition, thickness, and deposition method. By systematically investigating the various parameters, it was concluded that the sol-gel method, Ti-rich PZT film, and rapid thermal processing (RTP) are optimal for preparing reliable ferroelectic films. The details of our ferroelectric film have been reported elsewhere [3]. For the appropriate bottom electrode, platinum (Pt) is used, because a ferroelectric film with high (111) directionality can easily be grown on top of the Pt electrode. As for the top electrode, either IrO_2 or Ir/IrO_2 is used, because the metal oxide top electrode greatly improves the resistance of the ferroelectric capacitor against fatigue [4].

2.2 Vertical Shape Capacitor Etching Technology with Low Etching Damage

In addition to a high-quality ferroelectric film and proper electrodes, capacitor etching processes for the top electrode, the ferroelectric film, and the bottom electrode are very critical for maintaining good properties of ferroelectric capacitors, because severe etching damage is easily generated during the capacitor etching process [8]. These issues, which must be considered in seriously in FeRAM fabrication, are not relevant in the case of 1T-1C DRAM, because it is not necessary to etch the dielectric of the DRAM capacitor. In the 1T-1C DRAM capacitor, at first, each storage node (electrode) is separately formed, and then the dielectric is blanket-deposited. Plate electrode was blanket-deposited on the dielectric. The plate node is then defined as array by array, not cell by cell, so that it is not necessary to etch the capacitor dielectric cell by cell.

On the other hand, in FeRAM, the bottom metal electrode of the FeRAM capacitor is first blanket-deposited, and ferroelectric thin film is coated over the bottom metal electrode, which is then annealed at a high temperature for crystallization. After heat treatment for crystallization of the ferroelectric film, the top metal electrode for the plate node is also blanket-deposited over the crystallized ferroelectric film. After depositing all layers for the capacitor, each capacitor is formed by sequential etching of the top electrode, the ferroelectric thin film, and the bottom electrode. Although these three different layers, the top electrode, the ferroelectric film, and the bottom electrode, can be etched with an one-mask, two-mask, or three-mask process, it has been concluded that a three-mask process produces the smallest plasma-induced etching damage [9, 10]. The plasma-induced damage becomes much more pronounced as the number of masks decreases. On the contrary, the area penalty increases as the number of masks increases. Therefore, there is a trade-off between area penalty and plasma-induced damage for the current FeRAM capacitor technology. In order to make a small cell size for a 1T-1C FeRAM for a given CMOS technology, either a low-damage single-mask process or a novel process that is no longer based on ferroelectric film etching must be developed [10].

In order to etch a capacitor stack with one mask, it is very important to select an appropriate material for the hard etching mask. Thick titanium nitride (TiN) film is found to have excellent properties as a mask layer, such as good selectivity against the top electrode, the PZT film, and the bottom electrode. Even though the etching-induced damage might be generated at

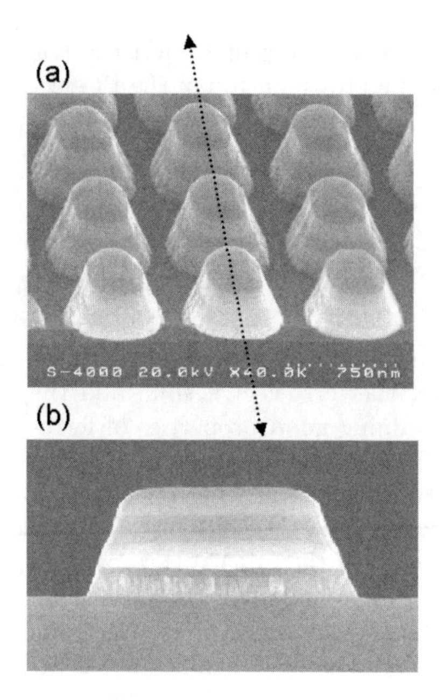

Fig. 4. The capacitor etching technology with one etching mask for 0.25 µm FeRAM technology. (**a**) The etching profile of the capacitor array, where the capacitor stack height is 400 nm. (**b**) The etching profile of an individual capacitor whose angle is 72°

the side-wall of the PZT film, it is easily removed by wet treatment with special chemistry [8]. The one-mask etching process can be relatively easily achieved by decreasing the stack height of the capacitor. Compared to a thick stack height, a small stack height requires less etching time, resulting in a small amount of etching-induced damage. In scaling down the capacitor stack height, it is crucial to scale down the ferroelectric film. In spite of the degraded performance of the ferroelectric capacitor with a reducing thickness of both the PZT and the bottom Pt electrode [11], we can maintain high remanent polarization, even in a capacitor with a total thickness of 400 nm, by optimizing the PZT process and adjusting the Pt electrode thickness. Figure 4 shows the etching profile of a ferroelectric capacitor with a total thickness of 400 nm.

2.3 Plug Technology

The small cell size factor for 1T-1C FeRAM must be implemented into the COB cell structure, because the memory cell capacitor is located directly over the array transistor and the bit line, thus eliminating the unwanted area penalty that cannot be avoided in the CUB cell structure. In order to produce the COB cell, it is necessary to develop a novel technology to connect between the diffusion layer of the array transistor and the bottom electrode of the ferroelectric capacitor. The connection technology is typically called plug technology, and is quite different from that of the 1T-1C COB DRAM. The difference comes from the fact that the FeRAM plug technology exists between the bottom electrode of the ferroelectric capacitor and the source/drain (S/D) junction of the array transistor, and it must provide an ohmic connection at both the plug–S/D junction and the plug–bottom electrode interfaces. Furthermore, plug technology should not have any detrimental influence on the ferroelectric capacitor. It is well known that PZT film is grown by heterogeneous nucleation, which means that PZT film properties are strongly dependent on the substrate materials [12].

In order to meet both requirements, FeRAM plug technology typically consists of the plug itself, the ohmic layer on top of the plug, and oxygen barrier layers that are placed between the ohmic layer and the bottom electrode. Among several candidates for oxygen blocking layers, a binary-stacked layer of Ir/IrO_2 or a triple-stacked layer of $Ir/IrO_2/TiAlN$ show strong resistance to oxygen penetration. With oxygen blocking layers, the plug and ohmic layer can be protected from oxidation during the high-temperature crystallization process in ambient oxygen.

For the plug material, both polysilicon and the W-plug are quite suitable, because the polysilicon plug is well matured technology, as manifested in current DRAM technology, and the W-plug is the standard process in the metal contact process of logic technology. When FeRAM is merged with logic technology, the use of FeRAM with the W-plug will be greatly beneficial. In the case of the polysilicon plug, a thin ohmic layer of cobalt silicide ($CoSi_2$) is deposited on top of the polysilicon plug prior to formation of the oxygen blocking layer, for good ohmic contact between the polysilicon plug and the bottom electrode [13]. The optimized plug technology can lead to high-quality ferreoelectric film being grown on top of the Pt bottom electrode while maintaining good contact resistance. On the other hand, in the case of the W-plug, a thin ohmic layer may not be necessary. Figure 5 shows the evaluation of the W-plug technology for the 0.25 μm FeRAM technology generation.

2.4 Encapsulated Hydrogen Barrier Technology

It is well known that a ferroelectric capacitor is easily degraded by hydrogen-induced damage [9]. It is also generally considered that a proper hydrogen barrier layer, which encapsulates the ferroelectric capacitor, is the most suitable solution for advanced CMOS technology, where various hydrogen compounds are heavily used in either deposition or etching processes as well as sintering processes. The encapsulated hydrogen barrier must have ideal barrier properties, such as a good stopping power to block hydrogen penetration, conformal deposition, good matching with the ferroelectric capacitor without causing its properties to deteriorate, and good compatibility with current CMOS technology. Among many candidates, a thin Al_2O_3 technology is found to be most adequate for a reliable hydrogen barrier, because it can satisfy the stringent requirements of encapsulated hydrogen barrier technology [14]. By using thin Al_2O_3 hydrogen barrier layer technology, we

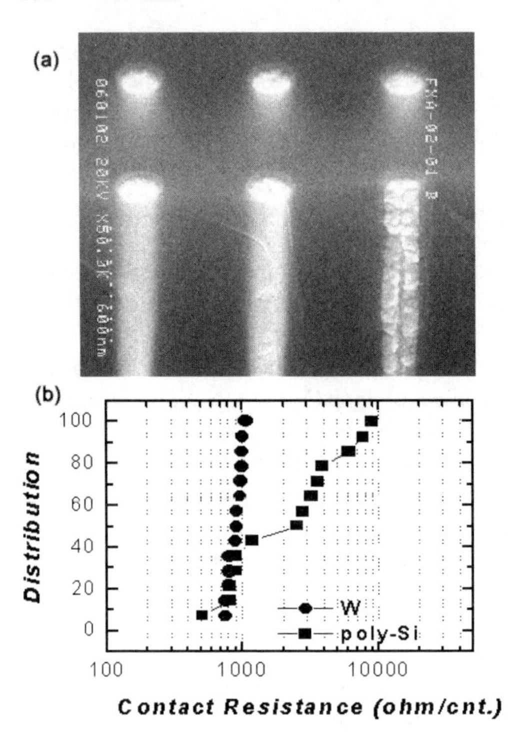

Fig. 5. The W-plug technology for 0.25 μm FeRAM technology. (**a**) A SEM image of the W-plug used for a 0.25 μm 32 Mb 1T-1C COB FeRAM. (**b**) The contact resistance of the W-plug between the bottom electrode and the diffusion junction of an array transistor of the 0.25 μm FeRAM technology, where the contact size is 0.25 μm × 0.25 μm

can eliminate hydrogen damage from hydrogen-rich CMOS processes, so that a reliable ferroelectric capacitor can be maintained even after full integration.

2.5 Novel Integration Technology

In order to realize the 15 F^2 cell size factor of 32 Mb FeRAM, it is required to develop novel cell integration technologies, in addition to key process technologies, as previously discussed. The novel cell structure has been invented by merging two adjacent plate line contacts into one common plate line contact. Thus, in the novel integration technology used for 32 Mb FeRAM, every adjacent pair of cells shares one common plate line [15]. It should be noted that every cell has its own separate plate line in conventional 1T-1C COB FeRAM, such as 25 F^2 4 Mb FeRAM. The common plate line technology consists of four steps: the formation of the capacitor contact, an additional local plate line, a local common via for two adjacent cells, and a global common plate line. These process steps are depicted in Fig. 6. The brief process sequences of common plate line technology are as follows. In order to fabricate the common plate line, it is indispensable to have an inter-layer dielectric (ILD) planarization process. For the purpose of ILD planarization, undoped silicate glass (USG) oxide is deposited on patterned substrates, and is followed by spin-coating of spin-on-glass (SOG) oxide films. After the SOG oxide film

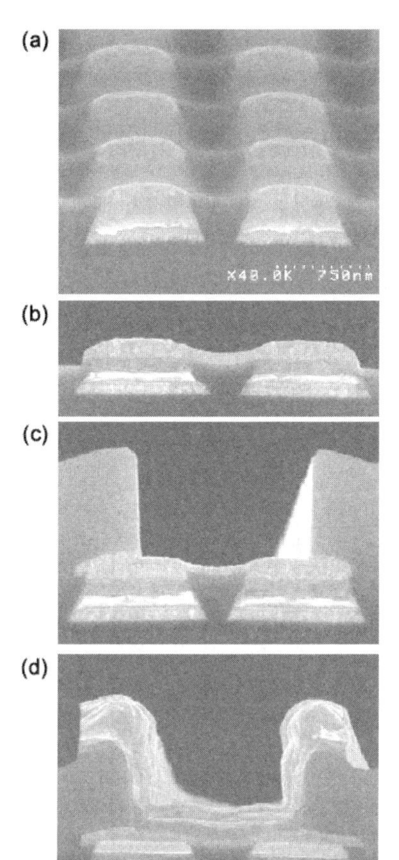

Fig. 6. The brief process sequence for common plate line technology. (**a**) Formation of the capacitor contact without a via. (**b**) Formation of additional a local plate line. (**c**) Formation of a common via for two adjacent cells. (**d**) Formation of a common plate line

coating process, the ILD oxide is then etched back until the top electrode is exposed (Fig. 6a). After the etch-back process, a landing pad with Ir/IrO_2 for an additional local plate line is formed directly over the top electrode, and is then shared by two adjacent cells (Fig. 6b). After that, thick ILD is deposited and a local common via is defined for two adjacent cells as shown in Fig. 6c, and then final recovery annealing process is carried out. Finally, Al/TiN metal for the plate line is deposited and patterned (Fig. 6d). Figure 7 shows the hysteresis loop and the Q–V curves of novel capacitors of 15 F^2 32 Mb FeRAM after completion of the full integration process.

2.6 Sub-10 F^2 Future FeRAM Technology

The key technologies, such as reliable ferroelectric capacitor technology, low-damage and vertical RIE etching technology, plug technology, encapsulated

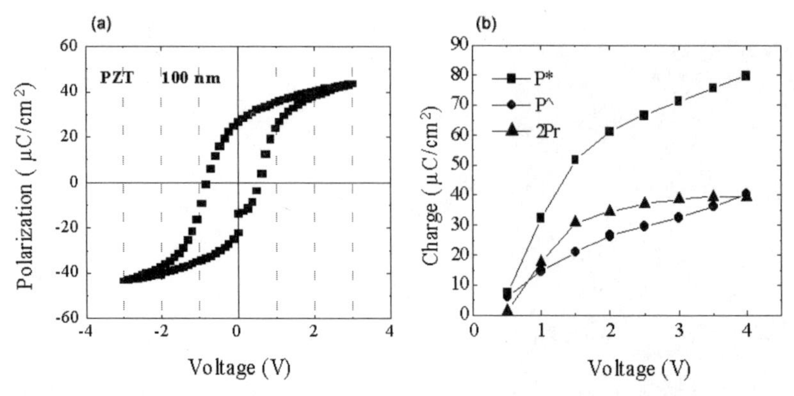

Fig. 7. Capacitor hysteresis properties and the Q–V curve of a novel cell for 32 Mb 1T-1C COB FeRAM. (**a**) The hysteresis curve. (**b**) The Q–V curve

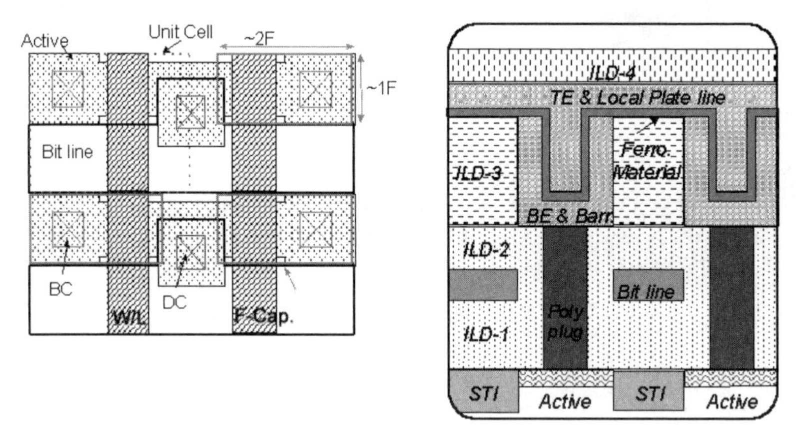

Fig. 8. A schematic of a future 6 F^2 1T-1C COB FeRAM. (**a**) The cell layout for a 6 F^2 1T-1C FeRAM. (**b**) The vertical structure of a 6 F^2 1T-1C FeRAM

hydrogen barrier technology, and common plate line technology, make it possible to have a 15 F^2 cell size factor, thereby improving the cost-effectiveness of FeRAM. However, the cell size factor of the state-of-the art FeRAM is still bigger than that of the current DRAM and Flash. In order to further improve the cost-effectiveness of FeRAM, technology innovations for the memory cell should be accomplished. In principle, an 1T-1C cell can be fabricated with a 6 F^2 cell size in open bit line cell architecture or an 8 F^2 cell size in folded bit line cell architecture. In order to realize either the 6 F^2 or the 8 F^2 cell size, it is essential to reduce the capacitor area. The available capacitor areas for given 6 F^2 and 8 F^2 cell are only allowed to be 2 F^2 and 3 F^2, respectively. In order to have enough signal charge from the limited capacitor area, we have to increase the capacitor surface area by forming a three-dimensional capacitor.

The proposed cell scheme, as shown in Fig. 8, is very similar to those of the most advanced MIM capacitor of DRAM technology. The fabrication process and issues of the proposed cell scheme are also strongly correlated with those of the MIM capacitor. Therefore, the key technologies for future FeRAM will be the CVD processes of metal electrodes and metal oxides, such as Pt, Ir, and IrO_2, barrier layer processes as well as CVD ferroelectric processes.

3 Summary

The key technologies of 1T-1C FeRAM, especially 1T-1C COB FeRAM for high-density memory, have been discussed. Novel integration technologies, such as common plate technology together with a high-quality ferroelectric capacitor, low-damage vertical capacitor etching technology, plug technology, and encapsulated hydrogen barrier technology, are decisive factors for producing a small cell factor of $15\,F^2$ for 32 Mb 1T-1C COB FeRAM. For a sub-$10\,F^2$ cell size factor, it is necessary to decrease the capacitor area down to 2–$3\,F^2$. When FeRAM technology advances toward the sub-quarter micron technology generation, the available charge will suffer because of such a limited area. Therefore, a three-dimensional capacitor is inevitable to increase the capacitor area, and this will not be possibly accomplished without CVD processes for ferroelectrics and electrodes.

References

1. J. T. Evans, R. Womack: IEEE J. Solid-State Circuits **23**, 1171 (1988)
2. K. Kim: Microelectron. Reliability **40**, 191 (2000)
3. K. Kim: Integr. Ferroelectr. **25**, 149 (1999)
4. D. J. Jung, S. Y. Lee, B. J. Koo, Y. S. Hwang, D. W. Shin, J. W. Lee, Y. S. Chun, S. H. Shin, M. H. Lee, H. B. Park, S. I. Lee, K. N. Kim, J. G. Lee: Tech. Dig. Symp. VLSI Technol. (1998) p. 122
5. S. Y. Lee, D. J. Jung, Y. J. Song, B. J. Koo, S. O. Park, H. J. Cho, S. J. Oh, D. S. Hwang, S. I. Lee, J. K. Lee, Y. S. Park, I. S. Jung, K. Kim: Tech. Dig. Symp. VLSI Technol. (1999) p. 141
6. M. K. Choi, B. G. Jeon, N. K. Jang, B. J. Min, K. Kim: Tech. Dig. Int. Solid-State Circuits Conf. (2002) No. 9.6
7. K. Kim: Integr. Ferroelectr. **36**, 21 (2001)
8. J. K. Lee, Y. Park, I. Chung, S. J. Oh, D. J. Jung, Y. S. Song, B. J. Koo, S. Y. Lee, K. Kim, S. B. Desu: J. Appl. Phys. **86**, 6376 (1999)
9. D. J. Jung, B. G. Jeon, H. H. Kim, Y. J. Song, B. J. Koo, S. Y. Lee, S. O. Park, Y. W. Park, K. Kim: Tech. Dig. Int. Electron Devices Meeting (1999) p. 279
10. N. W. Jang, Y. J. Song, H. H. Kim, D. J. Jung, S. Y. Lee, S. H. Joo, K. M. Lee, K. Kim: Tech. Dig. Symp. VLSI Technol. (2000) p. 34
11. Y. J. Song, N. W. Jang, D. J. Jung, H. H. Kim, H. J. Joo, S. Y. Lee, K. M. Lee, S. H. Joo, S. O. Park, K. Kim: Ext. Abstracts Solid-State Devices Materials, Tokyo (2001) p. 522

12. Y. J. Song, Y. Zhu, S. B. Desu: Appl. Phys. Lett. **72**, 2686 (1998)
13. H. H. Kim, D. J. Jung, S. Y. Lee, Y. J. Song, N. W. Jang, J. K. Lee, K. Kim: Integr. Ferroelectr. **39**, 239 (2001)
14. Y. T. Lee, H. J. Cho, S. J. Oh, S. H. Joo, J. J. Lee, K. M. Lee, S. O. Park, Y. W. Park, S. I. Lee: Ext. Abstracts Solid-State Devices Materials, Tokyo (1999) p. 394
15. K. Kim: 1st Int. Meeting Ferroelectric Random Access Memories, Gotemba (2001) Ext. Abstracts, p. 11

Testing and Reliability

Yasuhiro Shimada

Semiconductor Device Research Center, Matsushita Electric Industrial Co. Ltd.,
1-1 Saiwai-cho, Takatsuki, Osaka 5691193, Japan
shimada.yasuhiro@jp.panasonic.com

Abstract. In ferroelectric random access memories (FeRAMs), the difference in the displacement charge between the two polarization states is fundamental to discriminate the binary states, "1" and "0". Therefore, the durability of the difference in the displacement charge and circuit techniques for sensing the difference are intimately related to the reliability of FeRAMs. In this chapter, we deal with electrical degradation phenomena in ferroelectric capacitors, such as resistance degradation, time-dependent dielectric breakdown (TDDB), ferroelectric fatigue, charge retention, and imprint, which affect the sensing of displacement charges from a memory cell all the way. In each topic, we refer to related testing techniques to characterize and qualify the performance and reliability of ferroelectric capacitors. These discussions include some fundamental considerations on aging mechanisms in ferroelectric capacitors, which may provide essential and meaningful feedback for materials, circuit design, and manufacturing process improvements.

1 Leakage Current

Since ferroelectrics are used as capacitor dielectrics in FeRAMs, leakage currents and electrical breakdown are potential concerns in determining the operating voltage and device performance. Leakage currents in insulators, including ferroelectrics, arise from carrier conduction under high fields and high temperatures. For this reason, a charge in a ferroelectric capacitor under an electric field cannot be stored definitely, and it ultimately leaks off. Leakage current measurements are made by applying a ramped potential across the ferroelectric capacitor at a given temperature. The current–voltage (J–V) curves are plotted on different scales in accordance with an adequate conduction mechanism.

Possible conduction mechanisms in insulators are briefly summarized in [1], which includes Schottky emission, Frenkel–Poole emission, Fowler–Nordheim tunneling, and space-charge-limited and ohmic currents. In particular, the Schottky emission and Frenkel–Poole emission mechanisms are responsible for the leakage currents at high electric fields in ferroelectric capacitors using $Ba_{1-x}Sr_xTaO_3$ (BST) [2,3,4,5,6,7], $PbZr_xTi_{1-x}O_3$ (PZT) [8,9,10], and $SrBi_2Ta_2O_9$ (SBT) [11,12]. Schottky emission occurs when carriers overcome an interfacial potential barrier by thermal activation, so that the current is limited by the interface and depends on the temperature, the applied field, and the barrier height. Frenkel–Poole emission occurs when carriers trapped

H. Ishiwara, M. Okuyama, Y. Arimoto (Eds.): Ferroelectric Random Access Memories,
Topics Appl. Phys. **93**, 177–194 (2004)
© Springer-Verlag Berlin Heidelberg 2004

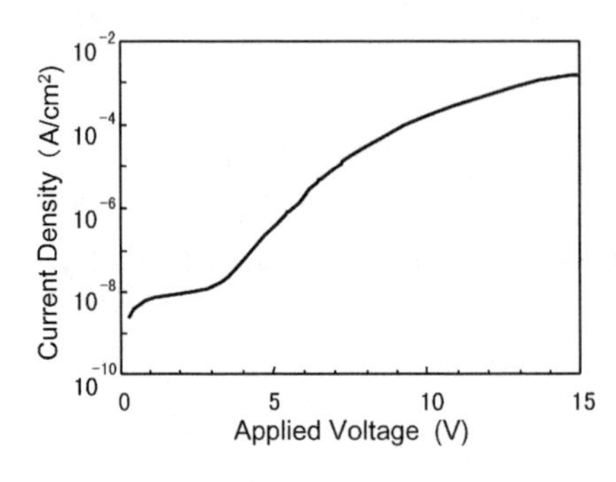

Fig. 1. The current–voltage characteristics of a $Ba_{0.7}Sr_{0.3}TiO_3$ film with a thickness of 140 nm

at defect levels in the bulk are emitted into the conduction band by an applied field that is high enough to change the defect potential, so that the current is limited by the distribution of defect levels in the bulk and depends on the temperature, the applied field, and the depth of the defect levels. When the J–V curve is plotted as log J versus $V^{1/2}$, the slope of the curve at high fields for Frenkel–Poole emission is twice as large as that for Schottky emission [5].

A typical current–voltage characteristic of a BST film with a thickness of 140 nm, prepared using a spin-on technique, is shown in Fig. 1 [2]. The leakage current in the BST film is 2×10^{-8} A/cm^2 at 3.3 V, which results in a charge loss of 0.2 fC per second for a capacitor area of 1 μm^2. Although a typical refresh cycle in DRAM is shorter than 100 ms, the charge loss caused from the leakage current in the BST capacitor is low enough to maintain the stored charge (~30 fC) during a refresh cycle. Thus BST is a promising material for high-density DRAM capacitor dielectrics. In the case of non-volatile FeRAMs, programmed ferroelectric capacitors are subjected to zero bias while the power is turned off, so that leakage currents do not contribute to charge retention. However, a significant leakage component causes a hysteresis deformation, resulting in malfunction as discussed later.

2 Electrical Breakdown

The time to electrical breakdown in a ferroelectric capacitor under a voltage stress is frequently defined as the time for the leakage current to reach a limited value above which the memory device does not work correctly. This electrical degradation is referred to as time-dependent dielectric breakdown (TDDB) and the time to electrical breakdown is referred to as the TDDB life. Thus the TDDB life is associated with resistance degradation over time.

The leakage current of a ferroelectric capacitor under a constant voltage stress shows a variety of time-dependent behaviors, depending on the materials, the film growth conditions, the electrode engineering, and the environmental conditions during the measurement. For instance, a gradual increase in leakage current associated with degradation of the insulation resistance is commonly observed when temperature and DC voltage stresses are applied

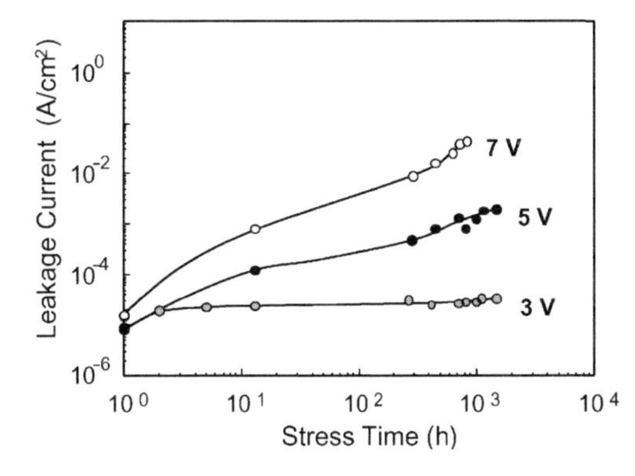

Fig. 2. Variations in the leakage current with stress time. The samples were subjected to stresses of positive applied voltages of 3, 5, and 7 V at 423 K for up to 1500 h. The leakage currents were measured at 3 V at room temperature

to ferroelectric capacitors for a certain period of time [3]. Even under less excessive temperature and voltage stresses, close to normal device operating conditions, and well below the critical onset of instant breakdown, a slight increase in the leakage current is still observed. Under excessive temperature and voltage stresses, close to the occurrence of instant dielectric breakdown, an anomalous increase in the leakage current followed by eventual breakdown is observed. These degradation phenomena in multi-layer ceramic capacitors were modeled quantitatively on the basis of defect chemistry [13]. Similar phenomena have been observed in $BaTiO_3$, BST, and PZT during the later part of the life terminated by the eventual breakdown [8, 10, 14, 15, 16, 17]. Figure 2 shows such leakage current behaviors observed in integrated BST capacitors stressed under different voltage conditions at 150 °C [14].

Many groups to date have pointed out the importance of the role of oxygen vacancies in the time evolution of the bulk-related leakage current of ferroelectric capacitors; that is, a field-driven redistribution of oxygen vacancies in the films would result in the formation of a forward-biased p–n junction [15, 17, 18]. However, their arguments are limited to the change of the leakage current in the bulk. When the leakage current under consideration is originally dominated by the interface-controlled Schottky emission, we must take account of the change in the leakage current at the interface as well as in the bulk.

If the interface barrier is formed by the effect of interface charge fixed at interface states rather than by that of a work function difference between the metal electrode and the ferroelectric, the barrier height depends on the built-in potential ϕ_b across the depletion region within the ferroelectric near the interface. The total charge in the depletion region can consist of ionized oxygen vacancies and can be attracted by the charge at the interface states. Then ϕ_b is given by

$$\phi_b = \frac{1}{2n}\frac{qN_d x_d^2}{\varepsilon\varepsilon_0}\,, \tag{1}$$

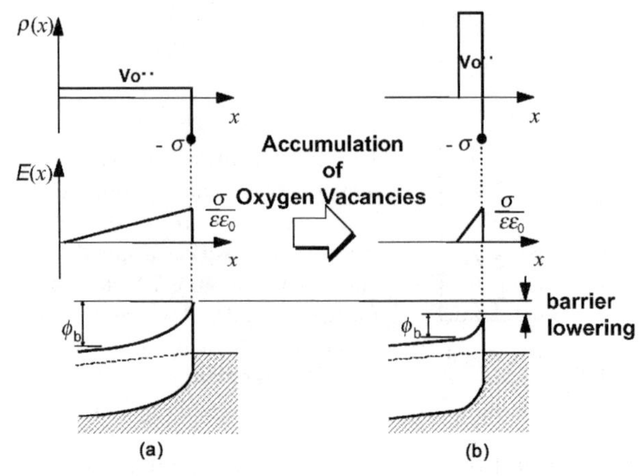

Fig. 3. A comparison of the charge distribution ρ, the electric field E, and energy band variation as a function of the direction x near the cathode of a BST capacitor (**a**) before and (**b**) after stressing

where q is the electronic charge, n is the average ionic valence of oxygen vacancies, N_d is the concentration of oxygen vacancies, x_d is the extent of the depletion region, ε is the dielectric constant of the ferroelectric, and ε_0 is the permittivity of vacuum.

When the temperature is elevated, oxygen vacancies initially distributed across the film are thermally activated and may begin to migrate toward the cathode under a high field. Finally, the oxygen vacancies accumulate near the cathode. Since the total charge in the depletion region $qN_\mathrm{d}x_\mathrm{d}$ remains constant in terms of the fixed interface charge, the extent of the depletion region x_d decreases as the concentration of oxygen vacancies N_d increases. Then the barrier height ϕ_b is lowered due to the accumulation of oxygen vacancies at the cathode, as illustrated in Fig. 3. As a result of the barrier lowering at the cathode, bulk-controlled conduction appears to be predominant. Then the space-charge-limited or Frenkel–Poole emission current will contribute to the total current.

When a lot of sample capacitors are tested, the TDDB lifetimes are distributed over a wide range of time durations. Then we need to assume a well-defined statistical distribution of the TDDB lifetimes. The Weibull distribution is one of the best descriptions for TDDB lifetimes [19]. Using a Weibull plot, for instance, we can determine the relation between voltage stress and TDDB lifetime for given ferroelectric capacitors. Figure 4a shows a Weibull plot of TDDB lifetimes for integrated BST capacitors in a microcontroller stressed under different field conditions at 125 °C. The time to 0.1 % cumulative failure is then expressed as a function of the field strength. Since the shape parameter of the distribution, m, is greater than unity, the failure mode of the integrated BST capacitors is intrinsic. From Fig. 4b, an estimated life-

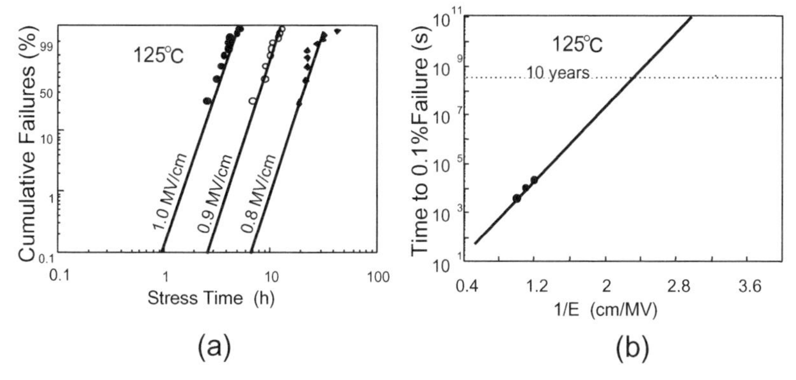

Fig. 4. Time-dependent dielectric breakdown (TDDB) characteristics for integrated BST capacitors with a thickness of 185 nm at 125 °C. (**a**) A Weibull distribution of cumulative failures versus the stress time. (**b**) The time to 0.1 % cumulative failure versus the reciprocal of the field strength

time to 0.1 % cumulative failure under a voltage stress of 5.5 V at 125 °C is longer than 10 years, which insures the reliable performance of the integrated BST capacitors [20].

The time to electrical breakdown should exceed the specified service life of a device, which is dominated by the major failure mechanism among different kinds of failure. For instance, ferroelectric fatigue is accepted as the dominant failure mechanism in FeRAMs. The number of read/write cycles, referred to as endurance, for commercialized FeRAMs is limited at present by fatigue and therefore specified as 10^{10} cycles. If the cycle time of the device is 100 ns, the cumulative time of DC voltage stress durations is only 1000 s. Then, a couple of hours is good enough for the required TDDB life. If the FeRAM should be an alternative to existing SRAMs and/or DRAMs, the TDDB life must be made to exceed 10 y. Such a challenge, for instance, is being made by protecting the FeRAM from environmental moisture using a SiN passivation layer [21]. Meanwhile, the fatigue issue should also be resolved to offer 10^{16} read/write cycles.

3 Hysteresis Measurement

Hysteresis measurements on ferroelectric capacitors are made using a Sawyer–Tower circuit [22], which has received a historical consensus as a standard technique for characterizing nonlinear capacitors. Figure 5a shows a typical circuit configuration. A ferroelectric capacitor C_f and a sense capacitor C_s are connected in series to a triangular AC voltage source, where C_f should be chosen to be much smaller than C_s so as to ignore the voltage drop across the sense capacitor. Since the voltage across the sense capacitor is given by Q/C_s, where Q is the charge stored in each capacitor, and is obtained by integrating

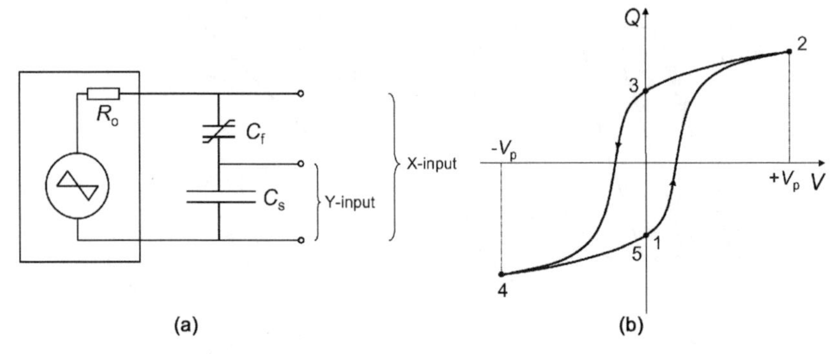

Fig. 5. (**a**) A Sawyer–Tower circuit for hysteresis measurements. (**b**) A typical hysteresis curve

the current flowing through the ferroelectric capacitor over a measurement cycle, a hysteresis curve is obtained, as illustrated in Fig. 5b, by connecting the voltage source and sense capacitor to the X and Y inputs, respectively, of an oscilloscope. Figure 6 shows the signal responses in accordance with the voltage V_f applied to the ferroelectric capacitor. For readers' convenience, each step is numbered in order in accordance with the numbered hysteresis excursion in Fig. 5b. Since the voltage generator charges up the ferroelectric capacitor through an output impedance of R_o, the measurement speed is limited by the time constant. In addition, impedance matching should be considered for higher-frequency measurements [23].

To measure the current response flowing through the ferroelectric capacitor, the sense capacitor is replaced with a sense resistor. A typical waveform of the response is illustrated in Fig. 6b. At higher frequencies this sense resistor in series with the output impedance will contribute to the limiting time constant of the system [24]. Then there will be a limitation on the capacitance of the ferroelectric capacitors that are measurable with such a system.

When the resistance of a ferroelectric capacitor is degraded to compete with its capacitive reactance, the leakage component will cause a deformation of the hysteresis curve along the Q-axis. Such a ferroelectric capacitor will burn out sooner or later, because the leakage component will be significantly large ($> 10^{-2}$ A/cm^2). Therefore, hysteresis measurements should be conducted with J–V measurements.

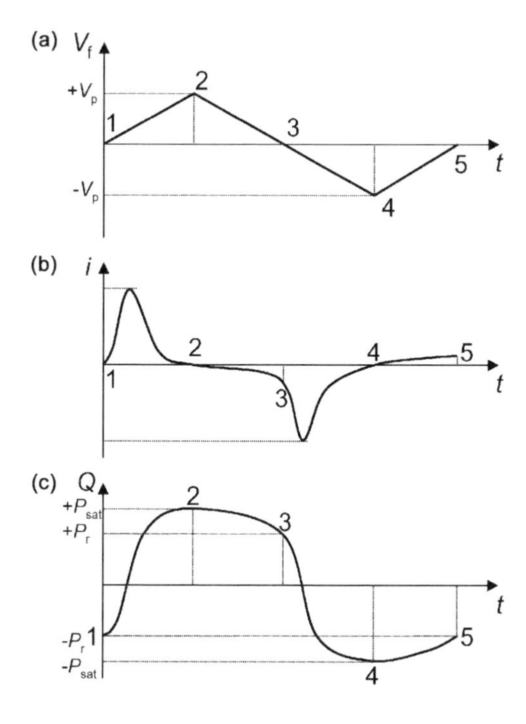

Fig. 6. Waveforms of (**a**) the voltage applied across, (**b**) current flowing through, and (**c**) charge stored in a ferroelectric capacitor during a hysteresis measurement cycle, where V_p is the amplitude of the cycling voltage across the ferroelectric capacitor, P_r is the remanent polarization, and P_{sat} is the saturation polarization

4 Ferroelectric Fatigue

Fatigue of ferroelectric capacitors brought about by voltage cycling must be minimized to prevent unintended memory malfunction during the specified lifetime. The degree of fatigue is characterized by the decrease in switchable polarization with respect to the number of voltage cycles. When PZT is used in nonvolatile ferroelectric memory capacitors, fatigue becomes very significant after billions of read/write cycles. It is believed that the presence of space charges at Pt electrode interfaces is responsible for fatigue in PZT capacitors [25, 26]. In order to mitigate the interface effect, several electrode materials have been examined. For instance, PZT films combined with RuO_2, LaSrCoO, or IrO_2 electrodes have been reported to exhibit no significant fatigue even after 10^{12} cycles [27, 28, 29, 30].

Following these efforts, an outstanding breakthrough in designing materials for nonvolatile ferroelectric memories has been made by considering the atomic-level microstructure of ferroelectrics [31, 32]. These materials exhibit a lack of fatigue even with simple metal electrodes, such as Pt. The structure of these materials has the general formula of multi-layer interstitial ferroelectric compounds:

$$(\mathrm{Bi_2O_2})^{2+}(A_{m-1}B_m\mathrm{O}_{3m+1})^{2-}\,, \tag{2}$$

where metal ion A can be Ca^{2+}, Sr^{2+}, Ba^{2+}, Pb^{2+}, Bi^{3+}, La^{3+}, and so on, and metal ion B can be Ti^{4+}, Ta^{5+}, or Nb^{5+}. For $m = 2$, for instance,

Bi_2O_2 layers are periodically alternated with perovskite-like AB_2O_7 layers with double oxygen octahedral units. Thus m is the number of octahedral units in between the Bi_2O_2 layers. The direction of the spontaneous polarization is parallel to the a–b plane of the layers, as this plane contains the periodic O–B–O chains that are known to be effective in the perovskite ferroelectric. In contrast, there is no spontaneous polarization along the c-axis because of the lack of periodic symmetry in the O–B–O chain. The periodic lattice parameters for stoichiometric SBT are $a = 5.52237$ Å, $b = 5.52408$ Å, and $c = 25.02641$ Å, exhibiting an orthorhombic distortion [33]. The choice of Sr for the metal A site is advantageous in suppressing the metal deficiency that causes degradations in ferroelectricity, compared to the use of volatile Pb in PZT. In addition, the stacked Bi_2O_2 layers can be an oxygen source for compensation of oxygen vacancies induced in perovskite-like octahedra, because the oxygen ions in the perovskite-like layers are much more stable than those in the Bi_2O_2 layers [34]. Thus oxygen vacancies generated in the perovskite-like layers to cause fatigue are readily exchanged with oxygen ions in the Bi_2O_2 layers [35]. Therefore the bismuth oxide materials, such as SBT, are hardly degradable in ferroelectricity, characterized by remanent polarization and coercive field, even if the deoxidation of the ferroelectric occurs under the influence of Pt electrodes.

Fatigue testing is coordinated by making hysteresis measurements on fatigued ferroelectric capacitors after a number of voltage cycles. In order to predict the reliability of FeRAMs, accelerated fatigue testing is required. In common accelerated tests for semiconductor devices, a number of devices are subjected to failure-causing stresses such as temperature and voltage bias, each related to a different acceleration factor that is needed to describe a relationship between failure times in a use condition and the accelerated condition. However, there is no definite temperature dependence of fatigue in SBT capacitors [36]. Recently, an empirical voltage acceleration model based on the Eyring model was proposed for fatigue in SBT capacitors [37]. The number of voltage cycles to fatigue failure is well described by the following expression:

$$\log(\text{cycles}) = C + \frac{E_\mathrm{a}}{kT} - n^* \log V_\mathrm{f} \,, \tag{3}$$

where E_a is the activation energy, k is the Boltzmann constant, T is the absolute temperature, V_f is the cycling voltage applied to the ferroelectric capacitor, C is a constant, and n^* is a function of temperature. Although the majority of fatigue data exhibit a logarithmic decrease in switching polarization with the number of voltage cycles, and there should be a meaningful relation between the motions of domain walls and defects, confident physical insights into the underlying mechanism have not yet been established.

5 Pulse Polarization Measurement

The macroscopic electric field in a poled ferroelectric capacitor in which a remanent polarization is preserved is substantially zero when both the capacitor

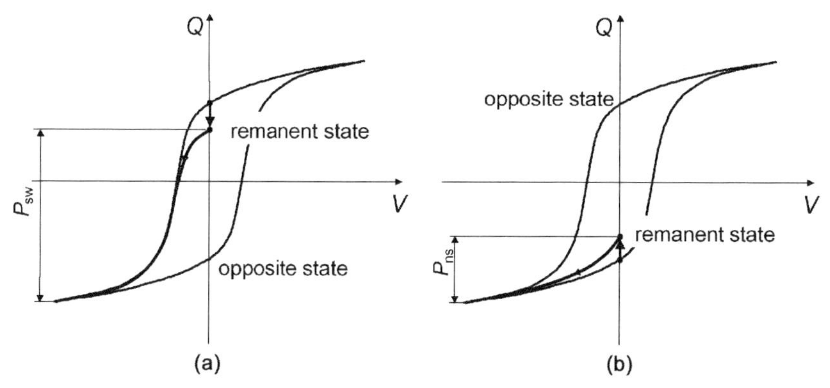

Fig. 7. A typical hysteresis loop, showing charge losses after the storage for a certain period of time. (**a**) P_{sw}: the switched polarization charge extracted from a capacitor in the up state. (**b**) P_{ns}: the nonswitched polarization charge extracted from a capacitor in the down state

electrodes are grounded after poling, so that the ferroelectric capacitor can be regarded as being under an equilibrium condition. Nevertheless, ferroelectric capacitors exhibit an aging effect characterized by the decay in remanent polarization during high-temperature storage (retention testing) [38,39,40,41]. This suggests that the aging effect is a manifestation of microscopic changes in the ferroelectric capacitor. Another aspect of the aging effect referred to is the loss of switchable polarization from the preliminarily poled remanent state to the opposite (complementary) state, resulting from the establishment of a preference for the remanent state over the opposite state during storage or unipolar voltage pulsing [42,43,44,45]. This phenomenon is termed imprint.

Both the retention and imprint measurements are based on the pulse polarization measurement technique, which yields P_{sw} or P_{ns} by pulsing the ferroelectric capacitor in accordance with whether a binary "1" or "0" is stored. To give a definite discrimination between "1" and "0", there should be a certain difference between P_{sw} and P_{ns}, which can be defined as

$$P_{\mathrm{nv}} = P_{\mathrm{sw}} - P_{\mathrm{ns}} \,. \tag{4}$$

Then P_{nv} is a function of the time interval between pulse polarization measurements. Following this definition, retention is characterized by making pulse polarization measurements for the same states as those preliminarily poled, as shown in Fig. 7. Imprint, on the other hand, is characterized by making pulse polarization measurements for the opposite states to those preliminarily poled. In either case, P_{nv} is an important measure for evaluating the nonvolatility of FeRAMs when the power is turned off.

6 Retention

Figure 8 shows a plot of P_{nv} versus retention time for $\mathrm{SrBi_2(Ta,Nb)_2O_9}$ (SBTN) capacitors [46]. The data set for the first 530 ms after poling was

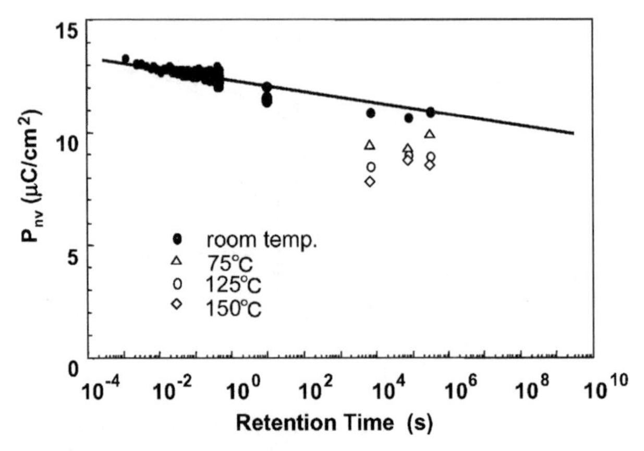

Fig. 8. The charge retention characteristics of SBTN capacitors integrated in a nonvolatile memory test structure for baking temperatures of 27, 75, 125, and 150 °C as functions of time, showing the change in the nonvolatile component P_{nv}

obtained from transient polarization measurements. The initial pulse polarization measurement was made at 10 s and the values of P_{nv} at 27 °C were approximately 12 μC/cm^2. After the initial pulse polarization measurement, test capacitors were baked at selected temperatures (27, 75, 125, and 150 °C) for 2, 24, and 100 h. Values of P_{nv} after storages at 27 °C showed good agreement with a straight fitting line extrapolated from data points in the transient regime. It is therefore reasonable to extend the fitting line to 100 h, and it is assumed that the decay in P_{nv} is governed by a single mechanism in the short to long time regime (10^{-3} to 10^5 s). The linear dependence of the decay curve on the logarithmic time yields a decay rate of 0.24 μC/cm^2 per decade at 27 °C. The logarithmic time dependence suggests that the polarization decay process is associated with a wide distribution over orders of magnitude in relaxation time, and is described by [39, 40]

$$P_{\mathrm{nv}}(t) = P_0 - m \log\left(\frac{t}{t_0}\right) , \tag{5}$$

where t is the time, t_0 is the characteristic time at which the linear behavior of $P_{\mathrm{nv}}(t)$ begins with respect to $\log t$, P_0 is the polarization at $t = t_0$, and m is the decay rate.

At elevated storage temperatures, a 2 h storage resulted in a significant decrease in P_{nv} by up to 4 μC/cm^2, depending on the storage temperature. However, measurements made after storage for another 24 and 100 h showed no pronounced change in P_{nv} within the measurement reproducibility. Thus the decrease in P_{nv} at high temperatures is supposed to occur within a short time, as seen in SBT [11]. To evaluate the temperature effect on P_{nv}, poled capacitors were baked at different temperatures from 27 to 220 °C only for 15 min, which are enough to give rise to such a decrease in P_{nv}. Supposing the

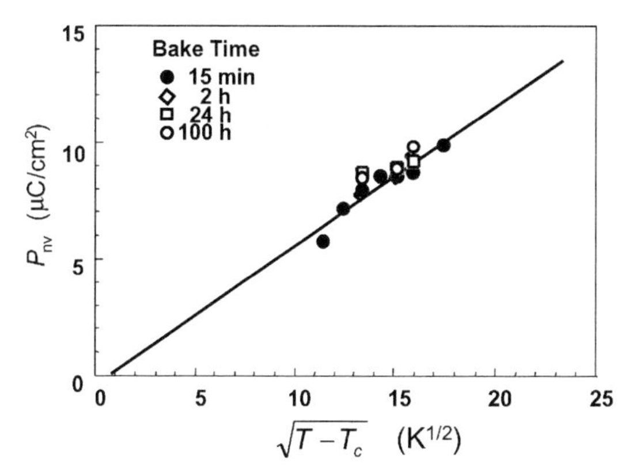

Fig. 9. The polarization charge difference P_{nv} versus the baking temperature T, where T_{C} is the Curie temperature. Polarization measurements were made at 27 °C after storage at temperatures from 27 to 220 °C for 15 min

second-order transition to be near the Curie temperature, T_{C}, the values of P_{nv} were plotted as a function of the baking temperature, T, so as to obtain the relation between P_{nv} and $(T_{\mathrm{C}}-T)^{1/2}$. As shown in Fig. 9, an extrapolation from this data set symbolized by closed circles indicates a collapse of P_{nv} at a temperature of 330 °C, which is close to an experimentally determined value of the Curie temperature for SBTN films [47].

To account for the temperature effect in the high-temperature storage test, values of P_{nv} in Fig. 8, whose measurements were made after storage for 2–100 h, were overlaid on the same plot (Fig. 9). Then we found a good agreement in the temperature dependence between the long-time storage values and the 15 min storage values. This indicates that the decrease in P_{nv} at high temperatures from its initial ($t_0 = 10$ s) value is primarily due to an instantaneous decrease in remanent polarization within 15 min, and that the temperature dependence is quite similar to that of spontaneous polarization in the vicinity of the second-order transition. The decrease in P_{nv} caused at high temperatures is explained by the loss of measurable charge in the capacitor electrodes, coupled with a polarization charge. Therefore, P_{nv} at high temperatures is quenched to room temperature.

Small changes in P_{nv} resulting from storage times of 2, 24, and 100 h storages at high temperatures lead to a supposition that P_{nv} at high temperatures has a logarithmic time dependence similar to that at 27 °C. When discussing the origin of the distributed relaxation time for the decay in remanent polarization, one can indicate two possible factors: (1) distribution in the coercive energy responsible for polarization reversal; and (2) distribution in the depth of the traps responsible for space charge transport. When the first mechanism is invoked, the polarization decay process is governed by the stochastic switching of thermally fluctuating displacement dipoles from one well to the other in the double-minimum potential with distributed coer-

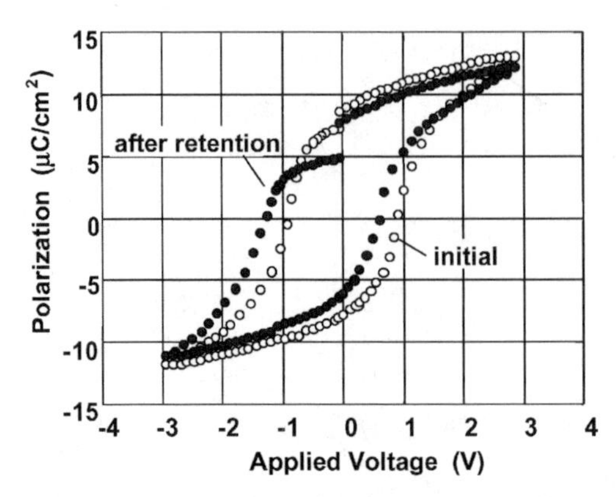

Fig. 10. Typical P–V hysteresis curves for positively poled SBTN capacitors before (○) and after (•) storage at 125 °C for 100 h, obtained by pulsed polarization measurements

cive potential barrier heights. The second mechanism accounts for the decay in the remanent polarization resulting from screening by space charges that are redistributed after poling. The decay time is then determined by the internal electric field and also the space charge concentration gradient in the capacitor.

Although either process leads to the distributed relaxation time, there is a substantial difference between them. The thermal fluctuation process is reversible if the coercive energy distribution does not change with time. In contrast, the space charge redistribution process is irreversible, and as a result the final polarization state is more stable than the initial state. Such a memory effect is referred to as imprint. This phenomenon was also obs erved in these SBTN samples, as shown in Fig. 10. Once imprint has occurred, the initial hysteresis loop is no longer restored even by polarization switching. This strongly indicates that the polarization decay process is accompanied by imprint caused by the space charge redistribution. This conclusion is consistent with the speculation that the decrease in polarization of SBT capacitors is independent of the reversal of domains [48]. The time required for emitting trapped charges into the conduction band may depend on the distribution in the depth of the traps, resulting in a logarithmic time dependence of the change in remanent polarization.

7 Imprint

Following a retention measurement, we can evaluate the deformation in the hysteresis curve of a ferroelectric capacitor by making a pulse polarization measurement for the opposite state to that preliminarily poled. In Fig. 11, a $\log P_{\text{nv}}$ versus $\log t$ plot is shown for the opposite remanent states in accordance with the following model [49]. Slow changes of the charge distribution

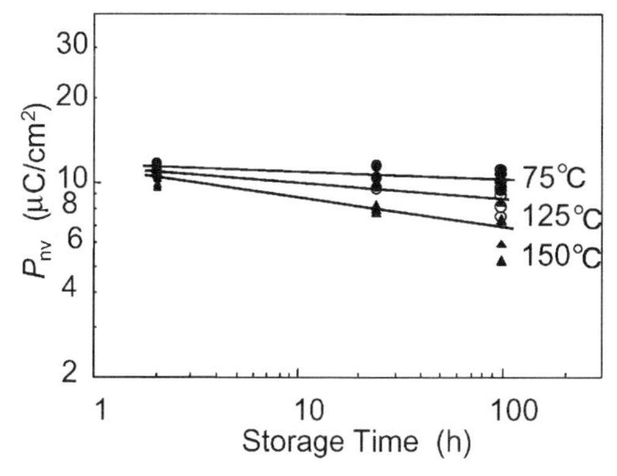

Fig. 11. The P_{nv} of $SrBi_2(Ta,Nb)_2O_9$ capacitors switched into the opposite remanent state after high-temperature storage

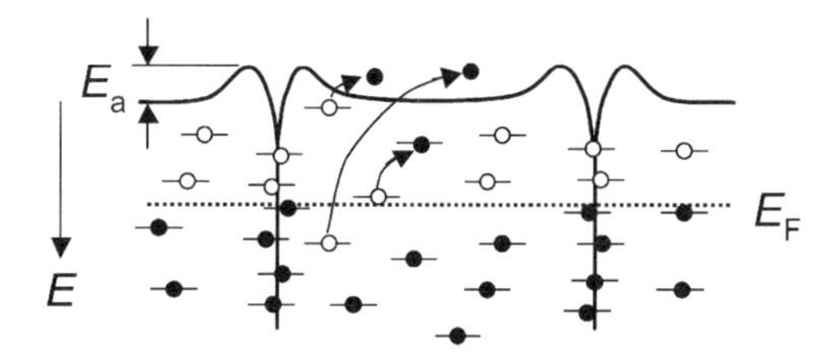

Fig. 12. An energy band diagram in the presence of distributed charged traps in a ferroelectric. A grain boundary or dislocation plane is a possible pinning center with a barrier potential of E_a

which take a long time have been attributed to carrier emission into the conduction band from localized traps distributed in the insulating energy gap [46]. For simplicity, we will consider the behavior of electrons for the negatively charged carriers. In the presence of charged traps in a ferroelectric, the situation may be illustrated by an energy band diagram, as shown in Fig. 12, in which most of the traps lying below the Fermi level E_F are occupied by electrons. In addition, some of the traps above E_F are also occupied by thermally activated electrons. The distribution of charged traps per unit volume from the band edge to the bottom of the energy gap can be described by the distribution function $g(E)$, so that the number of traps lying in a narrow range of energy between E and $E + \mathrm{d}E$ per unit volume is given by $g(E)\mathrm{d}E$. Thus we have

$$\int_0^\infty g(E)\mathrm{d}E = N\,, \tag{6}$$

where E is the energy depth from the bottom of the conduction band, and N is the total number of traps per unit volume. The number of electrons emitted from traps lying between E and $E + \mathrm{d}E$ during a short time interval between t and $t + \mathrm{d}t$ is proportional to the number of these traps; that is

$$r(t)\mathrm{d}t = \alpha g(E)\mathrm{d}E\,, \tag{7}$$

where $r(t)$ is the rate of electron emission from traps at an energy level of E per unit time and α is a constant. If the rate of electron emission from these traps into the conduction band depends on the energy depth, the waiting time t for the occurrence of electron emission from a trap at an energy depth of E is given by [50]

$$\frac{1}{t} = \nu_0 \exp\left(-\frac{E}{kT}\right)\,, \tag{8}$$

where ν_0 is the vibration frequency of trapped electrons, k is the Boltzmann constant, and T is the absolute temperature. From (7) and (8), we have

$$r(t) = \alpha k T g(E)\frac{1}{t}\,. \tag{9}$$

Then, the rate of electron emission from entire traps distributed in the energy gap $R(t)$ is given by

$$R(t) = \int_0^\infty r(t)\mathrm{d}E = \alpha N k T\frac{1}{t}\,. \tag{10}$$

It is immediately noticeable that $R(t)$ is independent of $g(E)$.

Here, we assume that domain pinning centers are distributed in the ferroelectric capacitor in a random manner and that the pinning centers capture encountered electrons at a constant rate. These pinning centers are preferably formed at grain boundaries or dislocation planes, at which many electrons are trapped densely (Fig. 12). Therefore, the band edges in the vicinity of grain boundaries are significantly distorted. In particular, the dense charged traps would give rise to barrier potentials surrounding the pinning centers with an average barrier height of E_{a}. Let us further assume that a certain fraction of switchable polarization is reduced by capturing an electron by a pinning center. Under this condition, the variation in P_{nv} with time follows

$$\frac{\mathrm{d}P_{\mathrm{nv}}(t)}{\mathrm{d}t} = -\beta R(t) P_{\mathrm{nv}}(t)\,, \tag{11}$$

where β is a proportionality constant associated with electron capture. Substituting (10) into (11), we have the solution of (11) for a long period of time between t_0 and t:

$$P_{\mathrm{nv}}(t) = P_0\left(\frac{t}{t_0}\right)^{-m^*}\,, \tag{12}$$

where $P_0 = P_{\mathrm{nv}}(t_0)$ and

$$m^* = \alpha\beta N k T\,. \tag{13}$$

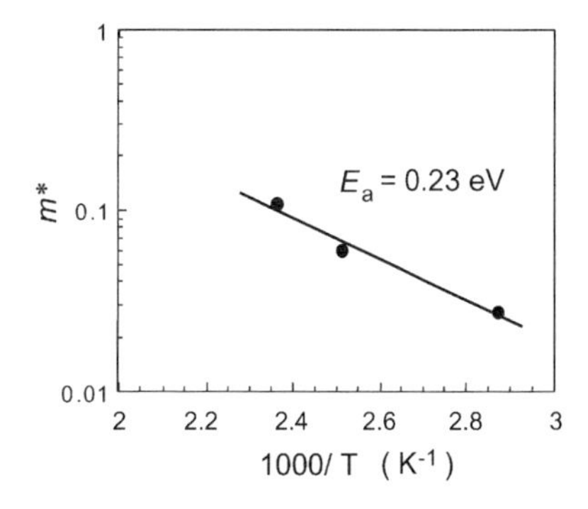

Fig. 13. The temperature dependence of the decay rate of switchable polarization for the opposite remanent state against a poled remanent state, showing an activation energy of 0.23 eV for thermal imprint

This deduction can be confirmed by straight lines in a log P_{nv} versus log t plot, as shown in Fig. 11. The slope of the fitting line for a given storage temperature is equal to the value of exponent m^* at the storage temperature under consideration. From the values of exponent m^* at different temperatures, 75, 125, and 150 °C, it turns out that the temperature dependence of m^* is exponential rather than linear (Fig. 13). This consequence is congruent with an assumption that an electron approaching a pinning center must overcome the barrier height E_{a} in order to be captured [45]. Then, the coefficient $\alpha\beta$ in (13) is proportional to $\exp(-E_{\mathrm{a}}/kT)$, because β in (11) is related to the electron capture rate at the pinning center. The coefficient kT in (13) is then considered to be constant as compared with the abrupt change in $\exp(-E_{\mathrm{a}}/kT)$ with temperature. Thus we have

$$m^* = \gamma N \exp\left(-\frac{E_{\mathrm{a}}}{kT}\right) , \tag{14}$$

where γ is a constant. Although the expression for P_{nv} in (12) is different form that in (5), the variation in P_{nv} with time can be written as

$$\frac{\mathrm{d}P_{\mathrm{nv}}}{\mathrm{d}t} \propto -\frac{1}{t^{m^*+1}} \tag{15}$$

for (12) and

$$\frac{\mathrm{d}P_{\mathrm{nv}}}{\mathrm{d}t} \propto -\frac{1}{t} \tag{16}$$

for (5). If m^* is much smaller than unity, P_{nv} following (12) will show the same behavior as that following (5). In fact, Fig. 13 indicates that the values of m^* are less than 0.1 for the examined temperatures and the exponential temperature dependence yields $E_{\mathrm{a}} = 0.23$ eV.

Similarly, if (5) is applied to describe the behavior of P_{nv}, m should have an exponential temperature dependence in the form of

$$m = m_0 \exp\left(-\frac{E_{\mathrm{a}}}{kT}\right) , \tag{17}$$

where m_0 is a proportionality constant. From the values of m obtained from $P_{\rm nv}$ versus $\log t$ plots for different temperatures, $E_{\rm a}$ is calculated be 0.19 eV. This coincidence leads us to believe that the decay in remanent polarization has the same origin as the decay in switchable polarization. A possible explanation is that emitted electrons are captured by preferred pinning centers, to screen or compensate for the polarization charges as well as to pin the domains. A similar argument can also be put forward for holes as the positively charged carriers. The above argument leads us to a comprehensive understanding of imprint as a result of the redistribution of emitted carriers, and this also agrees with the deduction in the previous discussion that the decay in remanent polarization is due to the redistribution of carriers rather than polarization reversal [48]. However, it should be noted that such redistribution of carriers must be highly localized, because the J–V characteristics before and after high-temperature storage indicate that there is no macroscopic change in the entire charge distribution throughout the capacitor during the thermal aging process [46].

References

1. S.M. Sze: *Physics of Semiconductor Devices*, 2nd ed. (Wiley, New York 1981)
2. M. Azuma, M. Scott, E. Fujii, T. Otsuki, G. Kano, C.A. Paz de Araujo: Abstracts 4th Int. Symp. Integr. Ferroelectr., Monterey (1992) p. 109
3. R. Waser, M. Klee: Abstracts 3rd Int. Symp. Integr. Ferroelectr., Colorado Springs (1991) p. 288
4. P. Bhattacharya, T. Komeda, K.-H. Park, Y. Nishioka: Jpn. J. Appl. Phys. **32**, 4103 (1993)
5. P. Li, T.M. Lu: Phys. Rev. B **43**, 14261 (1991)
6. Y. Shimada, A. Inoue, T. Nasu, K. Arita, Y. Nagano, A. Matsuda, Y. Uemoto, E. Fujii, M. Azuma, Y. Oishi, S. Hayashi, T. Otsuki: Jpn. J. Appl. Phys. **35**, 140 (1996)
7. Y. Fukuda, K. Numata, K. Aoki, A. Nishimura: Jpn. J. Appl. Phys. **35**, 5178 (1996)
8. X. Chen, A.I. Kingon, L. Mantese, O. Auciello, K.Y. Hsieh: Integr. Ferroelectr. **3**, 355 (1993)
9. H. Hu, S.B. Krupanidhi: J. Mater. Res. **9**, 1484 (1994)
10. X. Chen, A.I. Kingon, H.N. Al-Shareef, K.R. Bellur, K. Gifford, O. Auciello: Integr. Ferroelectr. **7**, 291 (1995)
11. D.J. Taylor, R.E. Jones, P. Zurcher, P. Chu, Y.T. Lii, B. Jiang, S.J. Gillespie: Appl. Phys. Lett. **68**, 2300 (1996)
12. K. Watanabe, A.J. Hartmann, R.N. Lamb, J.F. Scott: J. Appl. Phys. **84**, 2170 (1998)
13. R. Waser: J. Am. Ceram. Soc. **73**, 1645 (1990)
14. Y. Shimada, A. Inoue, T. Nasu, Y. Nagano, A. Matsuda, K. Arita, Y. Uemoto, E. Fujii, T. Otsuki: Jpn. J. Appl. Phys. **35**, 4919 (1996)
15. S.B. Desu, I.K. Yoo: J. Electrochem. Soc. **140**, L133 (1993)
16. K. Numata, Y. Fukuda, K. Aoki, A. Nishimura: Jpn. J. Appl. Phys. **34**, 5245 (1995)

17. D. J. Wouters, G. Willems, G. Groeseneken, H. E. Meas, K. Brooks: Integr. Ferroelectr. **7**, 173 (1995)
18. T. Baiatu, R. Waser, K.-H. Hardtl: J. Am. Ceram. Soc. **73**, 1663 (1990)
19. D. J. Klinger, Y. Nakada, M. Menendez (Eds.): *AT&T Reliability Manual* (Van Nostrand Reinhold, New York 1990)
20. E. Fujii, Y. Uemoto, S. Hayashi, T. Nasu, Y. Shimada, A. Matsuda, M. Kibe, M. Azuma, T. Otsuki, G. Kano, M. Scott, L. D. McMillan, C. A. Paz de Araujo: Tech. Dig. Int. Electron Devices Meeting, San Francisco (1992) p. 267
21. E. Fujii, T. Otsuki, Y. Judai, Y. Shimada, M. Azuma, Y. Uemoto, Y. Nagano, T. Nasu, Y. Izutsu, A. Matsuda, K. Nakao, K. Tanaka, K. Hirano, T. Ito, T. Mikawa, T. Kutsunai, L. D. McMillan, C. A. Paz de Araujo: Tech. Dig. Int. Electron Devices Meeting, Washington, DC (1997) p. 597
22. C. B. Sawyer, C. H. Tower: Phys. Rev. **35**, 296 (1930)
23. P. K. Larsen, G. L. M. Kampschoer, M. J. E. Ulenaers, G. A. C. M. Spierings, R. Cuppens: Appl. Phys. Lett. **29**, 611 (1991)
24. A. D. DeVilbiss, A. J. DeVilbiss: Integr. Ferroelectr. **26**, 285 (1999)
25. J. F. Scott, C. A. Araujo, B. M. Melnick, L. D. McMillan, R. Zuleeg: J. Appl. Phys. **70**, 382 (1991)
26. J. M. Benedetto, R. A. Moore, F. B. McLean: J. Appl. Phys. **75**, 460 (1994)
27. R. Ramesh, W. K. Chan, B. Wilkens, H. Gilchrist, T. Sands, J. M. Tarascon, V. G. Keramidas, D. K. Fork, J. Lee, A. Safari: Appl. Phys. Lett. **61**, 1537 (1992)
28. S. D. Bernstein, T. Y. Wong, Y. Kisler, R. W. Tustison: J. Mater. Res. **8**, 12 (1993)
29. I. K. Yoo, S. B. Desu, J. Xing: Mater. Res. Soc. Symp. Proc. **310**, 165 (1993)
30. T. Nakamura, Y. Nakao, A. Kamisawa, H. Takasu: Appl. Phys. Lett. **65**, 1522 (1994)
31. C. A. Paz de Araujo, J. D. Cuchiaro, L. D. McMillan, M. C. Scott, J. F. Scott: Nature **374**, 627 (1995)
32. T. Sumi, N. Moriwaki, G. Nakane, T. Nakakuma, Y. Judai, Y. Uemoto, Y. Nagano, S. Hayashi, M. Azuma, E. Fujii, S. Katsu, T. Otsuki, L. McMillan, C. A. Paz de Araujo, G. Kano: Tech. Dig. Int. Solid-State Circuits Conf., San Francisco (1994) p. 268
33. Y. Shimakawa, Y. Kubo, Y. Nakagawa, T. Kamiyama, H. Asano, F. Izumi: Appl. Phys. Lett. **74**, 1904 (1999)
34. B. H. Park, S. J. Hyun, S. D. Bu, T. W. Noh, J. Lee, H.-D. Kim, T. H. Kim, W. Jo: Appl. Phys. Lett. **74**, 1907 (1999)
35. O. Auciello, A. R. Krauss, J. Im, D. M. Gruen, E. A. Irene, R. P. H. Chang, G. E. McGuire: Appl. Phys. Lett. **69**, 2671 (1996)
36. T. Noguchi, T. Hase, Y. Miyasaka: Integr. Ferroelectr. **17**, 57 (1997)
37. V. Joshi, N. Solayappan, C. A. Paz de Araujo: Abstracts 12th Int. Symp. Integr. Ferroelectr., Aachen (2000) p. 96
38. A. Gregory, R. Zucca, S. Q. Wang, M. Brassington, N. Abt: IEEE Int. Reliability Phys. Symp., San Diego (1992) p. 91
39. R. Moazzami, N. Abt, Y. Nissan-Cohen, W. H. Shepherd, M. P. Brassington, C. Hu: Tech. Dig. Symp. VLSI Technol., Oiso (1991) p. 61
40. N. E. Abt: Abstracts 3rd Int. Symp. Integr. Ferroelectr., Colorado Springs (1991) p. 404
41. T. Mihara, H. Yoshimori, H. Watanabe, C. A. Paz de Araujo: Jpn. J. Appl. Phys. **34**, 2380 (1995)

42. J. M. Benedetto, M. L. Roush, I. K. Lloyd, R. Ramesh, B. Rychlik: Integr. Ferroelectr. **10**, 279 (1995)
43. W. L. Warren, D. Dimos, G. E. Pike, B. A. Tuttle, M. V. Raymond, R. Ramesh, J. T. Evans, Jr.: Appl. Phys. Lett. **67**, 866 (1995)
44. W. L. Warren, B. A. Tuttle, D. Dimos, G. E. Pike, H. N. Al-Shareef, R. Ramesh, J. T. Evans, Jr.: Jpn. J. Appl. Phys. **35**, 1521 (1996)
45. E. G. Lee, D. J. Wouters, G. Willems, H. E. Maes: Appl. Phys. Lett. **69**, 1223 (1996)
46. Y. Shimada, K. Nakao, A. Inoue, M. Azuma, Y. Uemoto, E. Fujii, T. Otsuki: Appl. Phys. Lett. **71**, 2538 (1997)
47. Y. Oishi, S. Hayashi, T. Otsuki, J. D. Cuchiaro, C. A. Paz de Araujo: Abstracts Mater. Res. Soc. Spring Meeting, San Francisco (1996) p. 336
48. K. Amanuma, T. Kunio: Jpn. J. Appl. Phys. **35**, 5229 (1996)
49. Y. Shimada, A. Noma, K. Nakao, T. Otsuki: Jpn. J. Appl. Phys. **38**, 2816 (1999)
50. N. F. Mott, R. W. Gurney: *Electronic Processes in Ionic Crystals*, 2nd ed. (Oxford University Press, London 1940)

Part IV

Advanced-Type Memories

Chain FeRAMs

Daisaburo Takashima and Yukihito Oowaki

SoC R & D Center, Toshiba Corp., Semiconductor Company,
2-5-1 Kasama, Sakae-ku, Yokohama 247-8585, Japan
daisaburo.takashima@toshiba.co.jp

Abstract. A chain FeRAM™ is a solution for future high-density and high-speed nonvolatile memory. One memory cell consists of one transistor and one ferroelectric capacitor connected in parallel, and one memory cell block consists of a member of cells in series. This configuration realizes a small memory cell of $4\,F^2$ size in the ideal case and a fast random access time. In this chapter, overview and design techniques for chain FeRAM are presented. Not only chain architecture but also several design techniques for realizing 1. high-speed, 2. high-density, and 3. low-voltage operation are discussed. 0.25 µm 8 Mb chain FeRAMs using these techniques are demonstrated.

1 Introduction

A ferroelectric RAM (FeRAM) has a great potential to create high-performance nonvolatile memory, because of its excellent characteristics, such as nonvolatility, a low power consumption, fast read/write operation, and a high endurance of over 10^{12} read/write cycles. Early papers on 512 b and 16 kb FeRAMs [1,2] with two-transistor and two-capacitor (2T/2C) cells were published in the late 1980s. The first 256 kb FeRAM using the one-transistor and one-capacitor (1T/1C) cell was demonstrated in 1994 [3]. On the other hand, recently, the market for portable electronic devices such as cellular phones and mobile computing systems has required megabit-class FeRAMs. To meet the above market demand, several megabit-class prototypes of 1 Mb FeRAMs [4, 5, 6] and 4 Mb FeRAMs [7, 8] have been demonstrated.

However, in previously reported 1Mb/4Mb FeRAMs [4, 5, 6, 7, 8], the chip sizes are too large, and the memory densities are too low to meet the above market demand, as shown in Fig. 1a. This large chip size is caused not only by the relatively large cell size, but also by a low cell area efficiency of 30–40 %, due to problems inherent in conventional FeRAMs, such as the large on-pitch plate driver. Moreover, the conventional FeRAM operates more slowly than DRAM, as shown in Fig. 1b. To overcome these problems, a chain FeRAM [9, 10, 11, 12, 13] was proposed in 1997. The chain FeRAM realizes a small cell size and a high cell efficiency thanks to small plate driver and sense amplifier areas, and fast random access.

In this chapter, an overview of chain FeRAM architecture and related circuit design techniques for realizing high-performance chain FeRAMs are presented. In Sect. 2, the features and problems of the conventional FeRAM are

H. Ishiwara, M. Okuyama, Y. Arimoto (Eds.): Ferroelectric Random Access Memories,
Topics Appl. Phys. **93**, 197–213 (2004)
© Springer-Verlag Berlin Heidelberg 2004

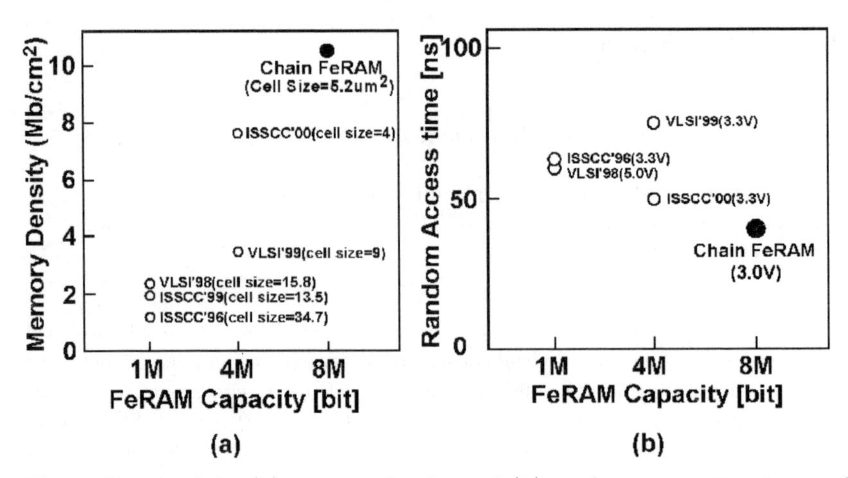

Fig. 1. Trends of the (**a**) memory density and (**b**) random access time in megabit-class FeRAMs

summarized. In Sect. 3, the concept and basic operation of chain FeRAMs are presented. In Sects. 4–6, the detailed circuit design techniques are presented for each subject. Section 4 presents the high-speed techniques, and simulated and measured performances of the 8 Mb chain FeRAM. In Sect. 5, first, the high-density advantage inherent in the chain architecture is discussed, and then three new techniques to reduce die size are introduced. After that, the 8 Mb chain FeRAM chip is demonstrated. Section 6 presents a low-voltage operational technique to enlarge the cell signal.

2 Conventional FeRAM

In the conventional 1T-1C FeRAM, a memory cell consists of one transistor and one ferroelectric capacitor connected in series, as shown in Fig. 2a. This is the same cell configuration as that of DRAM, except that ferroelectric materials, such as $PbZr_XTi_{1-X}O_3$(PZT) and $SrBi_2Ta_2O_9$(SBT), are used for the FeRAM capacitor, and cell data are held using ferroelectric polarization. This 1T-1C FeRAM cell configuration is simple. However, it is subject to two major problems.

The first problem is the cell size limitation. For the conventional FeRAM cell, the same cell configuration is adopted as for DRAM. Although the current FeRAM cell size is still larger than the DRAM one, the stacked ferroelectric capacitor [14, 15], and fine process technologies will realize the same cell structure and size as for DRAM, as shown in Fig. 2c. Therefore, when adopting the folded-bitline scheme [16], which is widely used in DRAMs, the memory cell is placed at one of two intersections of wordlines and bitlines, as shown in Fig. 2b. As a result, the memory cell size is limited to $2\,F \times 4\,F$ – that is, $8\,F^2$ – where F means the minimum feature size.

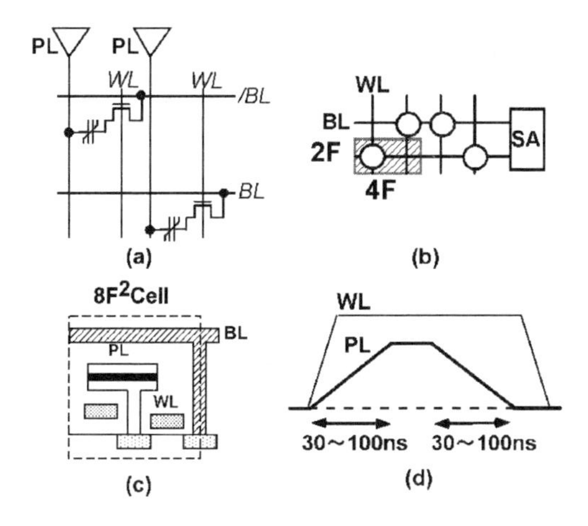

Fig. 2. The conventional FeRAM memory cell: (**a**) the cell array circuit diagram, (**b**) the minimum cell layout, (**c**) a cross-section in an ideal case and (**d**) the operation timing

The second problem is that access and cycle times are slower than those of DRAM due to the slow cell-plateline drive. In FeRAM operation, a cell-plateline (PL) must be driven to switch the polarization and read out cell data to the bitline [3], as shown in Fig. 2d. This causes a large cell-plateline delay, as shown in Fig. 2d. The attachment of many cells to one cell-plateline causes a heavy capacitive load of the cell-plateline, of the order of 10 pF. In FeRAM, a cell-plateline and its driver must be divided for each cell to avoid unwanted bias to unselected cells, and must be placed in a tight wordline pitch. This results in insufficient drivability of the plate driver. Moreover, in the present FeRAM memory cell, a highly resistive material such as Pt is used for the cell-plateline, resulting in a high resistance of the cell-plateline. This large *RC* delay causes a slow cell-plateline drive of 30–100 ns [3,4], and slow chip access and cycle times, whereas a large plate driver design to realize fast operation causes a huge chip size.

3 The Concept of Chain FeRAM

3.1 The Basic Structure

To overcome these problems, the concept of chain FeRAM has been proposed [9]. The chain FeRAM allows small $4\,F^2$ size memory cells using a planar cell transistor, and realizes fast random access. Figure 3a shows the circuit diagram of the chain cell block. One memory cell consists of one transistor and one ferroelectric capacitor connected in parallel, and one cell block consists of a number of memory cells connected in series. One terminal of the cell block is connected to the bitline (BL) via the block-selecting transistor, and the other terminal is connected to the cell-plateline (PL). Figure 3b shows the memory cell structure in the ideal case. Each ferroelectric capacitor is placed just above each cell transistor. Each contact between top/bottom capacitor electrode and source/drain terminal is shared by two neighboring cells. This

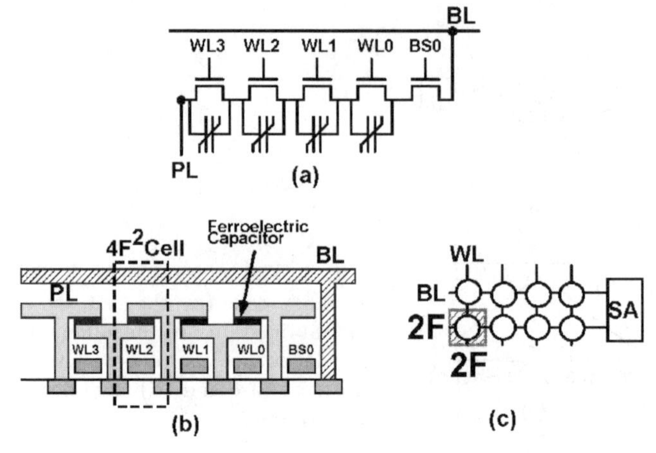

Fig. 3. The chain FeRAM memory cell block concept: (**a**) the circuit diagram, (**b**) a cross-section in an ideal case, and (**c**) the minimum cell layout

cell configuration allows a small $4\,\mathrm{F}^2$ size memory cell, since the minimum cell size is determined by the wordline pitch (2 F) and the bitline pitch (2 F), as shown in Fig. 3c.

3.2 Two Basic Operations

Figure 4 shows the basic operation of a chain FeRAM, using a driven cell-plate scheme [10, 11]. During the standby cycle, as shown in Fig. 4a, all wordlines (WL0–WL3) are boosted to high voltage, the block-selecting signal (BS) and cell-plateline (PL) are set at ground, and the bitline (BL) is also precharged at ground. So all ferroelectric capacitors are short-circuited by turned-on cell transistors. Therefore, as shown in the hysteresis loop of Fig. 4a, both "1" and "0" data are held steadily at zero bias points without noise to the capacitor electrodes. During the active cycle, as shown in Fig. 4b, when accessing the cell data of capacitor C_1, the selected wordline (WL2) is pulled down, and the block-selecting signal (BS) is pulled up. After that, the cell-plateline (PL) is pulled up to the V_{dd} level. Therefore, full the V_{dd} bias is only applied to the selected ferroelectric capacitor (C_1), and the cell data is read out to the bitline (BL), as shown in the loci of the hysteresis loop of Fig. 4b. When "1" data is read, a large charge is read out to the bitline by polarization switching. When "0" data is read, a small charge is read out to the bitline by nonpolarization switching. A sense amplifier amplifies the readout signal. On the other hand, cell data of unselected cells are protected by short-circuiting capacitors. Thus, completely random access is realized.

In order to realize fast FeRAM operation, a nondriven half-V_{dd} cell-plate scheme is also applicable to both conventional FeRAM [4, 17] and chain FeRAM [9], as shown in Fig. 5. When accessing a memory cell, a half-V_{dd}

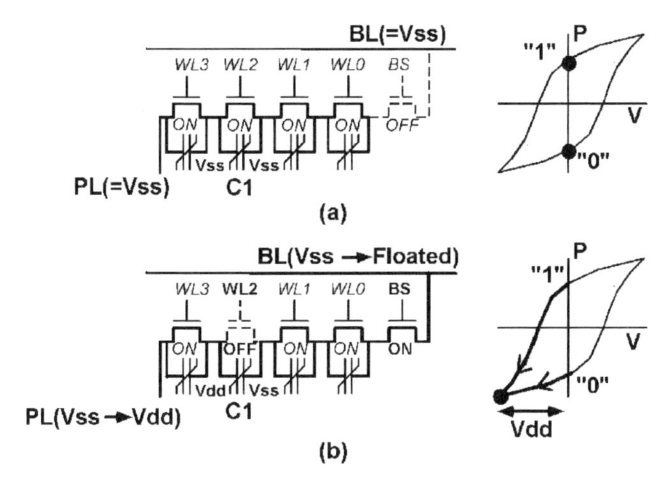

Fig. 4. The basic operation of a driven cell-plate scheme during (**a**) the standby cycle and (**b**) the active cycle

Operation Scheme / Memory cell Configuration	(1) Driven-Plate Scheme (PL: Vss → Vdd → Vss)		(2) 1/2Vdd-Plate Scheme (PL: 1/2Vdd)	
	Speed	Operation Voltage	Speed	Operation Voltage
(a) Conv. FeRAM 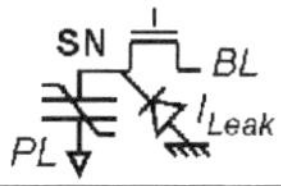 (SN, BL, I_{Leak}, PL)	[3] Slow	Low	[4,17] Fast	High (Need for Refresh)
(b) Chain FeRAM 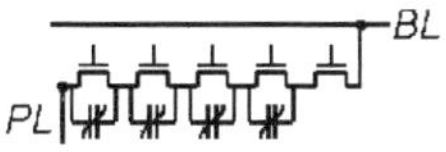 (BL, PL)	[10,11] Fast	Low	[9] Fast	High (No Refresh)

Fig. 5. A summary of operation schemes and their performance in conventional FeRAMs and chain FeRAMs

bias is applied to the ferroelectric capacitor without a driving cell-plateline, because the cell-plateline is fixed at half-V_{dd} and the bitline is precharged at ground. This eliminates cell-plateline drive delay. However, the conventional FeRAM requires a refresh operation [4, 17], because the cell storage node (SN) in the conventional FeRAM cell in Fig. 5 is floated during the standby cycle, and this causes cell data loss of "1" data as a result of the falling SN voltage to ground due to junction leakage. On the other hand, the half-V_{dd} scheme is applicable to chain FeRAM without any are held by short-circuiting ferroelectric capacitors.

Thus, the half-V_{dd} cell-plate scheme realizes fast operation. However, only half-V_{dd} is applied to ferroelectric capacitor. This imposes relatively high supply voltage operation on FeRAMs, because the present ferroelectric film still requires a relatively large bias of about 3 V to switch the cell polarization. Moreover, the small bias to the ferroelectric capacitor causes a small readout cell signal. Therefore, the driven cell-plate scheme is a solution for present FeRAMs. A practical array architecture and driven cell-plate scheme to realize fast chain FeRAM operation are discussed in Sect. 4.

4 The High-Speed Technique

Figure 6a shows the practical array architecture with a new fast driven cell-plate scheme [10] for an 8 Mb chain FeRAM chip. In chain FeRAM of this generation, eight cells are connected in series in a chain block. In practical design, in order to realize a folded-bitline configuration, two kinds of block-selecting signals (BS0 and BS1) and cell-platelines ($\overline{\mathrm{PL}}$ and PL) are introduced. For example, when accessing the cell data of capacitor C1 in the upper chain cell block, the selected wordline (WL6) is pulled down, and the block-selecting signal (BS0) of upper cell block is pulled up, as shown in Fig. 6b. After that, the cell-plateline ($\overline{\mathrm{PL}}$) connected to the upper block is pulled up to the array operation voltage (V_{aa}). Therefore, full V_{aa} bias is only applied to the selected ferroelectric capacitor (C_1), and the cell data is only read out to the upper bitline ($\overline{\mathrm{BL}}$). The lower bitline (BL) is used as a reference bitline. A sense amplifier (SA) amplifies the voltage difference between the bitline pair.

In this architecture, to overcome the problem of a slow cell-plateline drive, two new techniques are introduced. First, a metal M1 wiring cell-plateline is introduced, without any process penalty. The reason why there is no process penalty is discussed latter. This reduces the cell-plateline resistance to 1/25 of the conventional platinum (Pt) cell-plateline. Second, the cell-plateline and its plate driver are shared with 16 cells in two chain cell blocks, since the cell-plate drive of unselected cells does not cause cell data loss, because of the short-circuited ferroelectric capacitor. This scheme enables 16 times larger drivability of the plate driver. In the 8 Mb design, each driver is chosen to be 3.2 times larger, and each total driver area in a chip is 1/5 times larger. On the other hand, the increase in the cell-plateline capacitance due to the attachment of 16 cells is less than 20%, because only the selected ferroelectric capacitor contributes to the parasitic capacitance, since all unselected capacitors are short-circuited. These schemes achieve a high-speed cell-plateline drive of 10 ns. Figure 7 shows the simulated waveforms of the memory cell array of an 8 Mb chain FeRAM at 3.0 V V_{dd} and 70 ns cycle time. Due to the effect of the plate-driver being shared with 16 cells, this chip has a 3.2 times larger drivability of each plate-driver. Therefore, this chip realizes a fast cell-plateline ($\overline{\mathrm{PL}}$) drive of about 10 ns, as shown in Fig. 7. Figure 8 shows the

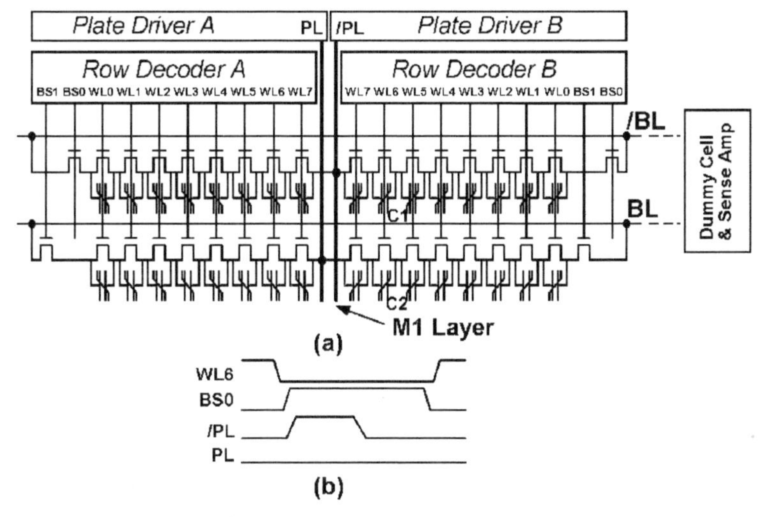

Fig. 6. A chain FeRAM array architecture with a fast and compact driven cell-plate scheme: (**a**) the circuit diagram and (**b**) the timing chart

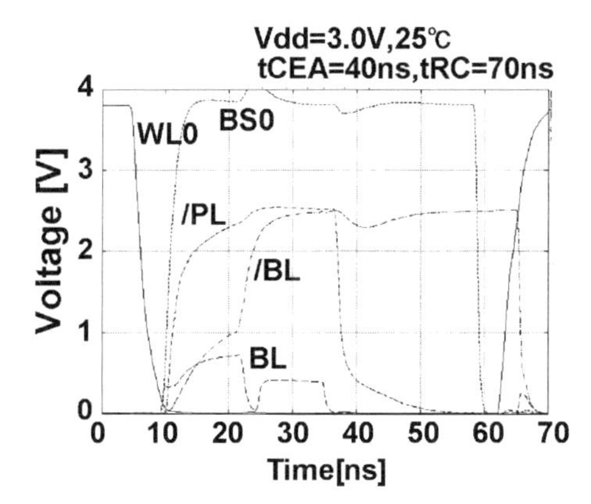

Fig. 7. Simulated waveforms of the 8 Mb chain FeRAM

measured access time Schmoo plot. This chip has achieved a 40 ns random access time (t_{CEA}) and a 70 ns read/write cycle time (t_{RC}) at $3.0\,\mathrm{V} \pm 0.3\,\mathrm{V}$ V_{dd}. These values are the fastest reported for FeRAMs.

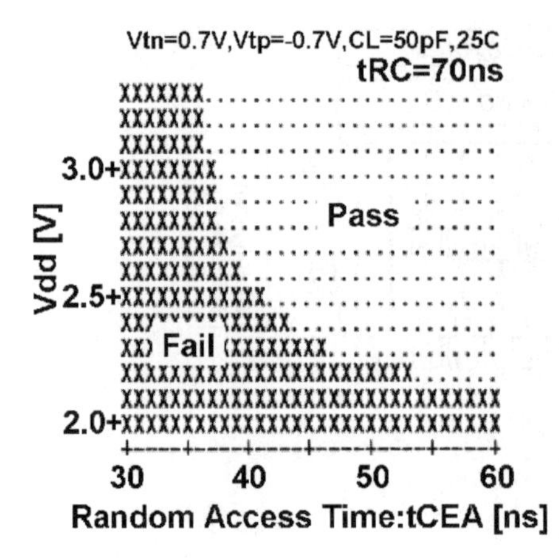

Fig. 8. A measured access time Schmoo plot

5 High-Density Techniques

In this section, first, the advantage of realizing the high density inherent in chain architecture [9,10] is discussed. Second, three new techniques [11] to reduce the chain FeRAM chip size further are presented. Third, after discussing the chip size shrink effect, an 8 Mb chain FeRAM chip is demonstrated.

5.1 The Advantage of Chain Architecture

The chain FeRAM enables a small memory cell of size $4\,F^2$. However, the current cell size is still as large as that of the conventional FeRAM cell, because of memory cell with its relatively large capacitor and without fine-stacked capacitor technology. However, other advantages of the chain FeRAM are as follows: 1. a small plate driver area and 2. a small sense amplifier area realizes a small die.

As discussed in Sect. 4, the chain architecture enables a fast cell-plateline drive, because the cell-plateline and its driver are shared by two chain cell blocks (16 cells), as shown in Fig. 6. This scheme also reduces the total plate driver area in a chip to 1/5. Eight cells in a chain block of Fig. 6 also share the bitline contact. This reduces the bitline capacitance per cell to about 1/3 of that of the conventional FeRAM. In our design, this advantage translates into sense amplifier area reduction to a half and a 1.5 times larger signal due to the small bitline capacitance.

5.2 The One-Pitch-Shift Cell

Figure 9 shows a new one-pitch-shift cell for the 8 Mb chain FeRAM. The chain FeRAM theoretically allows a small $4\,F^2$ size cell, thanks to the use of the cross-point type cell. However, the memory cell size of previously reported chain FeRAMs [10] was large. This is because the bottom electrode (BE) of

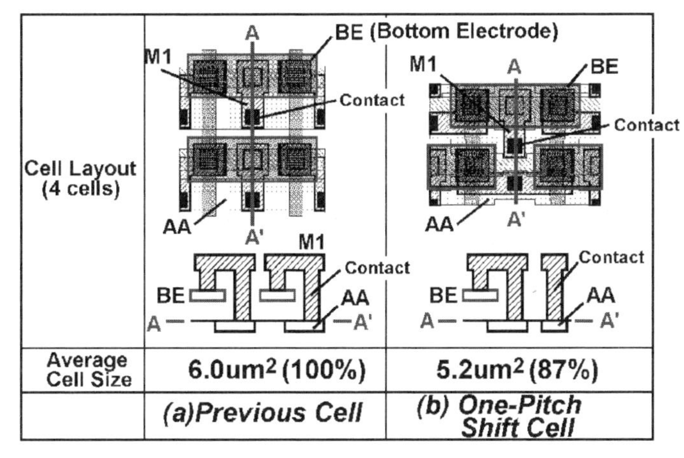

Fig. 9. (**a**) A previously reported memory cell and (**b**) the new one-pitch-shift cell for chain FeRAMs

the ferroelectric capacitor must be connected to the active area (AA) via a contact from the metal layer (M1), as shown in Fig. 9a, since, to insure easy fabrication, capacitor on a plug (stacked cell) technology is not introduced into the memory cell. Therefore, the contact between M1 and AA must be inserted between two BEs. This causes a waste of space between the BEs, when the space between the contact and the BE is large. On the other hand, in the new one-pitch-shift cell shown in Fig. 9b, the cell in the neighboring column is shifted by one wordline pitch. This cell arrangement saves this space, because two contacts are placed in a gap surrounded by four BEs. Therefore, by introducing the new memory cell, the average cell size, including the block-selecting transistor, is reduced to 87% of the previous cell size.

5.3 The Hierarchical Wordline Scheme without Extra Metal

Figure 10a shows an array block diagram of the 8Mb chain FeRAM. The magnified array block of Fig. 10a is shown in Fig. 10b. In this chip, a hierarchical wordline scheme is introduced to reduce the row-decoder and plate-driver areas, because the common decoder circuits for sub-row decoders and sub-plate drivers are placed in the main-row-decoder. This architecture reduces the die size by 5 mm^2. 8 Mb memory cells are divided into 32 kb sub-arrays. The 32 kb sub-array is organized into 256 WL's $\times$ 128 BL's. The main-row-decoder is placed at the chip top edge and controls 16 sub-row decoders and 16 sub-plate drivers. A main-block-selecting-signal (MBS) line from the main-row decoder, shown as a bold line in Fig. 10a,b, runs from the top to the bottom of the chip, and is used for selecting one of 32 row chain cell blocks.

One problem with the hierarchical wordline architecture shown in Fig. 10 is that this architecture requires an additional metal layer for the main-block-

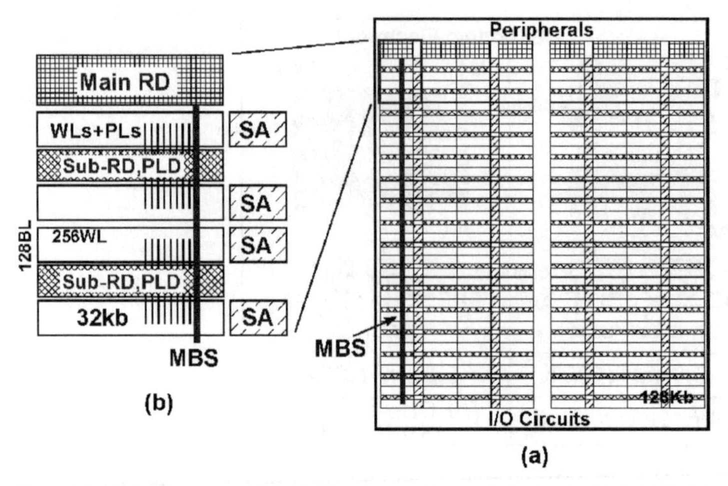

Fig. 10. (**a**) The array block diagram of an 8 Mb chain FeRAM chip, and (**b**) magnified array blocks of (**a**)

selecting-signal (MBS) line. The new chain cell block structure with an MBS line shown in Fig. 11 overcomes this problem. The M1 layer used for cell interconnections is also used for the main-block-selecting-signal (MBS) line, and this MBS line is placed in the otherwise unused space over the block-selecting transistors. Therefore, this architecture reduces the chip size by 5 mm^2 without an extra metal layer and area penalty for the chain cell block. Moreover, the M1 layer is also used for the cell-plateline ($\overline{PL}$ and PL). This enables a low-resistance M1 cell-plateline without any process penalty, and realizes a fast cell-plateline drive, as discussed in Sect. 4.

5.4 The Small-Area Dummy Cell Scheme

A key issue for realizing 1T-1C FeRAM operation is how to achieve a stable reference bitline (BL) voltage using small dummy cells, because FeRAM requires a relatively high reference BL voltage, and needs a large dummy capacitor for a dummy cell. In the FeRAM operation shown in Fig. 12a, when adopting a fast single-plate pulse sensing scheme [3, 5], after pulling up the cell-plateline (PL) to the array operation voltage (V_{aa}), the bitline (BL) voltages of the "0" data as well as those of the "1" data rise from ground level, by an amount determined by nonswitching and switching cell polarizations. Therefore, the reference BL voltage – the intermediate voltage between the "1" and "0" data – also becomes high. Moreover, considering the signal margin test, a high reference BL voltage of about a half V_{aa} is required, as shown in Fig. 12a.

In the conventional dummy cell [6] shown in Fig. 12b, the reference BL voltage is obtained by the charge of the dummy capacitor (C_d), by pulling up the dummy plateline (DPL). Therefore, in order to obtain a high reference BL

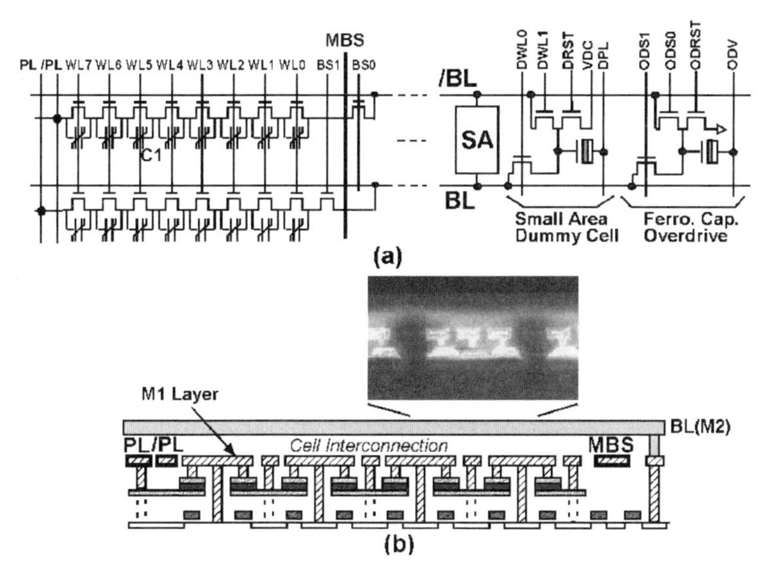

Fig. 11. (**a**) The array configuration of a chain FeRAM, and (**b**) a cross-section of the chain cell block and its SEM photograph

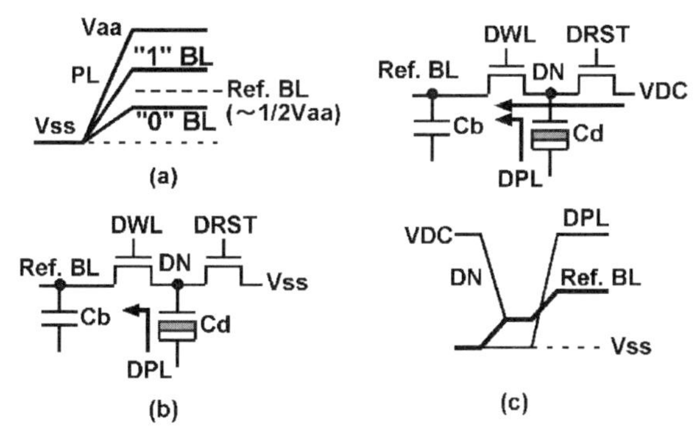

Fig. 12. (**a**) The required reference bitline (BL) voltage, (**b**) the conventional dummy cell, and (**c**) the new small area dummy cell and its timing chart

voltage of a half V_{aa}, a large dummy capacitor (C_{d}) of about 300 fF, which is comparable to the bitline capacitance (C_{b}), is required, as shown in Fig. 13a. One solution is to use a ferroelectric capacitor that has a high dielectric constant [5]. However, a fluctuation of polarization and a degradation of polarization due to fatigue and imprint vary the reference BL voltage. The new small-area dummy cell scheme of Fig. 12c overcomes this problem. In the new scheme, first, a stored charge ($C_{\mathrm{d}} \times V_{\mathrm{DC}}$) of the DN node precharged at the V_{DC} level is transferred to the reference BL by activating a dummy wordline (DWL). Second, an additional charge ($C_{\mathrm{d}} \times V_{\mathrm{DC}}$) by pulling up the dummy plateline (DPL) is also transferred to the reference BL. Therefore,

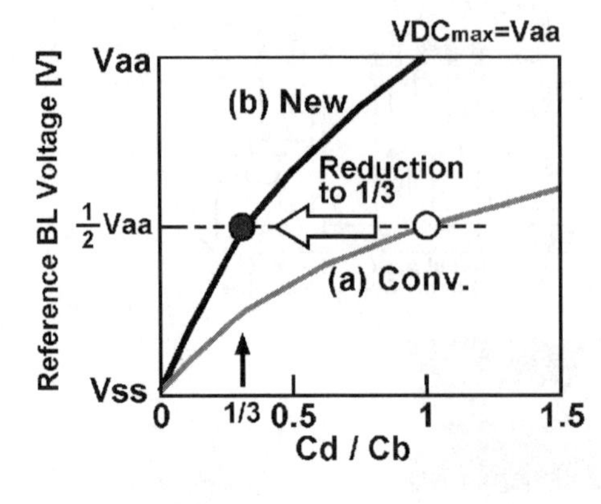

Fig. 13. The dependence of the reference bitline (BL) voltage on the ratio of the dummy capacitance (C_d) to the bitline capacitance (C_b) in (**a**) the conventional dummy cell scheme and (**b**) the proposed scheme

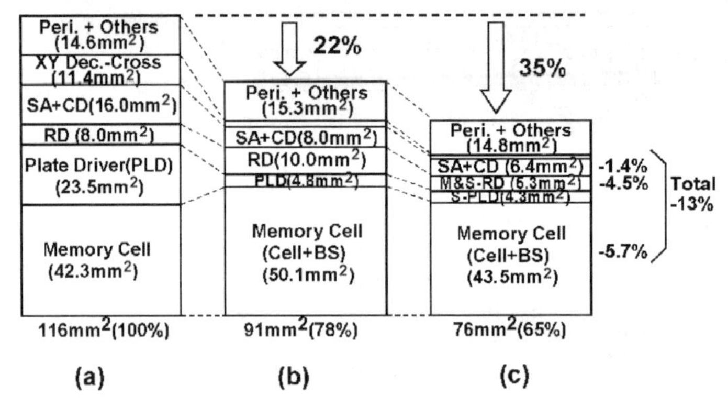

Fig. 14. A chip size comparison of 0.25 μm 8 Mb FeRAM in (**a**) conventional FeRAM, (**b**) chain FeRAM with a simple chain architecture, and (**c**) the chain FeRAM of this work

as shown in the two-step rises of the reference BL voltage of Fig. 12c, the new scheme can double the reference voltage, as shown in Fig. 13b. In other words, the dummy capacitor size is reduced to 1/3 to obtain the necessary reference BL voltage. The practical circuit of this scheme is shown in Fig. 11a. The dummy capacitor is shared with the bitline pair to reduce the dummy capacitor area further [6].

5.5 Chip Size Comparison

Figure 14 compares the 0.25 μm 8 Mb FeRAM chip size breakdown. The conventional FeRAM shown in Fig. 14a has a large die size due to its large plate driver and sense amplifier. On the other hand, in the chain FeRAM architecture shown in Fig. 14b, the small plate driver and sense amplifier reduce the die size by 14% compared with the conventional FeRAM [10]. These shrinks also reduce the large crossing-area between the X-decoder (the row-decoder and the plate-driver) and the Y-decoder (the sense amplifier and the column-decoder), and reduce the die size by 8%. Therefore, the chain architecture reduces the die size by 22%. Furthermore, in the 8 Mb chain FeRAM [11] shown in Fig. 14c, the new one-pitch-shift cell, the hierarchical wordline scheme, and the small-area dummy cell scheme reduce the die size by 5.7%, 4.5%, and 1.4%, respectively. The further shrinking of the crossing-area due to these techniques also reduces die size by 1.3%. The total chip area reduction by these new techniques is 13%. As a result, the 8 Mb chain FeRAM die size of this work is reduced to 65% of that of the conventional FeRAM.

A chain 8 Mb FeRAM was fabricated using 0.25 μm two-metal CMOS technology. A micrograph of the 8 Mb chain FeRAM is shown in Fig. 15. The capacitor size is 0.9 μm × 0.95 μm. The average cell size is 5.2 μm^2. A PZT film is used for the ferroelectric capacitor. The chip size is 76 mm^2. This chip achieved the highest memory density, 10.5 Mb/cm^2, and the highest average cell/chip area efficiency, 57.4%, reported in FeRAMs. The active current is 30 mA at a 70 ns cycle time. The features of the device are summarized in Table 1.

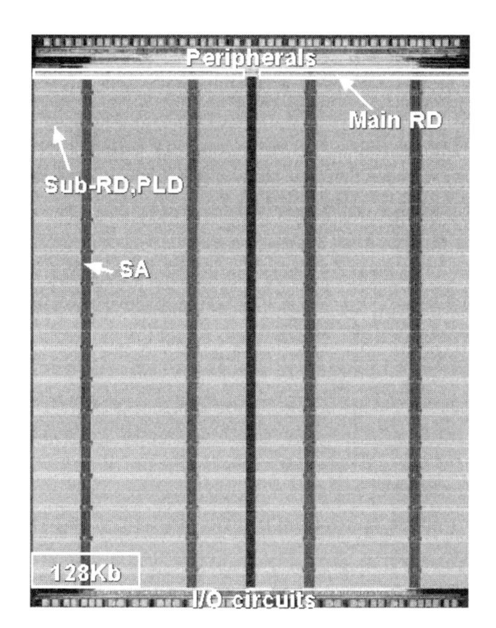

Fig. 15. A chip microphotograph of the 8 Mb chain FeRAM

Table 1. The device features of the 8 Mb chain FeRAM

Technology	0.25 μm-CMOS, 1-polycide, 2-metals
Organization	512 k × 16, 1 M × 8
Capacitor size	0.9 μm × 0.95 μm
Average cell size	5.2 μm^2
Chip size	76 mm^2
Cell efficiency	57.4%
Ferroelectric	PZT
Power supply	3.0 V, 2.5 V
Access time	40 ns at 3.0 V, 25 °C
Cycle time	70 ns at 3.0 V, 25 °C
Active current	30 mA at t_{RC} = 70 ns
Pin configuration	SRAM/Flash compatible
Functions	Pseudo-SRAM/static column

6 Low-Voltage Design

Another request regarding the application of ferroelectric memory to portable electronic devices is low-voltage operation, down to 2.5 V. However, low-voltage array operation degrades the cell signal readout to the bitline. Furthermore, as shown in Fig. 16a, in read operation, a plateline (PL) is pulled up to the array operation voltage (V_{aa}), and a bitline voltage rises from ground level even when "0" data is read out. This causes a small bias of less than V_{aa} between the cell-plateline (PL) and the bitline (BL); that is, a small bias applied to the ferroelectric capacitor. This causes signal loss of "1" data due to insufficient polarization switching, as shown in the locus of the hysteresis loop of Fig. 16a. One solution is to boost the cell-plateline voltage. But this causes an increase in the stress bias to the memory cell array.

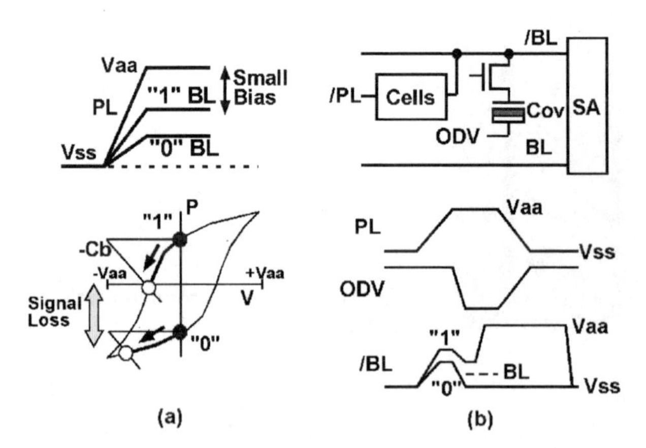

Fig. 16. (**a**) The timing chart and operating point analysis in conventional FeRAM operation. (**b**) The circuit diagram and its timing chart in the new ferroelectric capacitor overdrive scheme

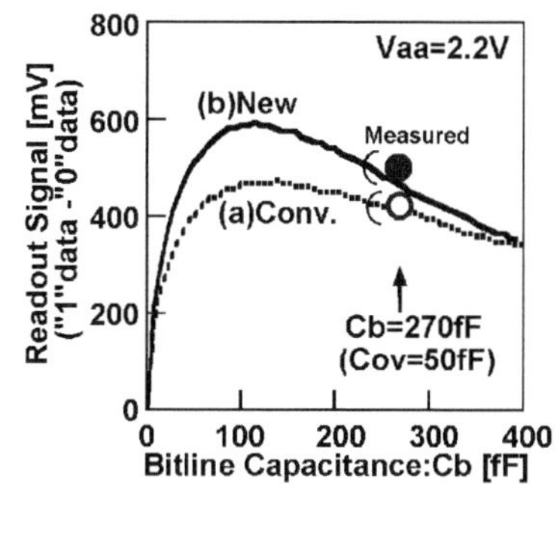

Fig. 17. The bitline capacitance dependence of the readout cell signal in (**a**) the conventional scheme and (**b**) the new overdrive scheme

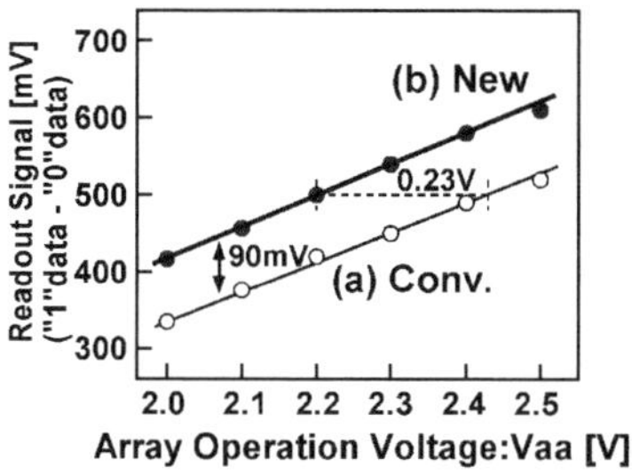

Fig. 18. The measured array operation voltage (V_{aa}) dependence of the readout cell signal in (**a**) the conventional scheme and (**b**) the new overdrive scheme

The ferroelectric capacitor overdrive scheme [11] of Fig. 16b overcomes this problem and enables enough polarization switching. In this scheme, the overdrive capacitor (C_{ov}) is introduced. A cell-plateline (PL) is pulled up to V_{aa}, and a bitline ($\overline{BL}$) rises from ground level. After that, the overdrive signal (ODV) is pulled down. By the coupling of capacitor (C_{ov}), the risen bitline is pulled down near ground, whereas the cell-plateline (PL) voltage is kept at V_{aa}. Therefore the full V_{aa} bias is applied to the ferroelectric capacitor, and the voltage swings of the cell-plateline (PL) and the bitline ($\overline{BL}$) are within the range of V_{aa}, as shown in Fig. 16b. This scheme realizes about a 0.4 V larger bias to the ferroelectric capacitor, and enables sufficient polarization switching, without over-bias to the cell-plateline and the bitline. In the practical design, the overdrive capacitor (C_{ov}) is also shared with the bitline pair to reduce the die size, as shown in Fig. 11a. The chip area penalty of this scheme is only 1%.

Figure 17 compares the readout cell signals of the conventional and proposed schemes for various bitline capacitances. In the conventional scheme, a small bitline capacitance causes a small signal due to small bias to the ferroelectric capacitor, and a large bitline capacitance also causes a small signal due to a large load capacitance. On the other hand, the proposed scheme has a larger signal than the conventional one, and especially so at the small bitline capacitance condition. This is due to both the small load capacitance and the large bias to the ferroelectric capacitor. The white and black circles show the measured signals of the conventional and proposed schemes, respectively. A 90 mV larger signal was observed by introducing the proposed scheme.

Figure 18 shows the measured array operation voltage (V_{aa}) dependence of the readout cell signal. These data were measured by sweeping the reference bitline voltage precisely. At any array operation voltages, the proposed scheme constantly has a signal that is 90 mV larger than that of the conventional scheme, thanks to the almost constant bias increase to the ferroelectric capacitor. This result implies that the signal-increasing ratio becomes larger in lower-voltage operation. The signal-increasing ratio was 26% at 2 V V_{aa}. Under the same cell signal conditions, the operation voltage of the proposed scheme is 0.23 V lower than that of the conventional scheme.

7 Summary

This chapter has reviewed the chain FeRAM architecture, and presented many circuit design techniques. The chain FeRAM allows a small memory cell of 4 F^2 size, thanks to the cross-point cell, and achieves random access. First, the concept and basic operation of chain FeRAM were presented. Second, a high-speed cell-plateline drive technique was presented. The chain FeRAM realized a 40 ns random-access time and a 70 ns read/write cycle time, which are the fastest reported in FeRAMs. Third, after discussing the merits of a high density of chain FeRAM, three new techniques to reduce the die further – 1. the one-pitch-shift cell, 2. the hierarchical wordline scheme without extra metal, and 3. the small-area dummy cell scheme – were presented. A chain architecture with these techniques reduced the die size to 65% of that of the conventional FeRAM, and realized a small 8 Mb chain FeRAM die of 76 mm^2, using a 0.25 μm two-metal CMOS. This chip achieved the highest memory density, 10.5 Mb/cm^2, and the highest average cell/chip area efficiency, 57.4%, reported in FeRAMs. Fourth, a new ferroelectric capacitor overdrive scheme to realize low-voltage operation was presented. A 90 mV larger signal was observed by introducing this overdrive scheme. This scheme reduces the array operation voltage by 0.23 V.

Acknowledgements

The authors thank J. Miyamoto and S. Sawada for their encouragement and support of this work. They are also grateful to members of the Toshiba-Infineon FeRAM Development Alliance (FDA) for chip design, device fabrication, and measurement.

References

1. J.T. Evans, R. Womack: IEEE J. Solid-State Circuits **23**, 1171 (1989)
2. R. Womack, D. Tolsch: Tech. Dig. Int. Solid-State Circuits Conf. (1989) p. 242
3. T. Sumi, N. Moriwaki, G. Nakane, T. Nakakuma, Y. Judai, Y. Uemoto, Y. Nagano, S. Hayashi, M. Azuma, E. Fujii, S. Katsu, T. Otsuki, L. McMillan, C. Araujo, G. Kano: Tech. Dig. Int. Solid-State Circuits Conf. (1994) p. 268

4. H. Koike, T. Otsuki, T. Kimura, M. Fukuma, Y. Hayashi, Y. Maejima, K. Amanuma, N. Tanane, T. Matsuki, S. Saito, T. Takeuchi, S. Kobayashi, T. Kunio, T. Hase, Y. Miyasaka, N. Shohata, M. Takada: IEEE J. Solid-State Circuits **31**, 1625 (1996)
5. W. Keaus, L. Lehman, D. Wilson, T. Yamazaki, C. Ohno, E. Nagi, H. Yamazaki, H. Suzuki: Tech. Dig. Symp. VLSI Circuits (1998) p. 242
6. T. Miyakawa, S. Tanaka, Y. Itoh, Y. Takeuchi, R. Ogiwara, S. M. Doumae, H. Takenaka, I. Kunishima, S. Shuto, O. Hidaka, O. Ohtsuki, S. Tanaka: Tech. Dig. Int. Solid-State Circuits Conf. (1999) p. 104
7. Y. Chung, M. K. Choi, K. Oh, B. G. Jeon, K. D. Suh: Dig. Symp. VLSI Circuits (1997) p. 97
8. B. G. Jeon, M. K. Choi, Y. Song, S. K. Oh, Y. Chung, K. D. Suh, K. Kim: Tech. Dig. Int. Solid-State Circuits Conf. (2000) p. 272
9. D. Takashima, I. Kunishima: IEEE J. Solid-State Circuits **33**, 787 (1998)
10. D. Takashima, S. Shuto, I. Kunishima, H. Takenaka, Y. Oowaki, S. Tanaka: IEEE J. Solid-State Circuits **34**, 1557 (1999)
11. D. Takashima, Y. Takeuchi, T. Miyakawa, Y. Itoh, R. Ogiwara, M. Kamoshida, K. Hoya, S. M. Doumae, T. Ozaki, H. Kanaya, M. Aoki, K. Yamakawa, I. Kunishima, Y. Oowaki: Tech. Dig. Int. Solid-State Circuits Conf. (2001) p. 40
12. D. Takashima: IEICE Trans. Electron. **E84-C**, 747 (2001)
13. D. Takashima, Y. Oowaki, I. Kunishima: Tech. Dig. Symp. VLSI Circuits (1999) p. 103
14. S. Onishi, K. Hamada, K. Ishihara, Y. Ito, S. Yokoyama, J. Kudo K. Sakiyama: Tech. Dig. Int. Electron. Device Meeting (1994) p. 843
15. K. Shoji, M. Moniwa, H. Yamashita, T. Kisu, T. Kaga, K. Torii, T. Kumihashi, T. Morimoto, H. Kawakami, Y. Gotoh, T. Itoga, T. Tanaka, N. Yokoyama, T. Kure, M. Ohkura, Y. Fujisaki, K. Sakata, K. Kimura: Tech. Dig. Symp. VLSI Circuits (1996) p. 28
16. H. Masuda , R. Hori, Y. Kamigaki, K. Itoh, H. Kawamoto, H. Katto: IEEE J. Solid-State Circuits **15**, 846 (1980)
17. K. Takeuchi, K. Matsuno, Y. Nakagome, M. Aoki: IEICE Trans. Electron. **E79-C**, 234 (1996)

Capacitor-on-Metal/Via-Stacked-Plug (CMVP) Memory Cell Technologies and Application to a Nonvolatile SRAM

Hiromitsu Hada[1], Kazushi Amanuma[2], Tohru Miwa[1], Sota Kobayashi[3], Toru Tatsumi[1], Yukihiko Maejima[3], Junichi Yamada[1], Hiroki Koike[1], Hideo Toyoshima[3], and Takemitsu Kunio[4]

[1] Silicon Systems Research Laboratories, NEC Corporation, 1120 Shimokuzawa, Sagamihara, Kanagawa 229-1198, Japan
h-hada@bp.jp.nec.com
[2] 1st Solutions Sales Operation Unit, Industrial Sales, NEC Corporation
[3] Advanced Technology Development Division, Manufacturing Operations Unit, NEC Electronics Corporation
[4] 4th System LSI Division, 2nd Business Development Operations Unit, NEC Electronics Corporation

Abstract. A capacitor-on-metal/via-stacked-plug (CMVP) memory cell was developed for the 0.25 μm CMOS logic embedded FeRAM. Using a 445 °C MOCVD $Pb(Zr,Ti)O_3$ process, a ferroelectric capacitor is formed after CMOS logic fabrication. Thus, FeRAM can be embedded without changing any logic devices and processes. Furthermore, this technology enables cell size reduction (3.2 μm^2 for 1T-1C), the minimum process damage on the ferroelectric capacitor, and a low manufacturing cost. This CMVP process technology permits a nonvolatile SRAM (NV-SRAM) cell consisting of a six-transistor ASIC SRAM cell and two backup ferroelectric capacitors stacked over the SRAM portion. The READ and WRITE operations in this cell are very similar to those of a standard SRAM. Because each memory cell can perform STORE and RECALL individually, both can execute massive-parallel operations. A $V_{dd}/2$ plate-line architecture makes READ/WRITE fatigue negligible.

1 Introduction

A conventional FeRAM memory cell has a ferroelectric capacitor connected to diffusion with a local interconnect or a poly-Si plug [1,2]. Embedding these memory cells on fine-scaled logic devices will present serious problems. First, backend processing causes the ferroelectric capacitor properties to deteriorate. In particular, exposure to a hydrogen-containing atmosphere, such as is forming gas annealing and tungsten chemical vapor deposition (W-CVD), substantially reduces the remanent polarization of a $Pb(Zr,Ti)O_3$ (PZT) capacitor [3]. Moreover, the addition of a FeRAM process must not alter the logic device parameters; otherwise, full custom design is required for FeRAM embedded logic devices, which will boost the design cost. In addition, there

H. Ishiwara, M. Okuyama, Y. Arimoto (Eds.): Ferroelectric Random Access Memories,
Topics Appl. Phys. **93**, 215–232 (2004)
© Springer-Verlag Berlin Heidelberg 2004

is great concem that ferroelectric and electrode materials may contaminate a transistor layer. In this chapter, in order to solve these problems a completely new FeRAM memory cell that has a Capacitor-on-Metal/Via-stacked-Plug (CMVP) structure is first presented [4,5]. This provides solutions for all the problems mentioned above and is applicable to 0.25 μm CMOS and beyond. Then, as an application of the CMVP cell, a nonvolatile SRAM with backup ferroelectric capacitors is proposed and fabricated.

2 The CMVP Memory Cell

2.1 The cell Structure and Process

Figures 1 and 2 illustrate the CMVP memory cell structure. A ferroelectric capacitor is formed above metal 2, and is connected to diffusion with a metal (Al) and via (W) stacked plug. This plug is formed at the same time with multi-layer metallization, and hence no additional process is necessary. Metal 1 is used for both a bit line and a stacked plug. Metal 3 is used only for a plate line. The cell size is limited by the metal 1 pitch, although the smallest cell size of 3.2 μm^2 is achieved.

The CMVP cell has the FeRAM process at the end of a standard CMOS process, as shown in Fig. 3. This gives plenty of benefits for its fabrication. First, there is no modification for CMOS logic devices and processes. Second, only passivation is necessary after the FeRAM process; backend process damage on the ferroelectric is minimized. Third, complete planarization is achieved for multi-level metallization. Fourth, most of the production line can be shared with other logic devices without any contamination, so that

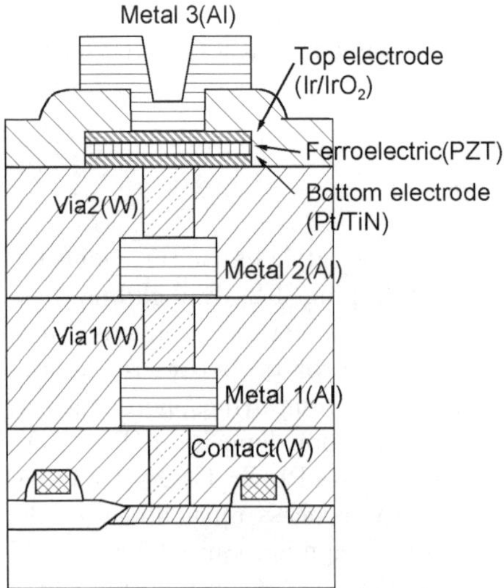

Fig. 1. A cross-section of the memory cell

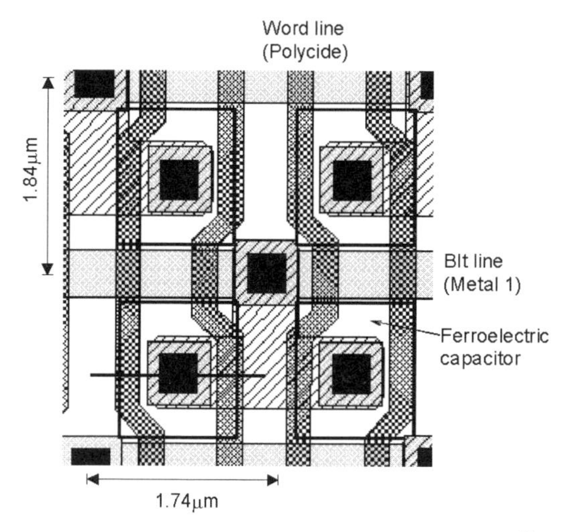

Fig. 2. The cell layout (1T-1C cell area: 3.2 μm^2). Metal 3 (plate lines) is not shown

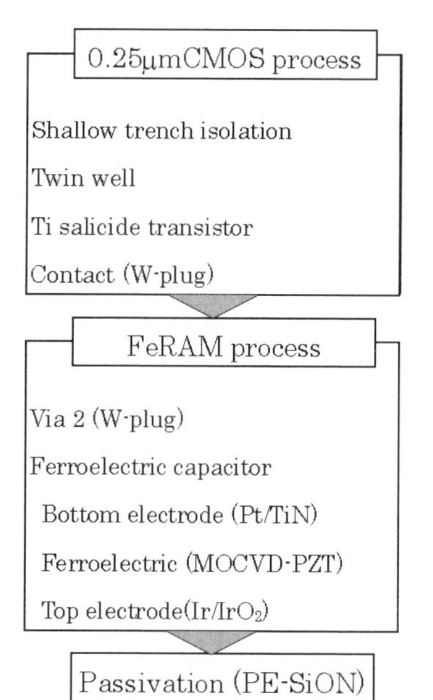

Fig. 3. The process flow of the CMVP memory cell

the FeRAM can be embedded at minimum cost. These aspects make CMVP the most suitable memory cell for embedded FeRAM. The devices and process below the second metal (M2) are exactly the same as for the 0.25 μm standard CMOS logic. The bit line is made of the first metal (M1), and the word line is Ti-silicide/polysilicon. The smallest cell size (1T-1C) is 3.2 μm^2.

After M2 formation, the following FeRAM process is added. The W-plug second via is formed and planarized by a chemical mechanical polishing

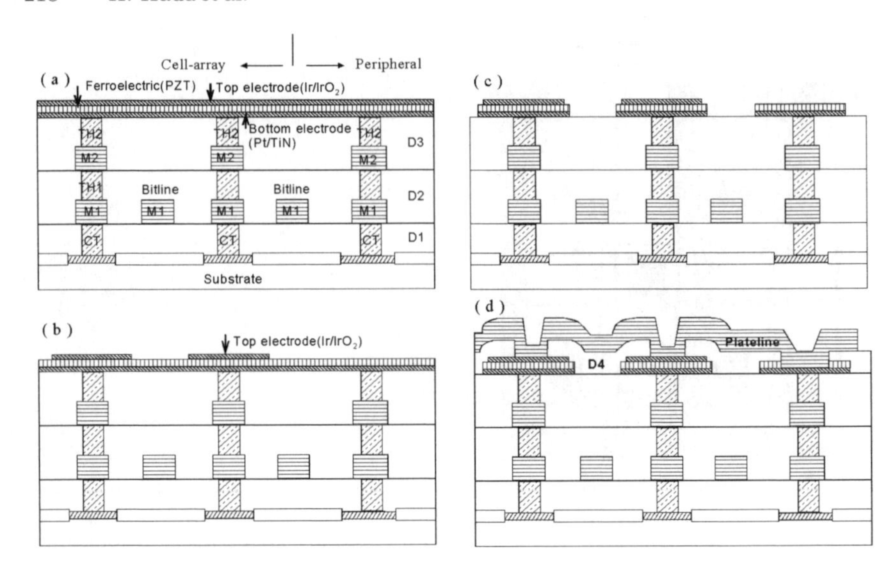

Fig. 4. The process scheme. (**a**) The top/bottom electrode and PZT are deposited on a planarized substrate. (**b**) The top electrode is etched by RIE. (**c**) The PZT and bottom electrode are etched. (**d**) The plateline is formed

Fig. 5. A cross-sectional SEM image after PZT deposition

(CMP) process. On this substrate, the bottom electrode (Pt/TiN), the ferroelectric (PZT), and the top electrode (Ir/IrO_2) are deposited as shown in Fig. 4a. PZT is grown using low-pressure metalorganic CVD (MOCVD) at 445 °C. Columnar grains grow homogeneously, and no effect of the W-plug second via is observed, as shown in Fig. 5. The top electrode is etched by reactive ion etching (RIE) with a resist mask in Ar/Cl_2 gas (Fig. 4b). The PZT and the bottom electrode are etched with another mask in $Ar/HBr/CF_4$

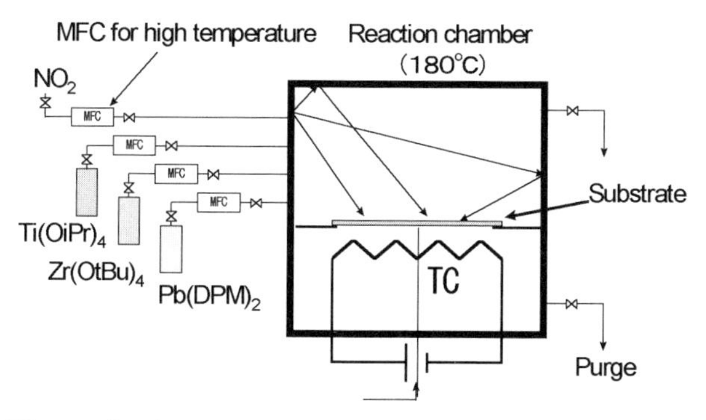

Fig. 6. A schematic diagram of the PZT MOCVD system

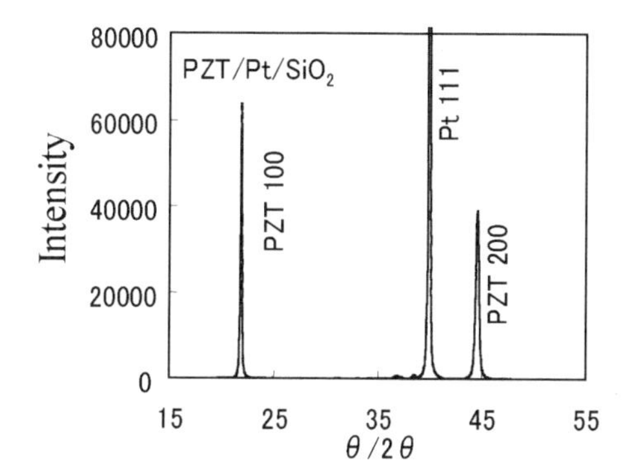

Fig. 7. The XRD pattern for the as-deposited PZT film

and Ar/Cl_2, respectively (Fig. 4c). A covering dielectric (D4) is deposited by O_3-TEOS CVD at 375 °C. Finally, a plate contact is opened by RIE and a plate line is formed (Fig. 4d). At the end of the cell array, the plate line is connected to M2 through the bottom electrode. This eliminates deep contact holes for the plate line, and thus the plate line can be connected to the plate-line driver with the same pitch as the memory cell. The combination of two-mask capacitor etching and plate-line contact through the bottom electrode gives good manufacturability.

Since a sol-gel or sputtering PZT deposition process requires 600 °C or above for PZT crystallization, a novel low-temperature MOCVD process has been developed. Figure 6 illustrates this MOCVD system. $Pb(DPM)_2$, $Zr(OtBu)_4$, $Ti(OiPr)_4$, and NO_2 oxidation gases are charged into a low-pressure (10^{-3} Torr) reaction chamber without a carrier gas. The reaction chamber is maintained at 180 °C, and the substrate is heated at 445 °C. A thin $PbTiO_3$ seeding layer is initially grown, followed by a PZT deposition at the same temperature. Without the seeding layer, the PZT crystalline phase was not obtained. X-ray diffraction (XRD) reveals that the as-deposited film is (100) oriented and no secondary phase exists, as shown in Fig. 7.

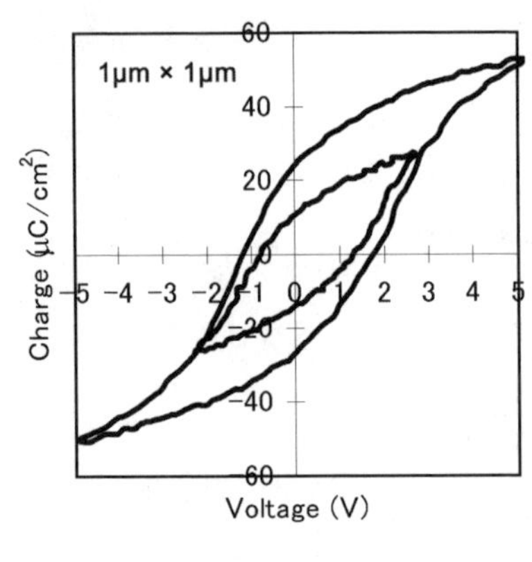

Fig. 8. Hysteresis curves for parallel-connected PZT capacitors (1 μm × 1 μm). The thickness of the PZT film was 200 nm. 100 kHz triangle waves of 2.5 V and 5 V were used

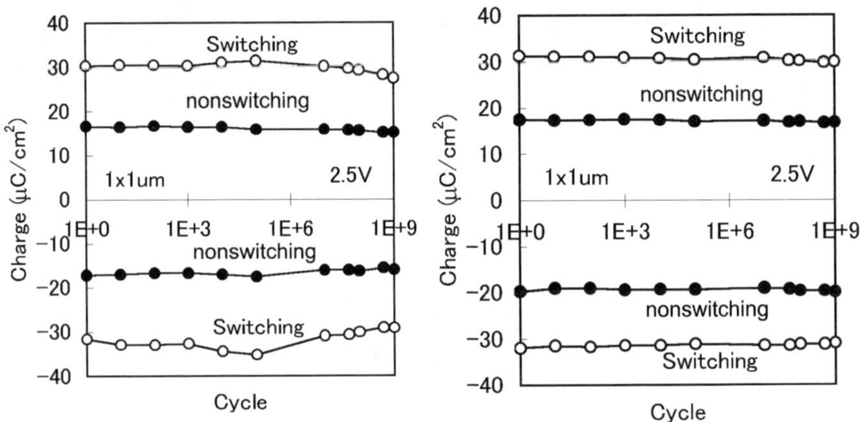

Fig. 9. (**a**) Fatigue and (**b**) imprint endurance for the 1 μm × 1 μm capacitors. The switching and nonswitching charges were measured by applying 2.5 V pulses

2.2 The Electrical Properties of the PZT Capacitor

MOCVD PZT showed good ferroelectric properties and no degradation was observed for a ferroelectric capacitor due to the low-temperature backend processing. After plate-line etching, capacitors show good hysteresis properties, as shown in Fig. 8. The PZT film thickness is 200 nm. More than 10 μC/cm^2 remanent polarization (P_r) was obtained for 1 μm × 1 μm capacitors at 2.5 V. The maximum polarization increased slightly as the capacitor size decreased from 3 μm × 3 μm to 1 μm × 1 μm, while P_r remained almost the same. This tendency was also observed when the ferroelectric was etched with a top electrode pattern. Therefore, it is not due to the fringe capacitor at the top electrode edge. 1 μm × 1 μm capacitors showed good fatigue and imprint properties. Figure 9 shows (a) fatigue and (b) imprint endurance for the 1 μm × 1 μm capacitors. As can be seen from the figure, the switching charge degradation during 10^9 switching cycles was less than 10 %, and little change was observed for switching or nonswitching charges due to 10^9 im-

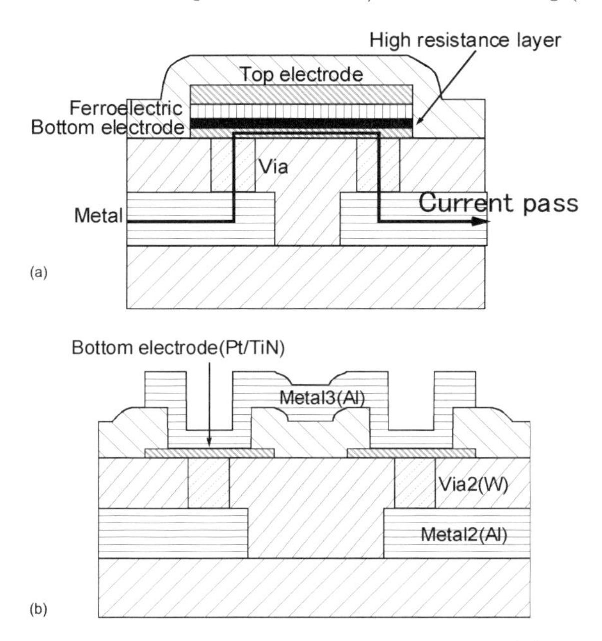

Fig. 10. A cross-section of the contact chain: (**a**) conventional; (**b**) this work. The top electrode and ferroelectric are removed by etching

print pulses. Furthermore, the I_d–V_g characteristics of the underlying n- and p-MOS transistors did not change after the FeRAM process.

2.3 Plug-Contact Resistance

One obstacle to a ferroelectric capacitor on a plug-contact is oxidation of the bottom electrode and/or the plug during ferroelectric deposition. This oxidation causes high plug-contact resistance. To prevent oxidation, several oxidation barriers have been reported [2]. In the case of the CMVP memory cell, a simple Pt/TiN bottom electrode gives a low plug-contact resistance because of the low-temperature FeRAM process. When a high-resistance layer exists beneath the bottom electrode, a conventional contact chain may present a low contact resistance, because the current can pass through as shown in Fig. 10a. To avoid this error, the contact chain shown in Fig. 10b was used in this study. The average contact resistance for the 0.4 μm × 0.4 μm via 2 was 20 Ω per contact, as shown in Fig. 11. It was also found from Auger depth profiling for the PZT/Pt/TiN/W/SiO_2/Si structure that neither TiN nor W under Pt were oxidized during PZT deposition. These results indicate that TiN works effectively as an oxygen barrier.

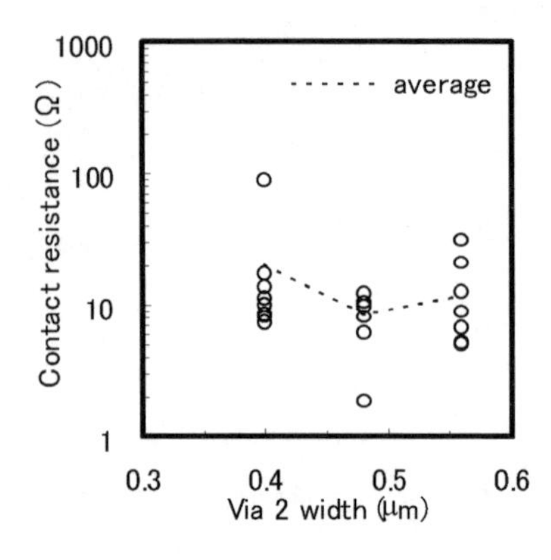

Fig. 11. The plug-contact resistance per contact

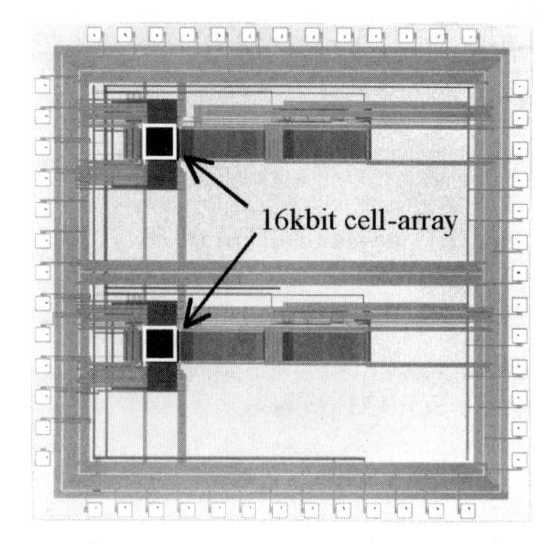

Fig. 12. A photograph of the cell-array monitor die. One chip has two 16 kb cell arrays

2.4 The Cell-Array Monitor

A cell-array monitor that has two 16 kb cell arrays was used to evaluate the cell designs. Figure 12 shows a photograph of this cell-array monitor. The peripheral circuits are 0.25 μm standard CMOS logic. Each cell array functions independently; thus various cell designs can be evaluated efficiently. For example, the yield for the 3.2 μm^2 cell with 0.8 μm × 0.8 μm top and 1.12 μm × 1.12 μm bottom electrodes was 9 % higher than that for the cell with 0.8 μm × 0.8 μm top and 1.08 μm × 1.08 μm bottom electrodes. Collecting these data leads to cell layout optimization.

3 A Nonvolatile SRAM with Backup Ferroelectric Capacitors (NV-SRAM)

On-chip rewritable nonvolatile memory is one of the key components in system LSIs. Especially in the IC-card field, nonvolatile memories are indispensable for storing identification codes and other personal information, so it is advantageous for them to be small and to execute PROGRAM/ERASE operations quickly [6,7,8,9]. In the near future, the trend toward short LSI production runs and greater product variety will make it more common to use rewritable nonvolatile memories for program code storage than mask ROMs.

Conventional on-chip nonvolatile memories are divided into two categories. These are FeRAM, which uses the remanent polarization of ferroelectric capacitors, and EEPROM and flash memory, which use V_t shifts in the floating-gate MOSFETs. FeRAMs appear promising for low-power on-chip use, because their operational voltage is low (about 2.5 V), and their WRITE endurance is high (over 10^8 cycles), offering improvements over on-chip EEPROMs and flash memories. Moreover, the high program voltage in floating gates is not suitable for advanced CMOS processes, and a micosecond-order programming delay prevents flexible use. FeRAMs' destructive READ operations, however, have limited their READ endurance to the same level as their WRITE endurance, preventing wider application.

There is demand in the market for on-chip nonvolatile memories, featuring low-voltage operation, high READ/WRITE endurance, and high-speed operation. However, no reported conventional programmable memories have offered all these features. We have developed a nonvolatile SRAM (NV-SRAM) technology based on shadow RAM [10] using ferroelectric capacitors. It offers three natural benefits. The first two are a lower operational voltage and a higher endurance than in floating-gate memories. These benefits stem from the characteristics of ferroelectric capacitors. The third is high-speed operation, which is inherited from SRAM.

A 512 byte NV-SRAM macro fabricated with 0.25 μm design rules has shown negligible READ/WRITE fatigue. It has an individual cell area of 18.6 μm^2 and a total macro area of 0.27 mm^2.

3.1 Conventional Shadow RAMs Using Ferroelectric Capacitors

Figure 13 shows the basic concept of a shadow RAM using ferroelectric capacitors [11,12,13]. A shadow-RAM cell consists of a volatile SRAM cell portion and a pair of ferroelectric capacitors. In a volatile state with an active power supply, data is held in the SRAM portion, and this data is READ/WRITE accessed in the same way as it is with a standard SRAM. Before voltage cutoff, the data in the SRAM portion is copied into the coupled nonvolatile capacitors in a STORE operation. When power is restored, this stored value is returned to the SRAM portion by a RECALL operation.

Figure 14a shows a circuit diagram for a typical conventional shadow-RAM cell. It consists of a six-transistor SRAM cell, two ferroelectric capacitors, and four extra transistors. When the power is on, two transistors gated with C1 isolate the capacitors from the SRAM portion, and two transistors gated with C2 discharge to the ground level. This cell occupies approximately twice the cell area of a six-transistor SRAM cell, and its STORE/RECALL

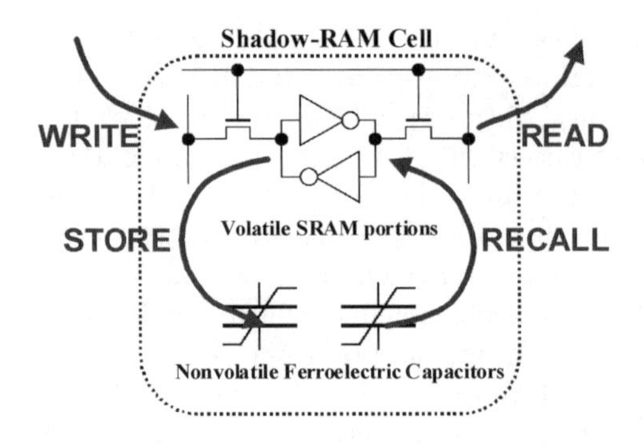

Fig. 13. Shadow RAM using ferroelectric capacitors

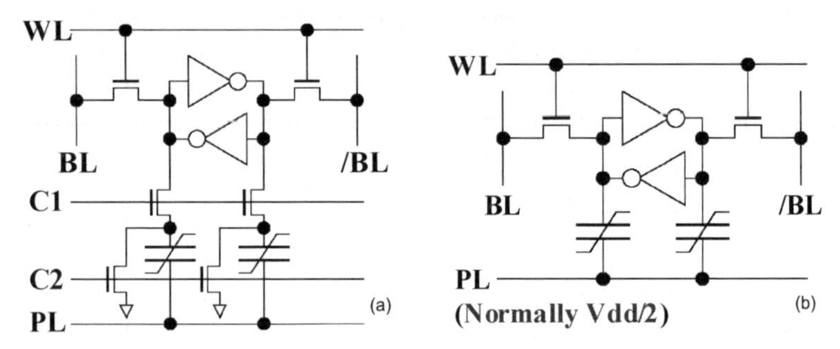

Fig. 14. Memory cell configurations. (**a**)A conventional shadow-RAM cell using ferroelectric capacitors and (**b**) a NV-SRAM cell

operation sequences are complex. The STORE operation is as follows. First, C1 is set to a high level, and C2 is set at the ground level to connect the capacitors to the SRAM-cell portion. Then, the plate line is driven from the ground level to V_{dd}. This is done to apply program voltages of plus and minus V_{dd} to each of the ferroelectric capacitors. In this conventional cell, C1 must be boosted over $V_{dd} + V_t$ to apply the full V_{dd} bias to the capacitor connected to the higher storage node.

A typical conventional shadow RAM using ferroelectric capacitors presents three design challenges. The first is concerned with density; designers must determine the minimum cell size possible. The second is concerned with the STORE and RECALL operations; massive-parallel operations are best in quick power-up/down sequences. The third challenge is concerned with the reliability of the stored data, which strongly depends on the program voltage. A sufficiently high program voltage must be combined with a low operational voltage.

3.2 The NV-SRAM Cell

3.2.1 The Cell Structure

Figure 14b shows the proposed NV-SRAM cell configuration. Each cell consists of a six-transistor SRAM cell and two backup capacitors. Each capacitor is directly connected to a SRAM-cell storage node. A unique feature of the proposed NV-SRAM cell is that the plate line of the NV-SRAM is set to $V_{dd}/2$ to keep the voltage bias across the capacitors low (between $-V_{dd}/2$ and $V_{dd}/2$). The coercive voltages of the capacitors are set to a value greater than $V_{dd}/2$ to eliminate polarization transitions. As a result, fatigue becomes negligible.

3.2.2 The READ Operation

To read out a stored value in an NV-SRAM cell, a READ operation similar to that in a standard SRAM is carried out. First, a pair of bit lines is precharged at a proper level (typically $V_{dd} - V_t$). Then a word line is selectively driven to an active level (typically V_{dd}), and one of the drive transistors in the SRAM portion pulls down one of the bit lines. Finally, a sense amplifier, connected to the bit line pair, amplifies the differential voltage between the bit lines. During this READ operation, the plate line is kept at $V_{dd}/2$.

In the READ operation, a ferroelectric capacitor on the lower storage node helps a drive transistor in the SRAM portion to discharge the connected bit line (Fig. 15a). As a result, the read access time in NV-SRAMs is shortened.

3.2.3 The WRITE Operation

To write a value in an NV-SRAM cell, a WRITE operation similar to that in a standard SRAM is carried out. First, a WRITE circuit drives one of the bit lines to V_{dd}, and the other to the ground level in accordance with the value to be written. Then, the word line is selectively driven to an active level, and the SRAM portion is forced to hold the value. During this WRITE operation, the plate line is kept at $V_{dd}/2$.

In contrast to the READ access time, WRITE cycle time can be extended by adding capacitors. To pull down a higher storage node to the ground level in a WRITE operation, the WRITE circuit also needs to discharge a ferroelectric capacitor on the storage node. The ferroelectric capacitor is about 100 fF and is comparable to the bit-line capacitance (Fig. 15b). It may require a time budget in the write cycle. Enhancing the WRITE circuit is an effective way of overcoming this disadvantage.

The static noise and WRITE margins of an NV-SRAM cell are naturally the same as those in an SRAM cell, which forms the SRAM cell portion of the NV-SRAM cell. This is because these margins define the DC characteristics of the memory cell. That is, they depend only on the resistance and conductance of the elements in the memory cells and are independent of the capacitance of the storage nodes.

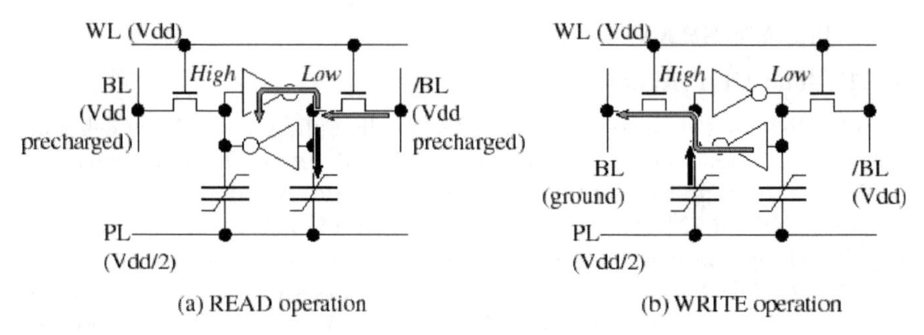

Fig. 15. The influence of capacitors on READ/WRITE operations (*arrows* show current paths). (**a**) READ operation. (**b**) WRITE operation

3.2.4 The STORE Operation

STORE operations must be carried out before voltage cutoff so that data is not lost. Figure 16 shows timing charts that indicate the polarization states of the ferroelectric capacitors at individual stages. In a volatile state, the plate line is set at $V_{\rm dd}/2$, and polarization is distributed randomly. During a STORE operation, the plate line is first driven to $V_{\rm dd}$, thus applying $V_{\rm dd}$ across the capacitor connected to the lower storage node (CapLow). This provides positive remanent polarization. By contrast, no bias is applied across the other capacitor (CapHigh) (1). The plate line is then discharged to the ground level, thus applying $-V_{\rm dd}$ across CapHigh. This provides negative remanent polarization, and removes the voltage bias from CapLow, and, by extension, retains positive remanent polarization (2). Finally, the voltage is cut off, and CapHigh retains negative and CapLow positive remanent polarizations. Because of the full $V_{\rm dd}$ biases during the STORE operations, the retention characteristics of the NV-SRAM are as high as those of standard FeRAMs.

3.2.5 The RECALL Operation

When power is next restored, an automatically invoked RECALL operation translates the states of the remanent polarization kept in the ferroelectric capacitors into voltage levels in their respective storage nodes in the SRAM portion. As the supplied voltage rises from the ground level to $V_{\rm dd}$, while the plate line is fixed at the ground level, the voltage at the storage node connected to CapHigh rises faster than at the other storage node (3). This is because, under these conditions, the differential capacitance of CapLow is higher than that of CapHigh because of the switching charge involved, and the node connected to CapLow has a greater coupling with the ground plate line. Subsequently, the latch function of the SRAM portion amplifies the differential potential of the complementary storage nodes and retains it. Consequently, the value stored in the ferroelectric capacitors by the previous

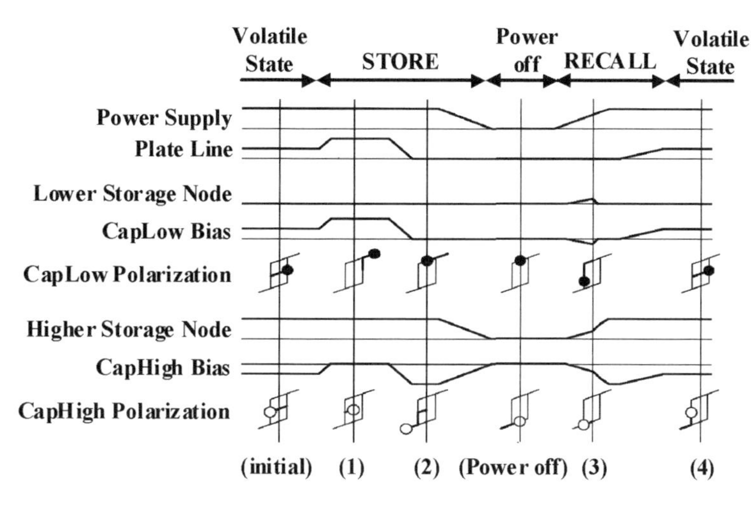

Fig. 16. The timing chart

STORE operation reappears in the SRAM portion. The plate line is finally driven to $V_{\mathrm{dd}}/2$ (4).

The STORE operation is carried out by simply driving the plate lines to V_{dd} and then to the ground level. The RECALL operation is only a result of the rise in the supply voltage. These sequence are very simple and can be completed within a short time. Note that each cell can perform both sequences individually and without the help of any peripheral circuit resources, such as sense amplifiers. As a result, these STORE and RECALL operations can execute massive-parallel operations. This means that if all of the NV-SRAM cells are connected to one plate-line node and they share a power supply, entire cell arrays can be stored and recalled simultaneously.

3.2.6 The Cell Layout

Figure 17a shows the memory-cell layout of an SRAM cell for a 0.25 μm standard ASIC macro, and Fig. 17b shows that of an NV-SRAM based on this SRAM cell. Figure 18 illustrates the three-dimensional (3-D) structure of an NV-SRAM cell. By using capacitor-on-metal/via-stacked plug (CMVP) process technologies [4,5], capacitors are formed above the SRAM portion after the standard CMOS process is complete. The topmost metal layer added forms plate lines. The bottom electrode of each capacitor is connected to an SRAM-portion storage node via a stacked plug, and the top electrodes in the cell array are connected to a plate line.

Using CMVP process technologies, circuit resources can be shared among LSIs with standard CMOS processes, and a ferroelectric memory macro can be easily embedded into a system LSI, because the transistor and wiring characteristics are fully compatible with the standard CMOS process. Our

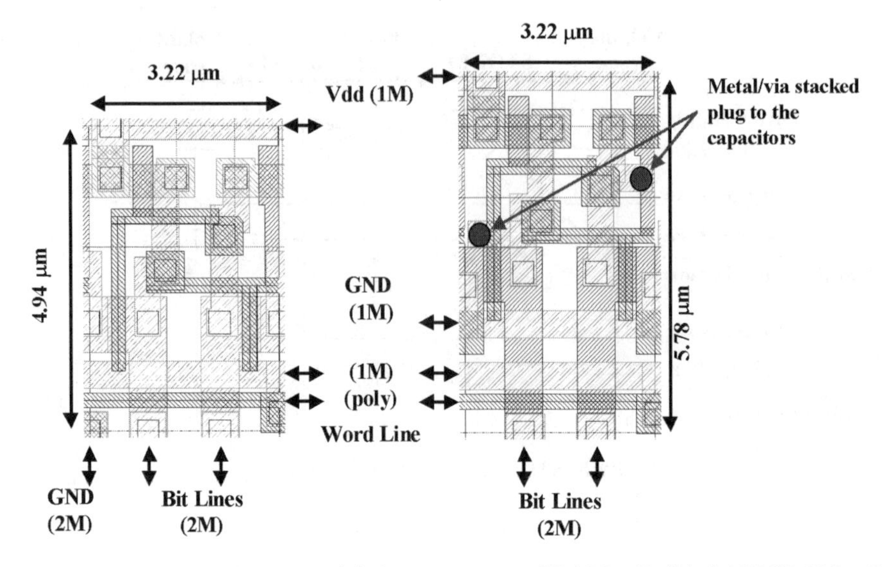

Fig. 17. Memory cell layouts. (**a**) A six-transistor SRAM cell. (**b**) A NV-SRAM cell

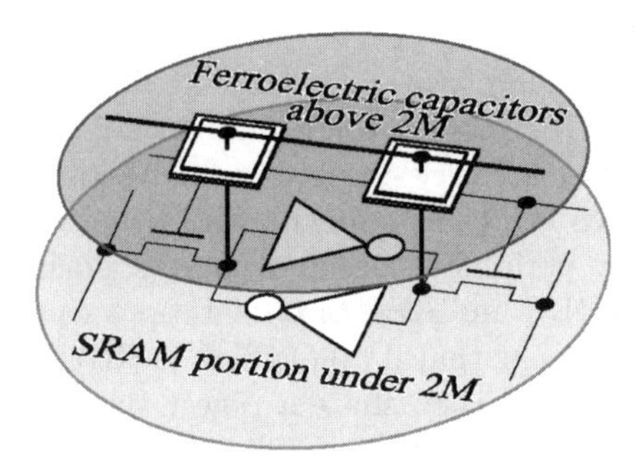

Fig. 18. The 3-D structure of the NV-SRAM cell

NV-SRAM cell is 1.17 times larger than an SRAM cell because of the area added by the stacked metal/via, which connects the storage nodes and the capacitors. The NV-SRAM cell size does not depend on the capacitor size, but on the wiring and transistor design rules. For 0.25 μm ASIC design rules, 1 μm^2 capacitors are used. These are large enough for stable operations. By using this simple structure, the size of the NV-SRAM cell can be reduced at the same rate as for the standard SRAM cell.

3.2.7 Other Advantages

In addition to their small cell size and their better resistance to READ/ WRITE fatigue, NV-SRAMs have two other advantages when compared to conventional nonvolatile memories: (1) they operate faster, and (2) they do

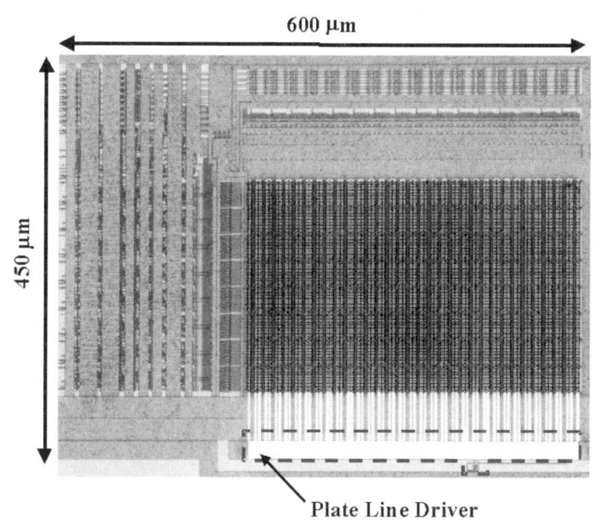

Fig. 19. The chip micrograph

not require boosting. In a conventional FeRAM, READ/WRITE operations are complicated because destructive-read cells must be written back in an operation cycle. By contrast, in an NV-SRAM, the READ operation is the same as in an SRAM. Thus, timing control is much simpler, and faster operation is possible. In a typical FeRAM and the conventional shadow RAM shown in Fig. 14a, in order to apply V_{dd} to the ferroelectric capacitors, both a selected word line and C1 must be boosted. By contrast, because the NV-SRAM is directly connected to capacitors, it can store nonvolatile data without boosting, which means that NV-SRAMs can be produced by using the standard CMOS process, without the additional steps needed for high-voltage transistors.

3.3 Chip Fabrication and Experiments

A micrograph of a 512 byte NV-SRAM macro is shown in Fig. 19. A 0.25 μm design rule double-metal-layer CMOS process with CMVP was used to produce it. In this process, ferroelectric capacitors were formed above the second metal layer. The ferroelectric material was MOCVD lead zirconate titanite. The bottom electrode is a multi-layer made of platinum and titanium, and the top electrode is a multi-layer made of iridium and iridium dioxide.

The design was based on a standard ASIC SRAM macro, with the original SRAM cell array simply being replaced by an NV-SRAM cell array. A plate-line driver was also added. The nonvolatile functions of such fabricated NV-SRAM macros have been observed to operate successfully. The ferroelectric capacitors do not affect the read access time – it is as fast as the ASIC SRAM macro. The bit-line discharge time during a WRITE cycle is, however, 1.2 ns slower than that in the ASIC SRAM macro because of the large ferroelectric capacitors, which are connected to the storage nodes. The chip characteristics are summarized in Table 1.

Table 1. Macro characteristics

Organization	512 byte
Supply voltage	2.5 V
Technology	0.25 μm double-metal CMOS with capacitor-on-metal/via-stacked-plug (CMVP)
Die size	$450 \times 600\ \mu m^2$
Cell size	$3.22 \times 5.78\ \mu m^2$
Capacitor size	1 μm × 1 μm × 2
Cycle time	6 ns at 2.5 V (estimated)
Active power	2 mW at 5 MHz

Figure 20 shows the measured fatigue characteristics of the ferroelectric capacitors for 2.5 V (V_{dd}) and 1.25 V ($V_{dd}/2$) cycling pulses. These were obtained from test devices consisting of parallel-connected $1 \times 1\ \mu m^2$ CMVP capacitors. The capacitors had an Ir/IrO_2/PZT (MOCVD-deposited, 200 nm thickness)/Pt/TiN structure and were located above the double-layer-metal wiring. The initial value of the switching charge was approximately 32 $\mu C/cm^2$ and that of the nonswitching charge was 15 $\mu C/cm^2$. The 2.5 V cycling in our test device corresponds to the READ/WRITE cycling of a normal 2.5 V FeRAM, and to the STORE/RECALL cycling of our 2.5 V NV-SRAM. The 1.25 V cycling corresponds to the WRITE cycling of the NV-SRAM. After 10^6 2.5 V switching cycles, the switching charge begins to decrease, and at

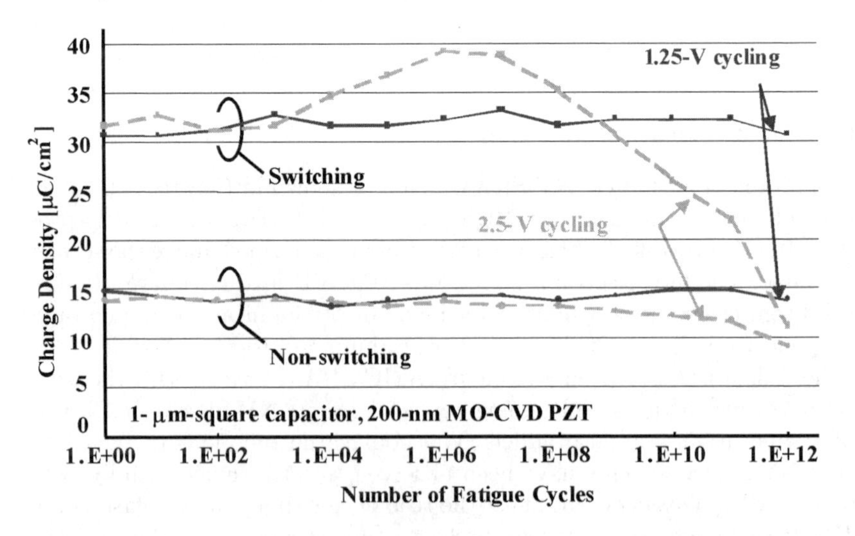

Fig. 20. Fatigue characteristics: 1.25 versus 2.5 V. The thickness of the PZT film was 200 nm

10^9 cycles it becomes lower than the initial value. At 10^{12} 1.25 V switching cycles, by contrast, the switching charge is hardly affected at all. A 2.5 V NV-SRAM has at least 10^4 times better READ/WRITE endurance than a 2.5 V FeRAM. It also has a power-up and power-down endurance of 10^8.

4 Summary

The CMVP memory cell was demonstrated as being a promising technology for embedded FeRAM. A PZT capacitor was formed on a metal (Al)/via (W) stacked plug using low-temperature MOCVD process. Two-mask capacitor etching and a plate-line contact through the bottom electrode were introduced to the CMVP process. 1 μm × 1 μm PZT capacitors showed P_r values of more than 10 μC/cm^2 at 2.5 V and good fatigue/imprint endurance. The 0.25 μm 16 kb cell-array monitor was fully functionable and proved effective for cell design optimization.

Then, the CMVP process was successfully applied to the fabrication of a 512 byte nonvolatile SRAM (NV-SRAM) test macro, in which a 0.25 μm double-metal-layer CMOS process was used. Its $V_{dd}/2$ plate-line architecture contributes to its negligible (more than 10^{12}) fatigue/imprint characteristics. Because of the CMVP 3-D structure, the memory cell occupies an area only 1.17 times larger than that of a standard ASIC SRAM cell produced with the same design rules. The cell has an area of 18.6 μm^2, and the total macro area is 0.27 mm^2. The READ/WRITE cycle time of the NV-SRAM is comparable with that of a standard SRAM macro, because its READ/WRITE operations are based on those of standard SRAMs.

References

1. N. Tanabe, T. Matsuki, S. Saitoh, T. Takeuchi, S. Kobayashi, T. Nakajima, Y. Maejima, Y. Hayashi, K. Amanuma, T. Hase, Y. Miyasaka, T. Kunio: Tech. Dig. Symp. VLSI Technol. (1995) p. 123
2. J. Kudo, Y. Ito, S. Mitarai, N. Ogata, S. Yamazaki, H. Urashima, A. Okutoh, M. Nagata, K. Ishihara: Tech. Dig. Int. Electron Device Meeting (1997) p. 609
3. T. Hase, T. Noguchi, Y. Miyasaka: Abstracts Integr. Ferroelectr. (1997) p. 29
4. K. Amanuma, T. Tatsumi, Y. Maejima, S. Takahashi, H. Hada, H. Okizaki, T. Kunio: Tech. Dig. Int. Electron Device Meeting (1998) p. 363
5. K. Amanuma, S. Kobayashi, T. Tatsumi, Y. Maejima, H. Hada, J. Yamada, T. Miwa, H. Koike, H. Toyoshima, T. Kunio: Ext. Abstracts Solid-State Devices Materials (1999) p. 384
6. T. Sumi, M. Azuma, T. Otsuki, J. Gregory, C. A. Paz de Araujo: Tech. Dig. Int. Solid-State Circuits Conf. (1995) p. 70
7. T. Fukushima, A. Kawahara, T. Namba, M. Matsumoto, T. Nishimoto, N. Ikeda, Y. Judai, T. Sumi, K. Arita, T. Otsuki: Tech. Dig. Symp. VLSI Circuits (1996) p. 46
8. T. Miwa, J. Yamada, Y. Okamoto, H. Koike, H. Toyoshima, H. Hada, Y. Hayashi, H. Okizaki, Y. Miyasaka, T. Kunio, H. Miyamoto, H. Gomi, H. Kitajima: Proc. Custom Integrated Circuits Conf. (1998) p. 439

9. J. Yamada, T. Miwa, H. Koike, H. Toyoshima, K. Amanuma, S. Kobayashi, T. Tatsumi, Y. Maejima, H. Hada, H. Mori, S. Takahashi, H. Takeuchi, T. Kunio: Tech. Dig. Int. Solid-State Circuits Conf. (2000) p. 270
10. J. Drori, S. Jewell-Larsen, R. Klevin, W. Owen, R. Simko, W. Tchon: Tech. Dig. Int. Solid-State Circuits Conf. (1981) p. 148
11. K. Dimmler, S. S. Eaton: Memory cell with volatile and nonvolatile portions having ferroelectric capacitors, US Patent No. 4,809,225 (1987)
12. S. S. Eaton, D. B. Butler, M. Parris, D. Wilson, H. McNeillie: Tech. Dig. Int. Solid-State Circuits Conf. (1998) p. 130
13. S. S. Eaton, M. Michael: SRAM with programmable capacitance divider, US Patent No. 4,918,654 (1990)

The FET-Type FeRAM

Hiroshi Ishiwara

Frontier Collaborative Research Center, Tokyo Institute of Technology,
4259 Nagatsuta, Midori-ku, Yokohama 226-8503, Japan
ishiwara@pi.titech.ac.jp

Abstract. The current status of the fabrication and integration of ferroelectric-gate field effect transistors (FETs) is reviewed. Novel applications of ferroelectric-gate FETs are first discussed, which include a single-transistor-cell-type digital memory, reconfigurable LSIs, and an analog memory for an artificial neural network. In the neural network application, the fabrication and basic operation of a pulse-frequency-modulation-type adaptive-learning neuron circuit are described, and also anSOI (silicon-on-insulator)-type FET array for synaptic connection is proposed.

Then, the data retention characteristics of ferroelectric-gate FETs with MFIS (M, metal; F, ferroelectric; I, insulator (buffer layer); S, semiconductor) and MFMIS structures are discussed. It is shown that the important factors in improving the data retention characteristics are (1) increasing the buffer layer capacitance, (2) decreasing the leakage current of both the ferroelectric film and the buffer layer, and (3) optimization of the area ratio between the MFM and MIS parts in the MFMIS structure. In the experiment, the properties of novel ferroelectrics such as $Sr_2(Ta,Nb)_2O_7$ and $Pb_5Ge_3O_{11}$ are discussed as well as those for the conventional ferroelectrics. Finally, a novel FET-type FeRAM with a 1T-2C structure is introduced and some experimental results are presented.

1 Introduction

So far, various kinds of ferroelectric memories have been discussed. However, the principle of the READ operation is the same in these memories. That is, the polarization reversal current is detected using a sense amplifier and thus a REWRITE operation is necessary after each READ operation. There is another type of ferroelectric memory, called an FET-type ferroelectric memory, which was proposed much earlier than the conventional capacitor-type memory. A key device in the FET-type memory is a ferroelectric-gate FET or an MFSFET (metal–ferroelectric–semiconductor FET), in which the gate insulator is composed of a ferroelectric material, as shown in Fig. 1. In this FET, the polarization direction of the ferroelectric film can be read out non-destructively using the drain current. The NDRO (nondestructive read-out) characteristic is one of the most important features of the FET-type FeRAM. The ferroelectric-gate FET can also be used as a nonvolatile analog memory if the polarization of the ferroelectric film is continuously controlled by applying pulse signals that are shorter than the polarization reversal time.

The original idea of the ferroelectric-gate FET was proposed by Bell Laboratories in 1955 [1] and many experimental studies were conducted in the

H. Ishiwara, M. Okuyama, Y. Arimoto (Eds.): Ferroelectric Random Access Memories,
Topics Appl. Phys. **93**, 233–251 (2004)
© Springer-Verlag Berlin Heidelberg 2004

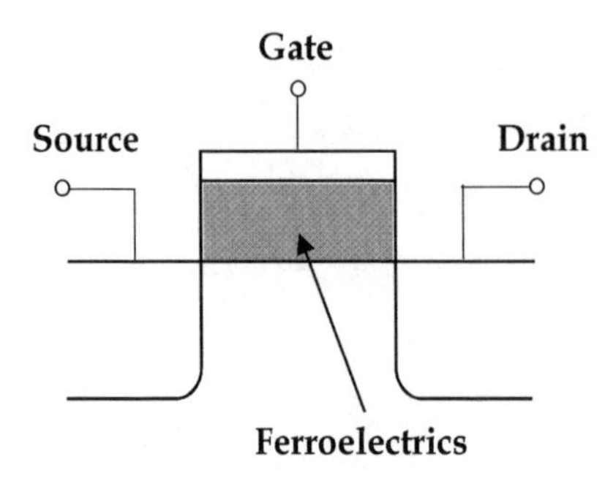

Fig. 1. The schematic of an MFSFET

1960s and 1970s. However, since it was very difficult to form a ferroelectric–semiconductor interface with good electrical properties, the studies were almost stopped in the late 1970s. Then, after the invention of the capacitor-type FeRAM, the importance of the ferroelectric-gate FET was again recognized and the studies are now getting popular. In this chapter, novel applications of ferroelectric-gate FETs are first reviewed, which include applications to a single-transistor-cell-type digital memory, reconfigurable LSIs, and a synaptic connection circuit in an artificial neural network. Next, it is described how a short data retention time is the present largest problem with ferroelectric-gate FETs and important factors to improve the retention characteristic are discussed. Then, the current status of the experimental studies on MF(MI)S (I, insulator) capacitors and FETs is described, and finally a novel 1T-2C-type NDRO cell with a good data retention characteristic is introduced and its basic operation is presented.

2 Novel Applications of Ferroelectric-Gate FETs

2.1 The Single-Transistor-Cell-Type Digital Memory

In order to increase the packing density of an FET-type FeRAM, it is desirable that each memory cell should be composed of a single ferroelectric-gate FET. However, in a conventional array of FETs, it is difficult to avoid the disturbance problem in the WRITE/READ operation and, thus, another MOSFET for cell selection is usually added in each cell. A proposed structure for a single-transistor-cell-type FeRAM is shown in Fig. 2, in which memory cells are fabricated using an SOI (silicon-on-insulator) structure [2]. In this figure, Si stripes with a lateral npn structure are placed on an insulating substrate: they are covered with a uniform ferroelectric film, and then metal stripes are placed on the film perpendicular to the Si stripes. Each Si stripe in the figure corresponds to a parallel connection of MFSFETs and each metal stripe to a common-gate electrode. In this structure, the packing density of the memory cells is expected to be very high, since no via hole across the ferroelectric film exists in the array area.

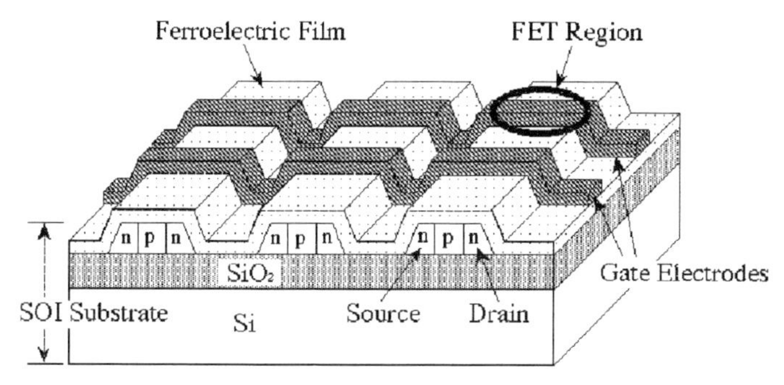

Fig. 2. The structure of a single-transistor-cell-type FeRAM

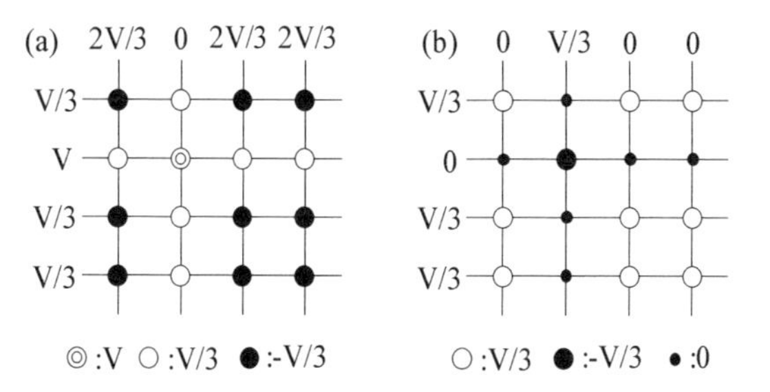

Fig. 3. The WRITE operation to a selected cell (**a**) and the compensation operation to minimize the data disturbance effect (**b**)

In order to write a datum "1" in a selected cell in the array, the so-called $V/3$ rule is used, in which V and $V/3$ are applied to the selected and nonselected metal electrodes, respectively, while 0 and $2V/3$ are applied to the selected and nonselected Si stripes, as shown in Fig. 3a. In this operation, the disturbance voltage in most nonselected cells is $-V/3$, and it is $V/3$ in the nonselected cells placed along the selected column and row. Although the disturbance voltage is 1/3 of the writing voltage, it has been found from an experimental result using ferroelectric capacitors that depolarization due to the disturbance voltage cannot be neglected in the worst case [2]. This problem has almost been solved by applying compensation voltage pulses in the next timing, as shown in Fig. 3b.

After the WRITE operation of "1", the Si surface approaches the inversion state because of the remanent polarization of the ferroelectric film, and the threshold voltage of the MFSFET is lowered. Thus, the stored data can be read out nondestructively by applying an appropriate gate voltage to a selected metal electrode and by measuring the current between the source and drain regions in a selected Si stripe. In a Si stripe with 10^3 parallel FETs, for example, it is desirable for an FET in the "1" state that the drain current increases by at least a factor of 10^4 when a READ pulse is applied to the gate, so that the READ current is 10 times larger than the total leakage current of all of the nonselected FETs. From this condition, the minimum READ voltage is determined to be 0.4 V, if the drain current of the MFSFET is increased by a factor of 10 in every 100 mV of applied gate voltage.

2.2 Reconfigurable LSIs

In the fabrication of a capacitor-type FeRAM, ferroelectric capacitors are formed on top of CMOS transistors which have been fabricated using a standard process. Thus, it is said that the integration of logic circuits and FeRAM is easier than that of logic circuits and DRAM. Furthermore, the former combination has the advantage that the stored data are retained even after the electricity is switched off. Because of these reasons, the integration of logic and memory functions on a chip (SOC; System-on-a-Chip) is one of the most promising applications of FeRAM. In this application, the role of FeRAM is storage of data, which is the same as that of DRAM.

On the other hand, if ferroelectric-gate FETs are added in a logic circuit as electrically programmable switches, it is possible to change the function of the circuit itself as well as to store data. An example of such a real-time reconfigurable logic circuit is the neuron-MOS circuit, where a neuron-MOSFET means a floating-gate MOSFET with several capacitively coupled input terminals [3]. When ferroelectric-gate FETs are implemented in a logic circuit, the functionality of the circuit is expected to be further improved because of their nonvolatile nature. Actually, an application to a DPGA (dynamic programmable gate array (DPGA)) has been proposed, in which SRAMs or anti-fuse devices in FPGA (field programmable gate array) are replaced with ferroelectric-gate FETs [4].

2.3 An Analog Memory for Storing Synaptic Weight in a Neural Network

A ferroelectric-gate FET is also useful as an analog memory. It can typically be used as an electrically modifiable synapse device for storing the weight value in an artificial neural network. In this application, it is also possible to give an adaptive-learning function to the FET [5]. That is, when the pulse width of the input signals is sufficiently narrow, polarization of the ferroelectric film is not completely reversed by the application of a single pulse, but it is gradually changed by the application of many pulses. This behavior can be regarded as an adaptive-learning function of the FET and the learned result can be read out using the drain current. In this case, the layout of synapse devices is essentially the same as that shown in Fig. 2 [6].

Recently, a PFM (pulse frequency modulation) type neuron circuit was actually fabricated and the adaptive-learning function was demonstrated [7]. A diagram of the circuit, in which ferroelectric-gate FETs are used as synapse devices, is shown in Fig. 4. The output pulse interval of this circuit is mainly

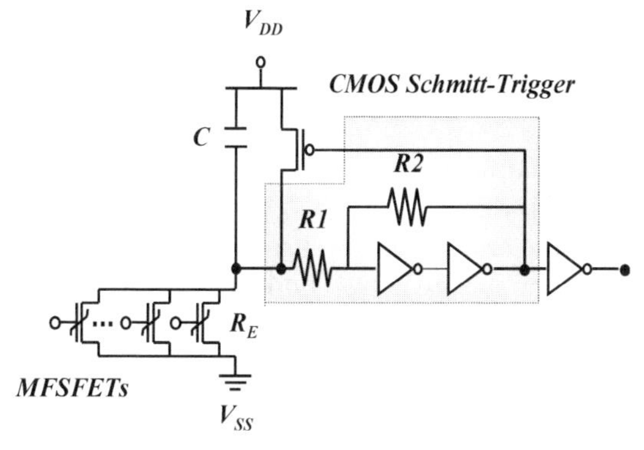

Fig. 4. An adaptive-learning neuron circuit composed of ferroelectric-gate FETs and a CMOS Schmitt-trigger oscillator

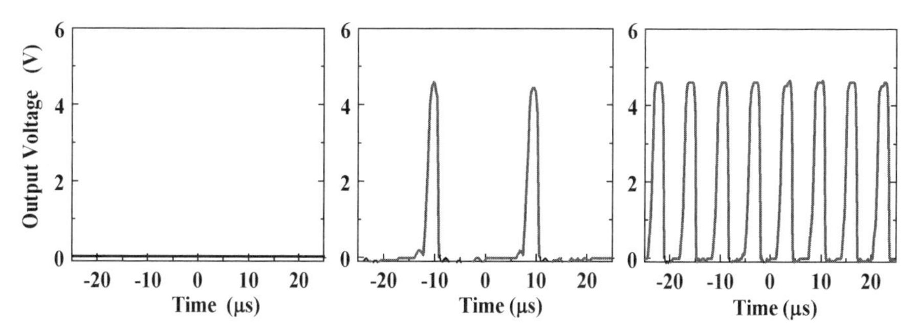

Fig. 5. The output waveforms of the neuron circuit

determined by a time constant of the capacitor C and the channel resistance R_{E} of the ferroelectric-gate FETs, and the charging and discharging operation of the capacitor C is controlled by the hysteretic transfer characteristic of a CMOS Schmitt-trigger circuit. A test circuit with a single synapse was fabricated on an SOI structure, using a 150-nm-thick SBT ($SrBi_2Ta_2O_9$).

In order to demonstrate the adaptive-learning function of the circuit, pulse input signals of 20 ns in width and 6 V in height were applied to the gate terminal of ferroelectric-gate FETs. The power supply voltage was 5 V and a DC bias voltage of 1.85 V was applied to the gate terminal. Under these conditions, the neuron circuit showed no oscillatory behavior before the first pulse was applied to the gate, as shown in the left-hand part of Fig. 5. The central and right-hand parts show typical output waveforms after a single pulse, and 60 pulses were applied to the gate. It can be seen from these figures that the circuit starts to oscillate by the application of a single pulse and the oscillation frequency increases as the number of input pulses increases. This behavior can be regarded as the adaptive-learning function of the circuit.

3 The Basic Operation of Ferroelectric-Gate FETs

3.1 Generation of the Depolarization Field

A typical sheet carrier density in the channel region of a MOSFET is about $2 \times 10^{12}\,\mathrm{cm}^{-2}$, assuming that $t_{\mathrm{OX}} = 30\,\mathrm{nm}$ and $V_{\mathrm{G}} - V_{\mathrm{T}} = 3\,\mathrm{V}$, where t_{OX} is the oxide thickness, V_{G} is the gate voltage, and V_{T} is the threshold voltage. This value corresponds to a charge density of $0.32\,\mu\mathrm{C/cm}^2$, and thus the remanent polarization of a ferroelectric film does not need to be large for the fabrication of ferroelectric-gate FETs, if it is effectively used to induce carriers on the semiconductor surface. However, if a ferroelectric film is directly deposited on a Si surface, it is generally difficult to form a good interface between them, since the constituent elements in both materials easily diffuse each other, and since a transition layer is formed at the interface. In order to improve the interface properties, a buffer layer is often inserted between a ferroelectric film and an Si substrate, which is composed of either a dielectric material (a MFIS structure, I, insulator) or a stacked structure of conductive and dielectric materials (a MFMIS structure).

In the MFIS or MFMIS structure, however, a new problem arises in that a depolarization field is generated in the ferroelectric film and it degrades the data retention characteristic of the FET significantly. The generation mechanism can be explained using a series connection model of ferroelectric and dielectric capacitors, as follows. When the power supply is turned off and the gate terminal of the FET is grounded, the top and bottom electrodes of the two capacitors are short-circuited. Under this condition, electrical charges $\pm Q$ appear on the electrodes of the both capacitors, due to the remanent polarization of the ferroelectric film and due to the charge neutrality condition at a node between the two capacitors (the FI interface in the MFIS structure or the floating gate electrode M in the MFMIS structure).

The Q–V relation for the buffer layer capacitor is $Q = CV$ and, thus, the relation in the ferroelectric capacitor becomes $Q = -CV$ under the short-circuited condition, where C is the capacitance of the buffer layer. That is, the direction of the electric field in the ferroelectric capacitor is opposite to that of the polarization of the film. This field is known as a depolarization field and it reduces the data retention time significantly, particularly when C is small.

3.2 Improvement of the Data Retention Characteristics

The above discussion shows that the buffer layer capacitance C must be as large as possible, so that the depolarization field becomes small. This condition means that a thin buffer layer with a high dielectric constant is preferable. So far, several buffer layers with high dielectric constants have been reported, including CeO_2 [8], $SrTiO_3$ [9], Y_2O_3 [10], Bi_2SiO_5 [11], $SrTa_2O_6$ [12], Si_3N_4 [13], $LaAlO_3$ [14], and so on. Their dielectric constants range from 7 to 100. The dielectric constant of $SrTiO_3$ is as high as 300, but when it is deposited on a Si substrate, the effective value reduces to about 70 because of the existence of a transition layer.

However, the dielectric constants of usual ferroelectric films are known to be much higher than those of the buffer layers. Thus, if both the ferroelectric and the buffer layer capacitors are made at the same size, most of the external voltage is applied to the buffer layer. In order to solve this mismatch problem of the dielectric constants, it is necessary to form a small-area ferroelectric capacitor with a thick film. However, if a too thick ferroelectric film is used, there is a drawback that the operational voltage of the FET becomes high.

The leakage current of both the ferroelectric film and the buffer layer is also important. If the charge neutrality at a node between the two capacitors is destroyed by the leakage current, electrical charges on the electrodes of the buffer layer capacitor disappear, which means that carriers on the semiconductor surface disappear and the stored data cannot be read out by the drain current of the FET, even if the polarization of the ferroelectric film is retained. Thus, it is very important to reduce the leakage current of both the ferroelectric film and the buffer layer.

Another important point is the matching of the induced charge between the buffer layer and the ferroelectric film. The remanent polarization values of PZT ($PbZr_XTi_{1-X}O_3$) and SBT are about $40\,\mu C/cm^2$ and $10\,\mu C/cm^2$, respectively, and they can induce the same density of positive and negative charges on the electrodes of a capacitor. These values are generally much larger than the maximum charge density induced by a dielectric film. For example, the maximum induced charge density of SiO_2 is about $3.5\,\mu C/cm^2$ for an electric field of $10\,MV/cm$ and the film breaks down for the higher electric field.

Thus, if a ferroelectric capacitor with a large remanent polarization is connected in series with a SiO_2 capacitor that has the same area, and if a sufficiently high voltage is applied across the two capacitors, the SiO_2 film breaks down before the polarization of the ferroelectric film is saturated. This situation is illustrated in Fig. 6a, for a combination of SBT and SiO_2. As shown in the figure, only a small hysteresis loop of SBT can be used under the condition that the SiO_2 buffer layer does not break down. It should be noted that this condition is independent of the film thickness of both capacitors.

In order to solve the induced charge mismatch problem, it is necessary to form an MFMIS structure and to optimize the area ratio between the ferroelectric capacitor and the buffer layer capacitor. That is, in order to use the saturation polarization of the ferroelectric film effectively, it is important to make the ferroelectric capacitor area small. If the area of the SBT capacitor is reduced to 1/5 of the SiO_2 capacitor, the vertical scale of the P–V (polarization versus voltage) characteristic equivalently becomes 1/5, as shown in Fig. 6b, and the saturation polarization curve can be drawn in the region in

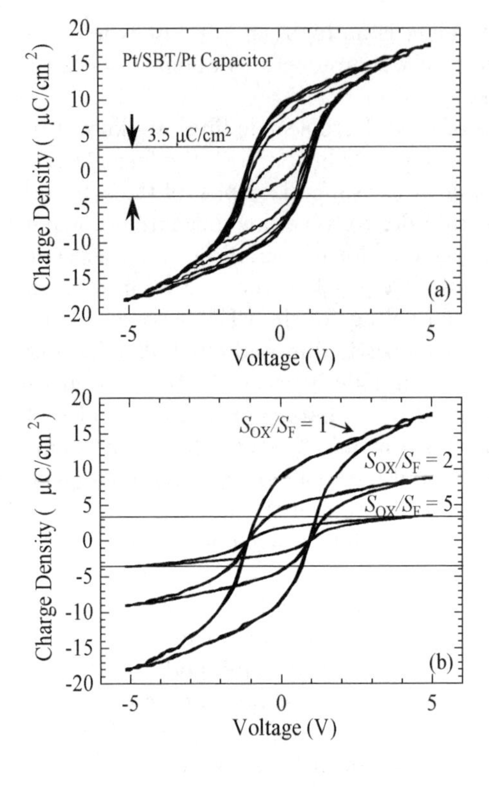

Fig. 6. The induced charge density mismatch between SBT and SiO_2 and (**b**) variation of the hysteresis loop by optimization of the capacitor area

which the polarization value does not exceed the maximum induced charge density of the SiO_2 ($\pm 3.5\,\mu C/cm^2$).

3.3 A Comparison of the MFIS and MFMIS Structures

The features of the MFIS structure are summarized as follows. (1) The gate structure is simple, which is particularly important in integrating FETs in a matrix form, as shown in Fig 2. That is, in the case of an MFMIS structure, it is necessary to form isolated floating gate electrodes before depositing a ferroelectric film. (2) The effect of the leakage current is localized around weak spots in the film, which is important in order to prolong the data retention time. In other words, in an MFMIS structure, the effect of the leakage current spreads out to the whole floating gate and the charge neutrality is completely destroyed in a short time.

On the other hand, the MFMIS structure has the following advantages. (1) Its roles as a diffusion barrier and a gate dielectric can be separated. For example, it is possible to use SiO_2 as the gate dielectric and to use an Ir/IrO_2 electrode as the barrier against diffusion of the constituent elements in PZT. (2) It is possible to optimize the area ratio between the ferroelectric capacitor and the buffer layer capacitor, so that the induced charges on both capacitors match. (3) It is possible to choose the electrode material so that a high-quality ferroelectric film is formed on the floating gate electrode.

4 Recent Experimental Results

4.1 The Relation between the Buffer Layer Thickness and the Area Ratio

In this subsection, the relation between the buffer layer thickness and the area ratio is first discussed and the validity of this relation is experimentally ascertained by fabricating MFMOS FETs composed of an SBT film and a SiO_2 buffer layer [15]. A typical value for the area ratio (S_{OX}/S_F) between the SiO_2 and SBT capacitors is 5, as shown in Fig. 6b. However, this value may be changed when the power supply voltage is fixed at a constant value. In what follows, the polarization characteristics of a ferroelectric film in an MFMOS structure are analyzed under various bias conditions, where the experimental data for various minor loops of an SBT capacitor are used.

The polarization hysteresis loops were measured for a 300-nm-thick SBT film, which was deposited by the conventional sol-gel spin-coating method and annealed for crystallization at 750 °C for 30 min in O_2 atmosphere. When a power supply voltage V_{SUP} is applied to the MFMOS structure, the relation $V_{SUP} = V_F + V_{OX}$ holds, where V_F is the voltage across the ferroelectric capacitor and V_{OX} is the voltage across the SiO_2 film. Then, the induced charge density Q_{OX} at the electrode of the MOS capacitor is expressed as $Q_{OX} = C_{OX}V_{OX} = C_{OX}(V_{SUP} - V_F)$, where C_{OX} is the oxide capacitance per unit area. Since Q_{OX} is also related to the polarization P in the form $PS_F = Q_{OX}S_{OX}$, the operational analysis of the MFMOS capacitor can be conducted graphically.

Figure 7 shows the Q_{OX}–V_F relationship calculated for different values of the SiO_2 thickness (t_{OX}) and the area ratio. The value of V_{SUP} was fixed at 5 V. The two hysteresis loops in the figure were selected from a set of experimental data so that their edges were in contact with the thick solid line corresponding to the case of $t_{OX} = 9$ nm. As can be seen in the figure, the polarization hysteresis loop is almost saturated for $S_{OX}/S_F = 10$, but not for $S_{OX}/S_F = 3$. Using the hysteresis loops in the figure, the degree of saturation can be defined as the ratio of the remanent polarization (P'_r) in the selected curve to that (P_r) in the fully saturated polarization curve. Then, the degree of saturation can be derived as a function of t_{OX} and the area ratio, and it can be expressed with a universal curve when the horizontal axis is normalized as shown in Fig. 8 [16]. As can be seen from the figure, a rough measure to obtain a well-saturated polarization curve under the assumed power supply voltage is $S_{OX} > S_F t_{OX}$ (nm); that is, area ratios larger than 5 and 10 are necessary for $t_{OX} = 5$ nm and $t_{OX} = 10$ nm, respectively.

Next, in order to ascertain the validity of the above relation, MFMOS-FETs were fabricated, in which the thickness values of the SiO_2 layer and

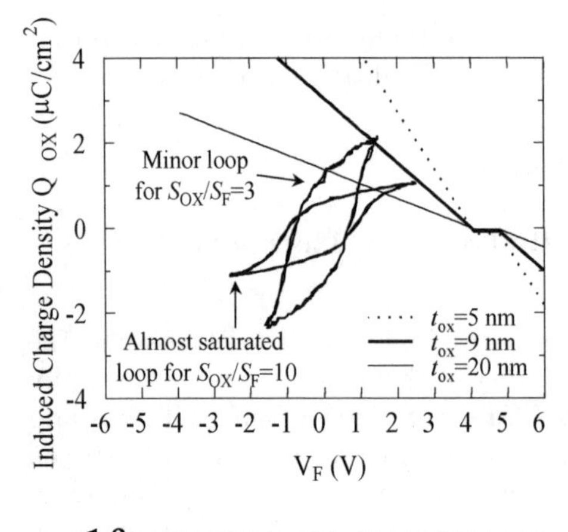

Fig. 7. The relationship between the induced charge density Q_{OX} and the voltage V_F across the ferroelectric capacitor

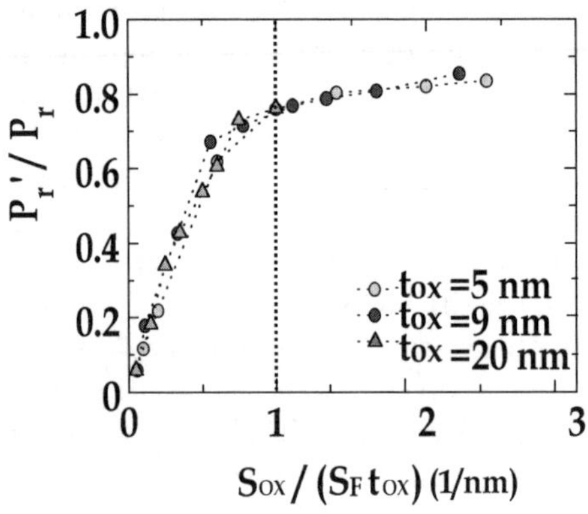

Fig. 8. The relationship between $S_{OX}/S_F t_{OX}$ (1/nm) and P_r'/P_r

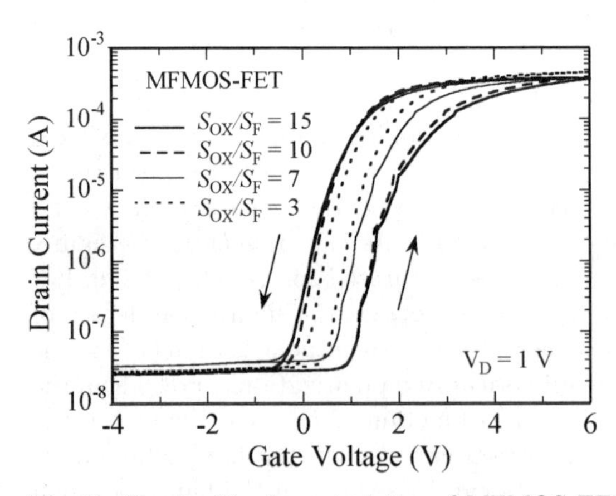

Fig. 9. The I_D–V_G characteristics of MFMOS FETs with S_{OX}/S_F values of 3, 7, 10, and 15

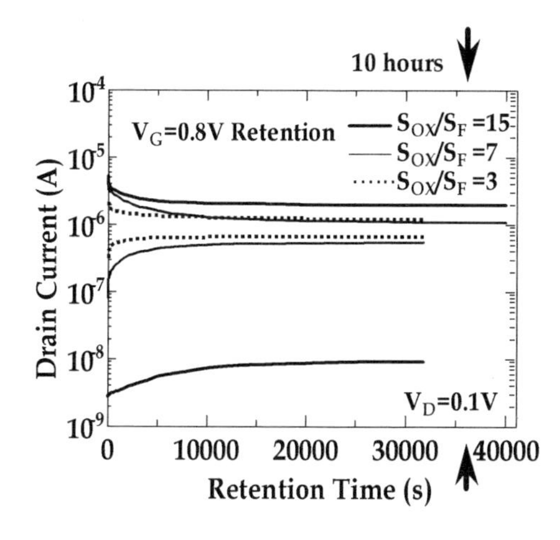

Fig. 10. The data retention characteristics of the fabricated MFMOS FETs

the SBT film were 9 nm and 150 nm, respectively. The memory operations of the fabricated MFMOS FETs were examined by measuring the I_D–V_G characteristics, as shown in Fig. 9. Counterclockwise hysteresis loops were obtained, as shown by arrows, which indicates that the threshold voltage is shifted due to the ferroelectric nature of the SBT film. As can be seen from the figure, the memory window increases from 0.35 V to 1.40 V as the S_{OX}/S_F ratio is changed from 3 to 15, and it is almost saturated at an S_{OX}/S_F ratio larger than 10. These results agree well with the estimation using Fig. 8.

Finally, a WRITE pulse of +6 V or −4 V was first applied for 100 ms to write "on" and "off" states, and then the drain current was continuously measured by maintaining V_G at 0.8 V and V_D at 0.1 V. Figure 10 shows the data retention estimation using the fabricated FETs. In the case in which $S_{OX}/S_F = 3$, the initial on/off ratio of the drain current is as small as 7.0, because of the narrow width of the memory window. On the other hand, in the case in which $S_{OX}/S_F = 15$, the drain current on/off ratio is larger than 1800 initially, and remains at a value larger than 200 even after about 10 hr has elapsed. It is concluded from a comparison of Figs. 8 and 10 that the use of a saturated polarization curve is very important for improving the data retention characteristics.

4.2 Control of the Crystal Orientation

As discussed in the previous section, when materials such as PZT and SBT are used in FETs, it is necessary to form an MFMIS structure and to optimize the area ratio between the MFM and MIS parts. However, this method is disadvantageous from the viewpoint of high-density integration. Thus, it is desirable to develop a new ferroelectric material, which has a small remanent polarization and a square-shaped hysteresis loop. In this section, *c*-axis oriented $(Ba,La)_4Ti_3O_{12}$ (BLT) is adopted as a promising candidate to satisfy

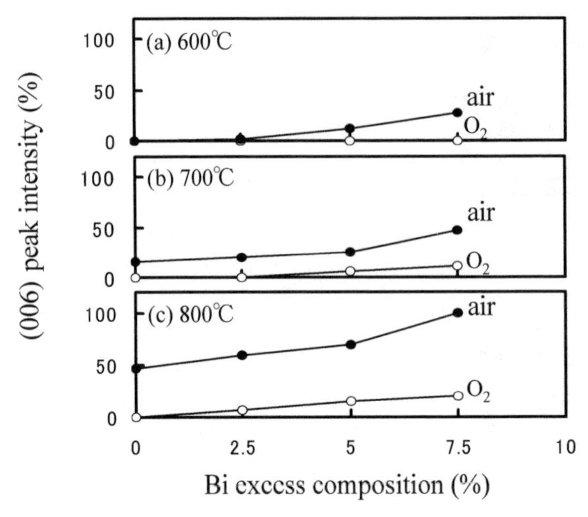

Fig. 11a–c. Relations between Bi-excess composition and XRD (006) peak intensity

the above conditions, and the crystallographical and electrical properties of the BLT films are investigated.

In the experiment, Si_3N_4 was chosen as a buffer layer and it was formed on Si(100) wafers using N or N_2 radicals [13]. BLT films were formed using a sol-gel spin-coating method, in which 2.5–7.5 % excess-Bi solutions were used and the film thickness was fixed at 100 nm [17]. Typical results are shown in Fig. 11, in which the intensity of the (006) peak is plotted as a measure of the preferred orientation of the *c*-axis. It can be seen from the figure that the peak intensity becomes stronger in air ambience, at the higher annealing temperature, and at the larger Bi composition. It is concluded from these results that BLT films are crystallized in air more easily than in O_2 ambience, regardless of the annealing temperature. It was also found that BLT films were almost perfectly oriented with the *c*-axis at an annealing temperature of 800 °C.

In order to ascertain the effectiveness of the *c*-axis-oriented film, C–V characteristics were measured for both the *c*-axis-oriented and the randomly oriented films. The BLT films were formed by spin-coating 10 % excess-Bi solution on n-type Si wafers covered with 3-nm-thick Si_3N_4 layers. The crystallization conditions were the same (at 800 °C in air) for both samples, but the pre-annealing temperature was changed to obtain the randomly oriented film.

The C–V characteristics for the Pt/BLT(100 nm)/Si_3N_4(3 nm)/Si structures are shown in Fig. 12. As can be seen from the figure, the memory window width in the *c*-axis-oriented film is about 1.2 V for a voltage scan of ±5 V. On the other hand, the hysteresis width of the film, which contains (117) crystallites, is very narrow for the same voltage scan amplitude. This

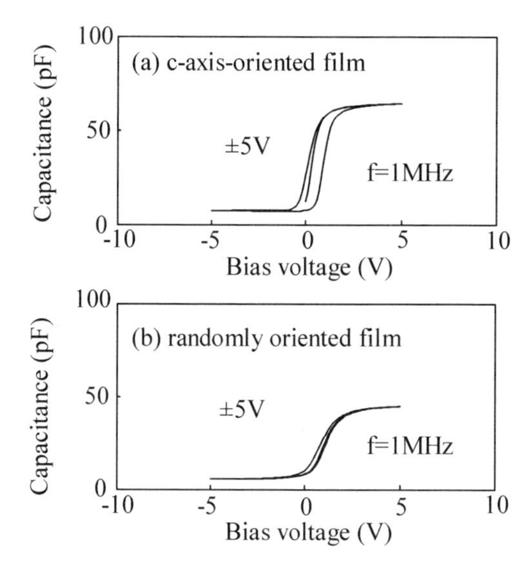

Fig. 12. The *C–V* characteristics of Pt/BLT/Si_3N_4/Si diodes. (**a**) *c*-axis-oriented film. (**b**) Randomly oriented film

difference can be explained by the difference in the remanent polarization values of the film. It was also found that the data retention characteristic of the *c*-axis-oriented film is much better than that of the randomly oriented film.

4.3 Other Important Buffer Layers and Ferroelectric Films

The important buffer layers so far reported are Bi_2SiO_5 and $SrTa_2O_6$. Bi_2SiO_5 (BSO) is a dielectric material with a dielectric constant of 30. Its crystal structure is orthorhombic and the *a*-plane of a BSO crystal is well lattice-matched to both the *a*-plane of Si and the *c*-plane of $Bi_4Ti_3O_{12}$(BIT). Thus, it has been reported that a BSO layer grows epitaxially on a Si(100) substrate, and that a BIT film grows on a BSO/Si structure using an MOCVD (metal-organic chemical vapor deposition) method [11]. The ferroelectric properties of an epitaxial BIT(100 nm)/BSO(30 nm)/Si structure were reasonably good and they were much improved by annealing at 800 °C for 2 h in an O_2 atmosphere. In the annealed sample, the capacitance values biased at the center of the *C–V* hysteresis loop of an MFIS diode did not change for 11 days.

The dielectric constant of $SrTa_2O_6$ is as high as 110. A 30-nm-thick $SrTa_2O_6$ layer was deposited by a sol-gel spin-coating method on a Si wafer, with a SiON layer formed by thermal nitridation [12]. crystallization annealing was conducted at 900 °C for 3 min. Then, a Pt floating gate electrode was formed by a vacuum evaporation method and a 400-nm-thick SBT sol-gel film was deposited. Using this MFMIS structure as a gate, FETs were fabricated. After optimizing the S_{OX}/S_F ratio at 5.9, the current on/off ratio was larger than five orders of magnitude initially, and it was still larger than 10 after 2 days had elapsed.

Concerning the ferroelectric films, important results have been obtained using $Sr_2(Ta,Nb)_2O_7$ and $Pb_5Ge_3O_{11}$. $Sr_2(Ta,Nb)_2O_7$ (STN) is a ferroelectric material with a tungsten–bronze crystal structure. For a film with a Ta/Nb ratio of 70/30, the remanent polarization and the dielectric constant have been measured to be 1.0 μC/cm^2 and 40, respectively [18]. MFMIS FETs with a S_{OX}/S_F ratio of 3.12 were fabricated using the gate structure of Pt/STN (150 nm)/Pt/IrO_2/poly-Si/SiO_2(13 nm)/Si, in which the memory window in the I_D–V_G characteristic was 3.8 V for the gate voltage scan of ±5 V. Concerning the retention characteristics, the variation of the high state capacitance of an MFMIS diode with the above structure was not large (from 6 pF to 5.4 pF) during the 2-week measurement period, while the low state capacitance did not change during the measurement.

$Pb_5Ge_3O_{11}$ (PGO) is a ferroelectric material with a dielectric constant of 30, a remanent polarization of 3 μC/cm^2, and a coercive field of 100 kV/cm. Using this material, MFMIS FETs with excellent FET characteristics have been fabricated [19]. The film was deposited by MOCVD and a damascene process was used for forming a gate structure composed of Pt/PGO(300 nm)/Ir/poly-Si/SiO_2(6 nm)/Si. In the fabricated FETs, the memory window in the I_D–V_G characteristic was as large as 3 V for the gate voltage scan of ±3 V. Furthermore, the current on/off ratio was larger than eight orders of magnitude and the shift in the threshold voltage was retained at least for 5 days.

5 The 1T-2C-Type Ferroelectric Memory

5.1 The Cell Structure and Basic Operation

As discussed above, the improvement of the data retention characteristic is the most important issue in commercialization of the ferroelectric-gate FET. However, since the leakage current due to the depolarization field is one of the main reasons for the short retention time, it is very difficult to prolong the retention time to 10 years. In order to solve this problem, a 1T-2C-type cell has been proposed, in which two ferroelectric capacitors with the same area are connected to the gate electrode of a MOSFET [20], and the basic operation of the cell has been demonstrated [21].

The cell structure and the polarization directions of the ferroelectric capacitors are shown in Fig. 13. In order to write a datum in this cell, a positive or negative pulse is applied between the terminals A and B, so that the two ferroelectric films are polarized oppositely with respect to the gate electrode of the MOSFET. In this case, the electrical charges induced at the electrodes of both capacitors cancel each other and no charge is induced at the gate electrode of the FET, which means that no depolarization field is generated in the ferroelectric film, and a retention time as long as that of the usual 1T1C-type memory can be expected. In the cell shown in Fig. 13, the ferroelectric film thickness of both capacitors is assumed to be the same. However,

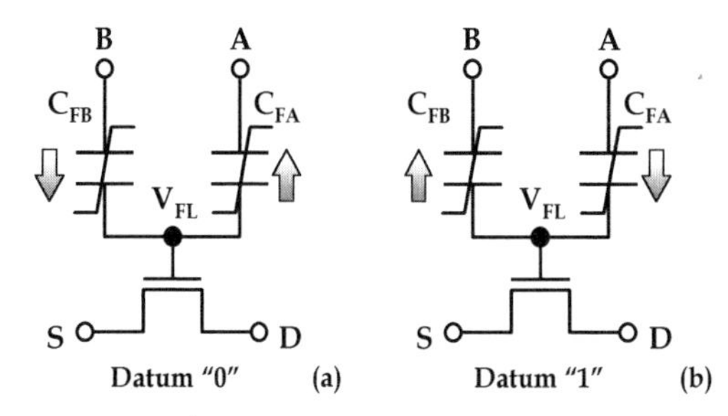

Fig. 13a,b. An equivalent circuit of the 1T-2C-type memory and the polarization directions of data "0" and "1"

it is not necessary for them to be the same if the polarization is well saturated in both ferroelectric films.

For the READ operation, positive voltage pulses are applied to terminal B, keeping terminal A open. In this operation, when the stored datum is "0", no polarization reversal occurs in the ferroelectric film and a small drain current flows through the MOSFET, assuming an n-channel FET. On the contrary, when the stored datum is "1", the polarization of C_{FB} is reversed and a large drain current flows. As discussed here, the application of READ pulses is essential to distinguish the "0" and "1" data in this cell. However, the pulse height does not need to be high enough to reverse the polarization of the ferroelectric film completely, but a partial change of the polarization is preferable from the viewpoint of avoiding the fatigue phenomenon and avoiding the data disturbing effect on nonselected cells.

5.2 Experimental Results

The 1T-2C ferroelectric memory cell was designed by using a 5 μm rule and it was fabricated on a silicon-on-insulator (SOI) structure. The gate MFMOS structure of a memory cell is as follows: Pt (60 nm)/SBT (150 nm)/Pt (60 nm)/Ti (10 nm)/SiO_2 (9 nm)/Si. The SBT film was deposited by a liquid source misted chemical deposition (LSMCD) method and annealed at 750 °C for 30 min in an O_2 atmosphere for crystallization. Figure 14 shows the difference in the drain current between the data "0" and "1". It can be seen from the figure that the current on/off ratio in the READ operation is as high as 430 and that the current level after the READ operation is almost the same as that before the operation. Then, the NDRO (nondestructive read-out) function was investigated by applying READ pulses repeatedly. Figure 15 shows the variation of the drain current for the datum "0" and the datum "1" state, in which the drain current of each memory state hardly changed even after 10^3 pulses were applied.

Finally, the data retention characteristics of the fabricated memory cell were examined by measuring the variation of the drain current with time for various polarization states, as shown in Fig. 16. The states (A) and (B) correspond to the data "1" and "0" for the 1T-2C memory cell, respectively. The state (C) corresponds to datum "1" of the conventional MFMOS FET,

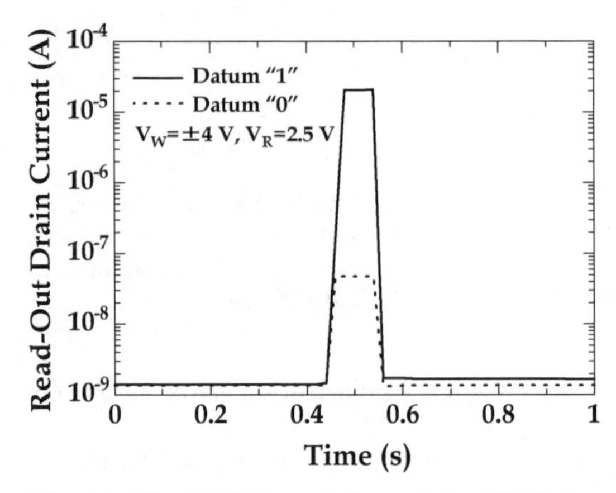

Fig. 14. The READ operations of the 1T-2C-type ferroelectric memory cell

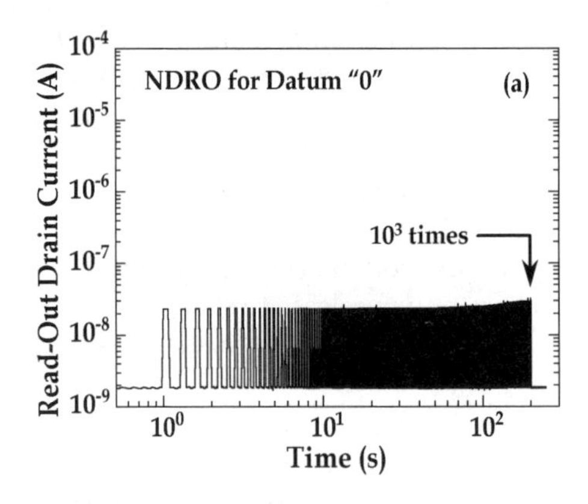

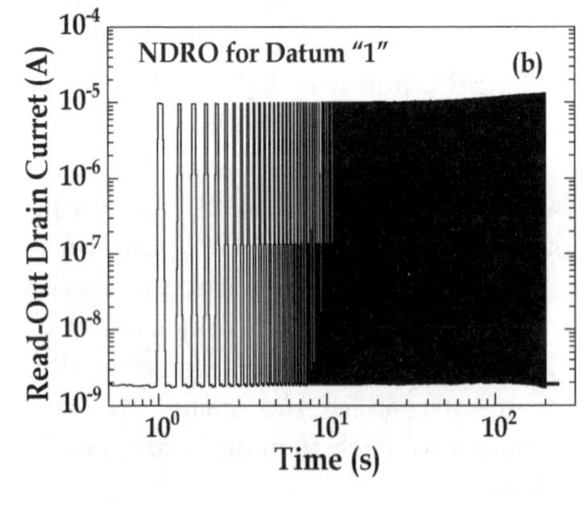

Fig. 15. NDRO operation for (**a**) datum "0" and (**b**) datum "1", in which 10^3 READ pulses were employed

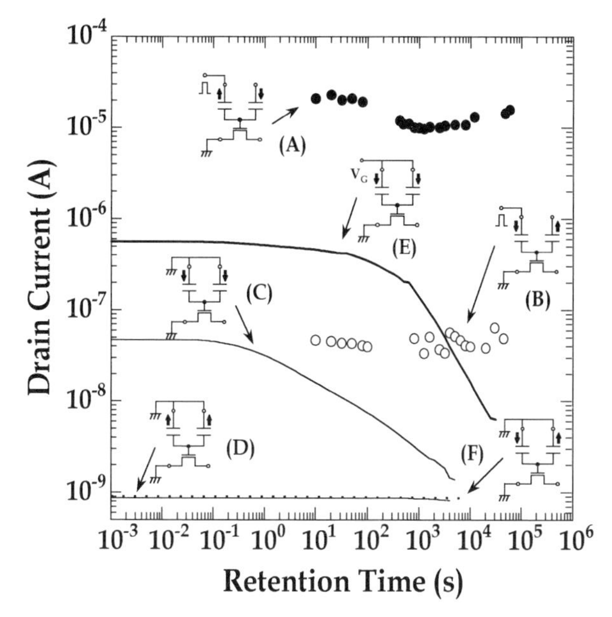

Fig. 16. The data retention characteristics of the fabricated 1T-2C memory cell at various polarization states of the ferroelectric capacitors. The polarization directions of the capacitors and the measurement conditions are as follows. (A) C_{FA}, down; C_{FB}, up. (B) C_{FA}, up; C_{FB}, down. (C) C_{FA}, down; C_{FB}, down; $V_A = V_B = 0\,V$. (D) C_{FA}, up; C_{FB}, up; $V_A = V_B = 0\,V$. (E) C_{FA}, down; C_{FB}, down; $V_A = V_B = 0.5\,V$. (F) (*dashed line*) C_{FA}, up; C_{FB}, down; $V_A = V_B = 0\,V$

since the polarization directions of both capacitors are downward. Similarly, the state (D) corresponds to datum "0" of the conventional MFMOS FET. In the state (E), an adequate V_G of 0.5 V was applied to both capacitors during measurement, so that the retention characteristics were improved. The state (F) was chosen to monitor the initial state of the 1T-2C memory cell with time.

When the polarization directions of both capacitors are parallel [states (C), (D), and (E)], the "1" and "0" data can be distinguished without any READ signals, but the drain current of the datum "1" drastically decreases within 10^4 s, due to the depolarization field. On the other hand, when the polarization directions of both capacitors are opposite [states (A) and (B)], the data can be distinguished only when READ pulses are applied, and the response current to READ pulses does not degrade at least up to 10^5 s, as shown in Fig. 16. It is concluded from these measurements that the effective suppression of the depolarization field is the origin of the excellent data retention characteristics of the 1T-2C memory cell.

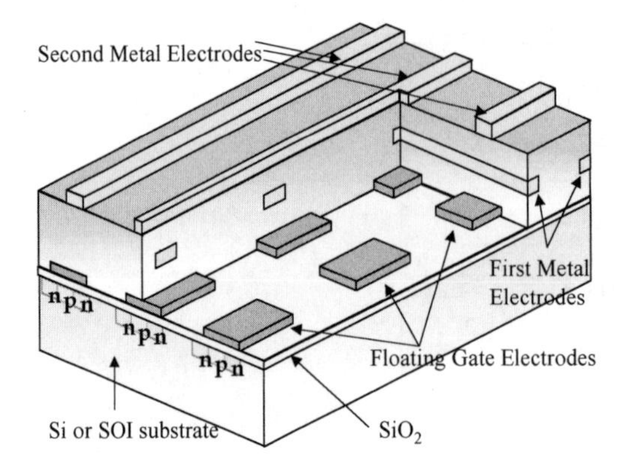

Fig. 17. Integration of the 1T-2C-type memory on an SOI structure

5.3 High-Density Integration

In order to realize a high-density memory, the cell structure shown in Fig. 17 is proposed [20], which is a simple extension of the single-transistor-cell-type memory shown in Fig. 2. In this structure, Si stripes, each of which acts as a MOSFET connected in parallel, are placed on an insulating substrate, floating gate electrodes are placed along the stripes on the gate SiO_2 layer, and double conductive stripes embedded in a ferroelectric film are placed along and perpendicular to the Si stripes. The size of the floating gate and the width of the double conductive stripes are so determined that the overlap areas between the floating gate and the first conductive stripe, and between the floating gate and the second stripe, are equal.

More recently, a planar $8F^2$ (where F is the minimum feature size) cell has been proposed, in which MOSFETs are connected in series and conductive stripes as the upper electrodes of ferroelectric capacitors are placed in parallel on a single-layer ferroelectric film [22].

References

1. I. M. Ross: US Patent No. 2791760 (1957)
2. H. Ishiwara, T. Shimamura, E. Tokumitsu: Jpn. J. Appl. Phys. **36**, 1655 (1997)
3. T. Shiata, T. Ohmi: IEEE Trans. Electron. Dev. **40**, 570 (1993)
4. H. Takasu, Y. Fujimori, T. Nakamura: Abstracts 196th Meeting Electrochem. Soc. Hawaii (1999) No. 1033
5. H. Ishiwara: Jpn. J. Appl. Phys. **32**, 442 (1993)
6. H. Ishiwara: Integr. Ferroelectr. **17**, 11 (1997)
7. S.-M. Yoon, E. Tokumitsu, H. Ishiwara: IEEE Trans. Electron. Dev. **47**, 1630 (2000)
8. T. Hirai, K. Teramoto, K. Nagashima, H. Koike, Y. Tarui: Jpn. J. Appl. Phys. **34**, 4163 (1995)
9. E. Tokumitsu, R. Nakamura, H. Ishiwara: IEEE Electron. Dev. Lett. **18**, 160 (1997)
10. B.-E. Park, E. Tokumitsu, H. Ishiwara: Jpn. J. Appl. Phys. **37**, 5145 (1998)

11. T. Kijima, H. Matsunaga: Jpn. J. Appl. Phys. **38**, 2281 (1999)
12. E. Tokumitsu, G. Fujii, H. Ishiwara: Appl. Phys. Lett. **75**, 575 (1999)
13. Y. Fujisaki, T. Kijima, H. Ishiwara: Appl. Phys. Lett. **78**, 1285 (2001)
14. B.-E. Park, H. Ishiwara: Appl. Phys. Lett. **79**, 806 (2001)
15. S.-M. Yoon, E. Tokumitsu, H. Ishiwara: Jpn. J. Appl. Phys. **39**, 2119 (2000)
16. H. Ishiwara: Proc. 12th IEEE Int. Symp. Applications of Ferroelectrics **1**, 331 (2000)
17. T. Kijima, Y. Fujisaki, H. Ishiwara: Jpn. J. Appl. Phys. **40**, 2977 (2001)
18. Y. Fujimori, T. Nakamura, A. Kamisawa: Jpn. J. Appl. Phys. **38**, 2285 (1999)
19. T. Li, S.-T. Hsu, B. D. Ulrich, L. Stecker, D. R. Evans, J. J. Lee: IEEE Electron. Dev. Lett. **23**, 339 (2002)
20. H. Ishiwara: Integr. Ferroelectr. **34**, 11 (2001)
21. S.-M. Yoon, H. Ishiwara: IEEE Trans. Electron. Dev. **48**, 2002 (2001)
22. S. Yamamoto, H. S. Kim, H. Ishiwara: Jpn. J. Appl. Phys. **42**, 2059 (2003)

Part V

Applications and Future Prospects

Ferroelectric Technologies for Portable Equipment

Yoshikazu Fujimori, Takashi Nakamura, and Hidemi Takasu

Semiconductor R & D Headquarters, ROHM Co., Ltd.,
21, Saiin Mizosaki-cho, Ukyo-ku, Kyoto 615-8585, Japan
yoshikazu.fujimori@dsn.rohm.co.jp

Abstract. Key technologies for portable equipment are low-voltage operation, low power consumption, and protocol-flexibility. Ferroelectric technology can contribute nonvolatility and flexibility to the portable systems. In this chapter, process technologies for low-voltage operation are first described. Then, nonvolatile logic circuits are introduced for low power consumption. Several nonvolatile latch circuits, which are basic elements of powering off LSIs, are shown. We also discuss reconfigurable architecture using ferroelectric technologies.

1 Introduction

Portable equipment, including the cellular phone and the personal digital assistant (PDA), is spreading rapidly today. The power consumption decrease demand for LSIs has risen along with it. The power consumption of a transistor is decreased by reducing the design rule. However, the power consumption increases due to the requirement for high integration and high performance, and an improvement by means of a review of the device structure or other methods is required. There is a proposal by which the power consumption of the transistor is reduced without degradation of the performance by using high-k materials for the gate insulator as the one expedient. Moreover, there is a method of decreasing the power consumption, which is called the multiple-threshold (MT) CMOS method, and involves stopping the power supply to the circuit blocks that do not operate.

The requirement for a power consumption decrease is also strong for the semiconductor memory. Especially, current nonvolatile memories such as EEPROM (electrically erasable programmable read-only memory) and FLASH consume a large amount of power during the writing of data. It is not especially low in power consumption, although new memories such as MRAM (magnetoresistance RAM) [1] and OUM (Ovonic unified memory) [2] are high in performance. The ferroelectric memory is more advantageous than MRAM and OUM with regard to low power consumption. MRAM and OUM are the types that use variable resistance and thus an applied voltage must be enlarged as the size decreases to obtain the same amount of signal. On the other hand, the ferroelectric memory does not have the necessity for applying a voltage greater than the polarization saturation voltage of the ferroelectric capacitor, and the operational voltage can be decreased by reducing the

H. Ishiwara, M. Okuyama, Y. Arimoto (Eds.): Ferroelectric Random Access Memories,
Topics Appl. Phys. **93**, 255–269 (2004)
© Springer-Verlag Berlin Heidelberg 2004

polarization saturation voltage. The polarization saturation voltage can be decreased by process improvements and by making a thinner film.

The ferroelectric memory is not only an ideal memory with clear advantages such as nonvolatility, low power consumption, high endurance, and high-speed writing, but is also the most suitable device for memory embedded applications [3]. Its manufacturing process makes it more compatible with the standard CMOS process than the traditional nonvolatile memory process, since it does not require high-voltage operation, and the ferroelectric process does not influence the characteristics of the CMOS devices used in logic cells, analog cells, and core cells. In the spreading of the intellectual property (IP) application for the LSI industry, this embedded application takes on a more important role.

In the near future, ferroelectric memory technology will be incorporated into reconfigurable devices as programmable interconnect switches besides being used as embedded memories. These ferroelectric memory based reconfigurable devices can be used as dynamic programmable gate arrays (DPGAs), which can be reconfigured from their original logic in a system under an operational mode.

New logic circuits will be operated with lower power consumption or a resume function by introducing ferroelectric gated transistors or ferroelectric capacitors. Even though ferroelectric memory technology has many advantages, it is not popular yet, since the conventional semiconductor process degrades the ferroelectric layer easily. The means of preventing degradation of ferroelectric films in the silicon wafer process will also be discussed.

2 Process Technologies for Low-Voltage Operation

2.1 Experimental and Characterization

$Pb(Zr,Ti)O_3$ (PZT) ferroelectric thin films were prepared by the conventional sol-gel method [4]. The sol-gel solution was made by Mitsubishi Materials Corp. and the solvent was 2-metoxyethanol. The Zr/Ti ratio was 52/48. Silicon p-type wafers were used as the substrates. Pt(175 nm)/IrO_2(65 nm) or IrO_2 (65 nm) was deposited as the bottom electrode by magnetron sputtering. Figure 1 shows the flow diagram for film deposition. After spin-coating of the PZT precursor, the films were baked at 200 °C to evaporate the solvent and pre-annealed at 400 °C in dry air to eliminate organic components. These processes were repeated several times, until the film thickness was about 300 nm. The deposited films were crystallized by low-pressure rapid thermal annealing (RTA) and/or the two-step annealing technique. The iridium system top electrode was deposited by sputtering and patterned by dry etching. The area of the top electrode was 50 μm × 50 μm.

Figure 2 shows a flow diagram of the ferroelectric film crystallization process. In the conventional crystallization process, the spin-coated films are crystallized by RTA (first annealing). Then, the top electrodes are deposited.

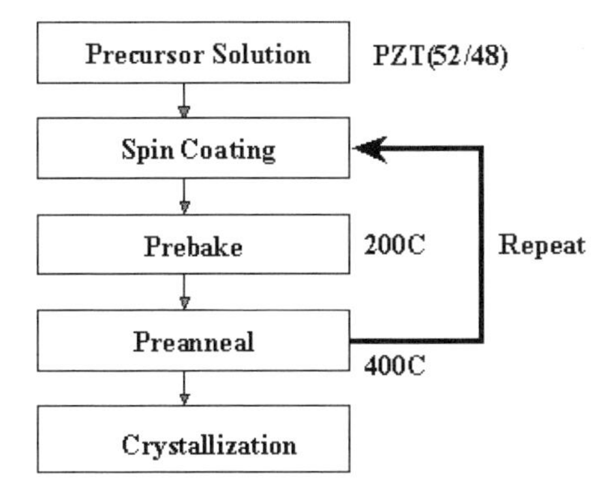

Fig. 1. The flow diagram for film deposition by the sol-gel method

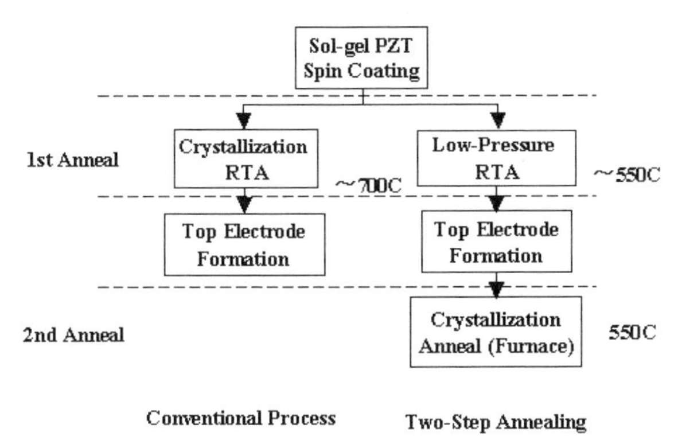

Fig. 2. The flow diagram for PZT film crystallization

In the case of the two-step annealing process, the films are annealed by RTA under low-pressure oxygen ambience, where typically the oxygen pressure is 50 Torr (first annealing). Then, the top electrodes are formed. Lastly, the crystallization annealing is carried out in a furnace at 550 °C for 20 min in oxygen gas [5].

The D–E hysteresis loops were measured using a conventional Sawer–Tower circuit. For measurement of the switching charge density of the fatigue and imprint measurements, −5 V and 5 V pulse trains were used; the pulse width was 1 μs and the pulse interval was 1 s.

2.2 Experimental Results

To investigate the effect of low-pressure annealing, the spin-coated films were crystallized by RTA under normal pressure (760 Torr) and low pressure (50 Torr) in the conventional single-step crystallization process. It was found in the normal pressure process that the annealing temperature should be above 700 °C to obtain sufficient remanent polarization. It was also found in the low-pressure process that the shapes of the hysteresis loops did not

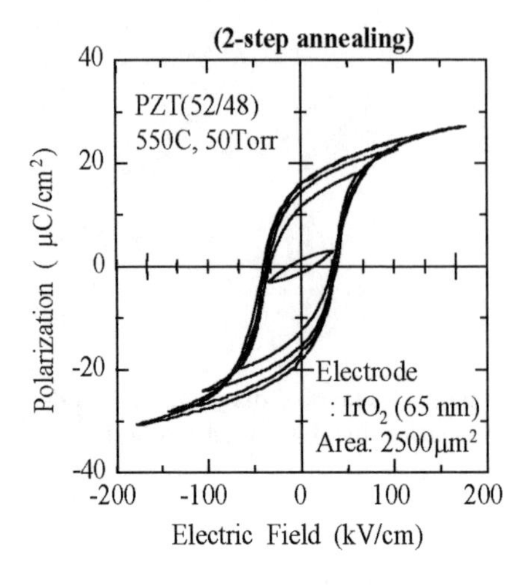

Fig. 3. The D–E hysteresis loops of PZT capacitors crystallized by two-step annealing

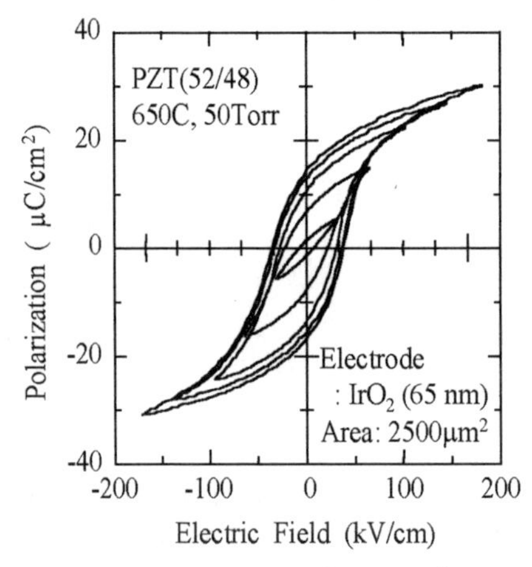

Fig. 4. The D–E hysteresis loops of PZT capacitors crystallized by low-presure annealing only

change down to 650 °C. However, the perovskite phase and ferroelectric hysteresis were not observed when the films were annealed at 550 °C at 50 Torr.

Figure 3 shows the D–E hysteresis loops of the two-step annealed capacitors. The films were annealed at 550 °C by low-pressure RTA (first annealing). Following the top electrode formation, the crystallization anneal was carried out at 550 °C in a normal-pressure furnace (second annealing). For comparison with the two-step annealing, D–E hysteresis loops of the PZT film which was crystallized at 650 °C by low-pressure RTA (first annealing) are shown in Fig. 4. The figure shows that the crystallization occurs at the first annealing. In the case of two-step annealing, well-saturated D–E hysteresis loops were obtained and the saturation characteristics were also good (Fig. 5).

The fatigue characteristics were measured by means of 5 V fatigue pulses. The two-step annealed film exhibited no fatigue up to 10^{10} cycles. Figure 6 shows the imprint characteristics. The imprint bake temperature was 150 °C

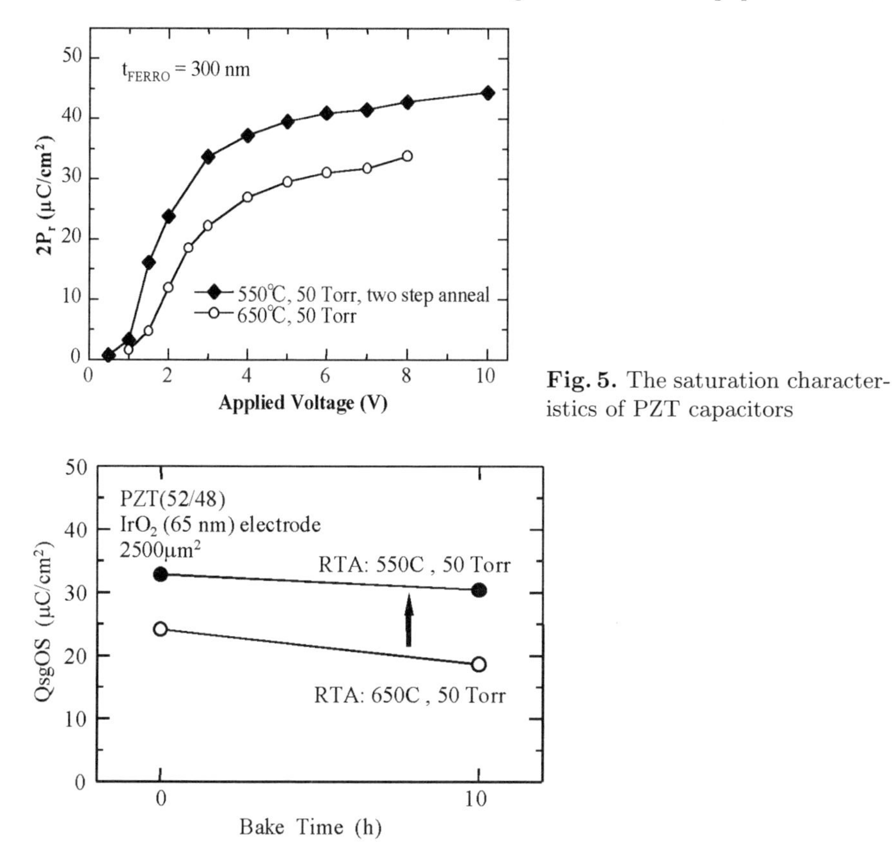

Fig. 5. The saturation characteristics of PZT capacitors

Fig. 6. The imprint after fatigue characteristics of PZT capacitors

and the baking time was 10 h. Imprint measurements were carried out after the fatigue pulses (5 V, 10^8 cycles) were applied. Using the two-step annealing process, the imprint characteristics was improved. The opposite state signal charge (Q_{sgOS}) was maintained above 30 μC/cm^2 after the imprint that followed the fatigue test. One of the reasons for lowering the crystallization temperature by low-pressure annealing is to remove the residual gases (e.g., acetone, CO_2, and H_2O) from the sol-gel films. In the case in which the first low-pressure annealing is not carried out, the gases escape from the films during crystallization, the top electrodes peel off, and the crystallized films contain many bubbles.

When the conventional crystallization process is used, we assume that nucleation occurs on the bottom electrode and the perovskite structure grows from the ferroelectric/bottom electrode interface to the surface. Thus, the composition deviation of the ferroelectric films is concentrated at the surface – but the deviation is small (less than a few weight percent) and very difficult to investigate precisely. That surface layer is considered to be the cause of

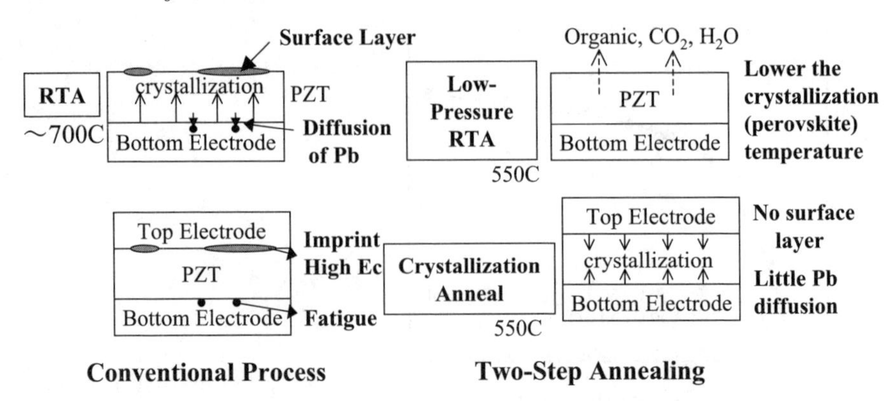

Fig. 7. A schematic drawing of the film crystallization process

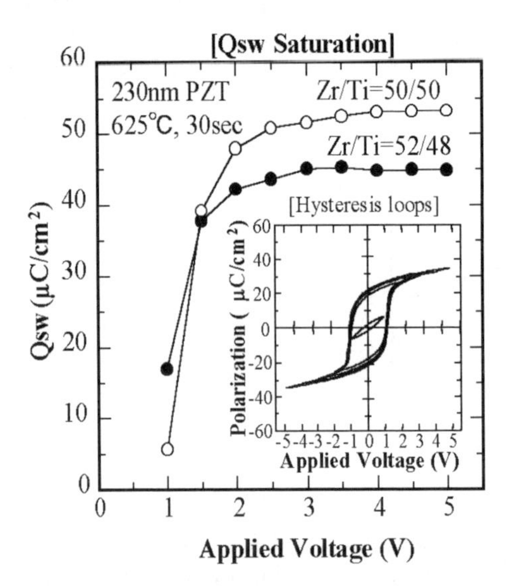

Fig. 8. The characteristics of PZT capacitors crystallized by two-step annealing deposited on a Pt/IrO_2 electrode

the fatigue and imprint characteristics. In the case of two-step annealing, it is assumed that the perovskite phase grows from the both sides, the surface layer is not formed, and the fatigue and imprint characteristics are improved (Fig. 7).

When PZT films were deposited on Pt/IrO_2 bottom electrodes, a larger switching charge and a smaller polarization saturation voltage were obtained, because the PZT films were oriented to (111). The polarization characteristic of a PZT film on a Pt/IrO_2 bottom electrode using two-step annealing is shown in Fig. 8. In the two-step annealed sample (closed circles), the polarization saturation voltage goes down to 1.65 V, while it is higher than 2 V in a film with the same thickness formed by a conventional method (open circles). These results show that a lower-voltage operation of the FeRAM is made possible by using this method. The operational voltage will be de-

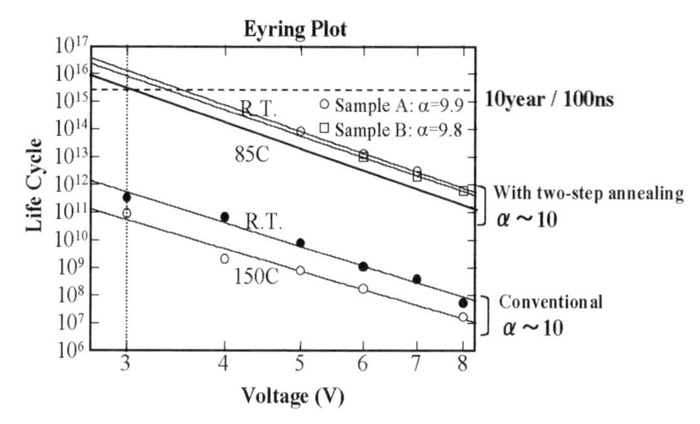

Fig. 9. Acceleration test in voltage stress

creased to less than 1 V in the figure by using thinner films. Recently, there has been a report about having obtained a polarization saturation voltage of 0.5 V by using thin films [6].

When this process is used, the reliability of the PZT films improves considerably. A fatigue characteristic accelerated by the voltage is shown in Fig. 9. As can be seen from the figure, PZT film formed by the new process has an endurance of more than 2×10^{15} times in 3 V. Ferroelectric film already has the ability that a read/write cycle is unlimited, and the read/write cycle limit of the FeRAM will be removed by the establishment of an evaluation method.

3 Nonvolatile Logic

The reduction of the power consumption is becoming more essential for portable systems. Several standby leakage reduction schemes have been reported. One is the MT-CMOS. This method, however, requires high $V_{\rm th}$ additional latches branched from the critical path, like a balloon latch [7]. The other is the variable-threshold (VT) CMOS [8]. This requires a triple-well structure and an additional circuit to control the substrate bias. According to the ITRS 2001 roadmap, the gate leakage for low standby will be comparable to source–drain leakage by 2005. The nonvolatile latch circuit disussed in this section is useful to reduce gate leakage like that of a MT-CMOS circuit.

3.1 The Ferroelectric Nonvolatile Latch

Figure 10 shows the concept of the MT-CMOS with a nonvolatile latch. By adding nonvolatility to a conventional latch, the latched data are kept during power off. Figure 11 shows a nonvolatile latch circuit. This circuit requires two additional ferroelectric capacitors (C1, C2) and one plate line (PL) to

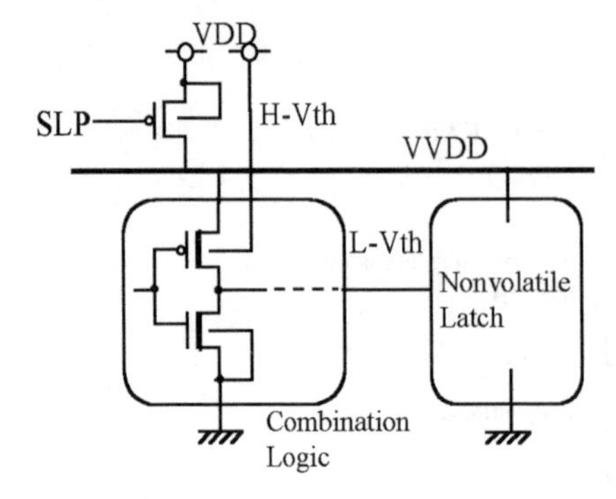

Fig. 10. The concept of MT-CMOS with a nonvolatile latch

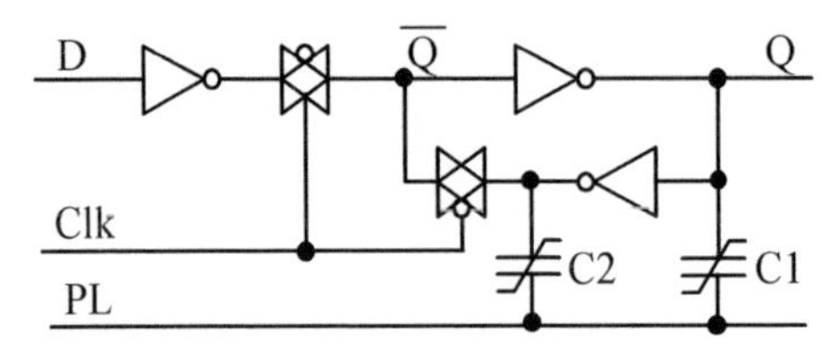

Fig. 11. The nonvolatile latch circuit

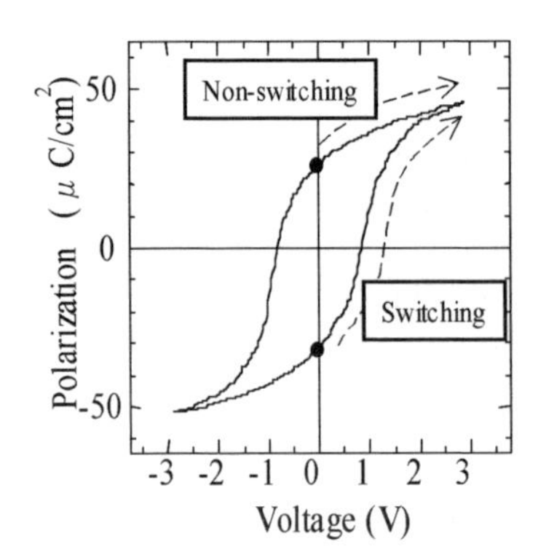

Fig. 12. The ferroelectric hysteresis loop

store and recall the nonvolatile latched data. Ferroelectric capacitors require three additional masks (in the case in which a FeRAM is used as an embedded memory, no additional masks are necessary); however, there is no area penalty in using stacked-type capacitors [9]. In an active state with a power supply, data are held in the inverter loop like a conventional latch. Before power is down, the latched data are transferred to the ferroelectric capacitors (the STORE operation). The stored data are recalled at the first step when the power is turned on (the RECALL operation).

The ferroelectric hysteresis loop is shown in Fig 12. Ferroelectric materials have two states at zero bias and show two different capacitances when a voltage is applied. The switching capacitance is larger than that of nonswitching, so we can know the stored data as ferroelectric polarizations. Figure 13 shows timing charts of the STORE and RECALL operations. In the active state, PL is fixed at the V_{DD} or GND level to prevent useless ferroelectric polarization

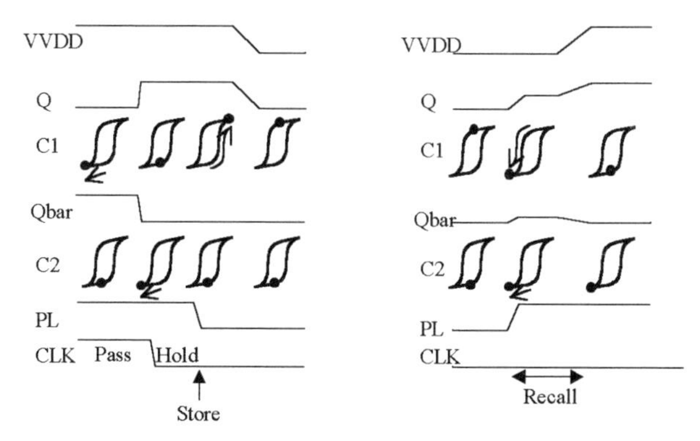

Fig. 13. The timing charts of the STORE and RECALL operations

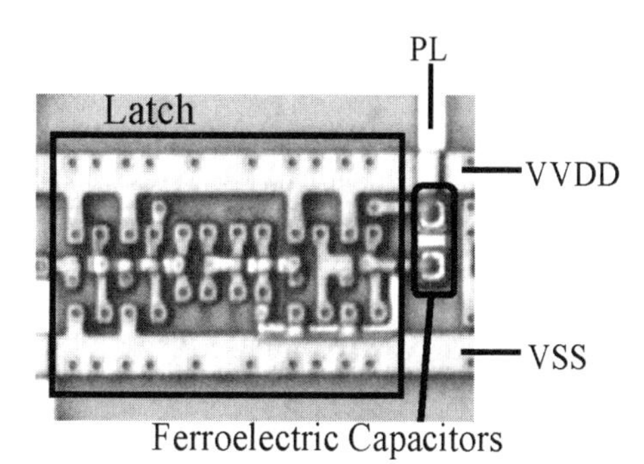

Fig. 14. A photograph of the test circuit

reversal. By changing the PL state, the ferroelectric capacitors are polarized in a complementary way. When the power is next restored, PL is first driven to the V_{DD} level. The voltages of Q and $\overline{\mathrm{Q}}$ rise to different levels according to the ferroelectric polarizations. After that, the inverter loop is activated by supplying the power.

3.2 Test Chip Measurement

We fabricated test circuits to confirm the operations of the nonvolatile latch, because a ferroelectric capacitor has no complete simulation model. Figure 14 shows the photomicrograph of a test chip, which was fabricated in a 0.6 μm CMOS processusing PZT (lead zirconate titanate) ferroelectric capacitor technology. Figure 15 shows the measured waveforms of the nonvolatile latch. The baking time of the data retention measurement is 15 h at 150 °C, which corresponds to 6 years at room temperature. We tested five chips at each state and all of the chips passed this retention test. Figure 16 shows simulated waveforms of the RECALL operation. The voltage difference between Q and $\overline{\mathrm{Q}}$ is larger than 1 V when the power ist restored. It is enough to restore the data. Figure 17 indicates the measured setup time of the nonvolatile latch as a function of the PL voltage during circuit operation.

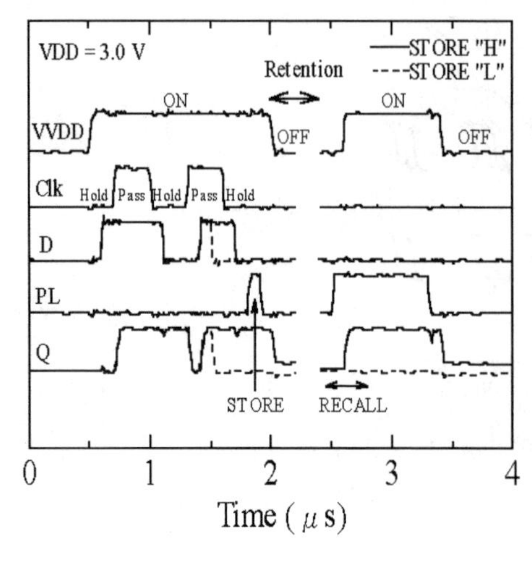

Fig. 15. The measured waveform of the nonvolatile latch

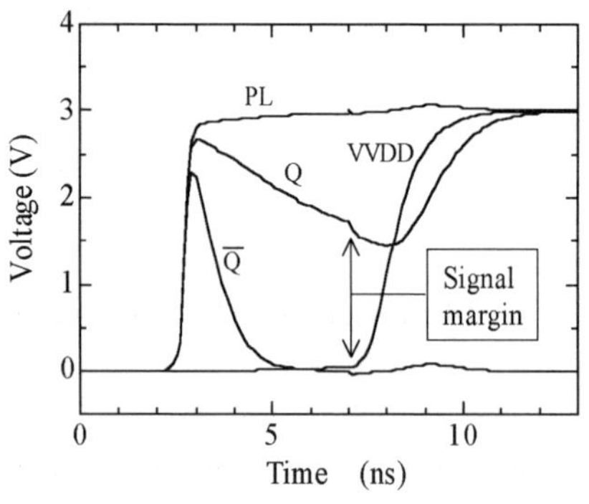

Fig. 16. The simulated waveform of the RECALL operation

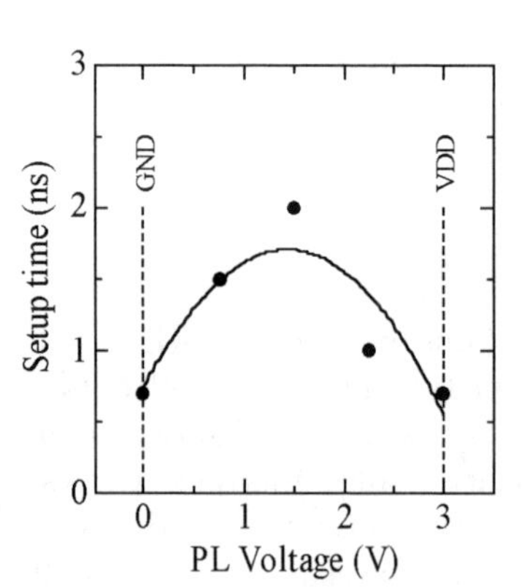

Fig. 17. The setup time as a function of the PL voltage

When the PL is kept at $V_{\mathrm{DD}}/2$, a reversal of the ferroelectric polarization occurs every time the Q state changes and the operation speed decreases. Therefore the PL should be fixed at the V_{DD} or GND level. These values for the setup time are comparable with those of a conventional latch circuit.

3.3 Scaling Issues

Table 1 shows scaling projections for the circuit parameters. We assume that ferroelectric capacitor size is $8\,F^2$ (F is the feature size) and that the ferroelectric thickness is proportional to V_{DD}. The ferroelectric capacitance (C_{ferro}) decreases 0.625 times per generation.The scaling factor of the delay caused by the ferroelectric capacitance is estimated at 0.4. This corresponding growth rate is 35% per year, which is much faster than that of a gate delay (14% per year). Therefore the nonvolatile latch can be scaled down without the operating speed being degraded.

Table 1. Scaling projections for the circuit parameters

Parameter (unit)	Values					Scaling Factor
Generation (nm)	0.6 μm (this work)	180	130	90	65	0.7×
V_{DD} (V)	3.0	1.5	1.2	1.1	0.9	0.8× (0.85×)
Ion (μA/μm)	200	250	300	400	500	1.25×
Capacitor size (μm^2)	2.8	0.26	0.14	0.07	0.03	0.5×
Ferroelectric thickness (nm)	240	130	100	85	70	0.8× (0.85×)
C_{ferro} (fF)	110	27	17	10	6.5	0.625×
$C_{ferro}V/I$ (1 at 180 nm)	–	1	0.4	0.16	0.064	0.4×

3.4 Other Nonvolatile Latch Circuits

Figure 18 shows a ferroelectric latch unit (FLU). The latch unit used in ferroelectric capacitors can keep its logic status even when the electrical power is lost. The previous latched data will be given when electrical power is supplied back to the circuit.

Figure 19 shows the operational principle of nonvolatile latches. Two identical ferroelectric capacitors, C1 and C2, are connected in series. During the WRITE operation, PL1 and PL2 are kept at a half V_{cc} and data are set on the center node IN. If the coercive voltage is smaller than half V_{cc}, both ferroelectric capacitors are polarized upward and downward. The remanent polarization stores the data while the power is down. Before the power is turned on, PL1 and PL2 are driven to V_{cc} and ground level, respectively. At this time, the IN node needs to be in the high-impedance state. The polarization reversal occurred to one of these two capacitors and the center node voltage changes out of half V_{cc}. After that, an inverter or latch is activated to amplify the node voltages and then the stored data are RECALLed.

Ferroelectric gated transistors or ferroelectric capacitors can be used to minimize the electrical power consumption and the logic circuits (Fig. 20). This type of FET is called a ferroelectric FET. It has the remarkable advantage of nondestructive readout.

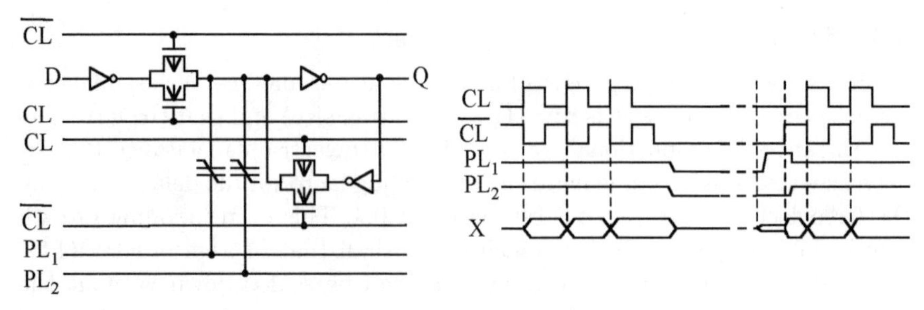

Fig. 18. The ferroelectric latch unit

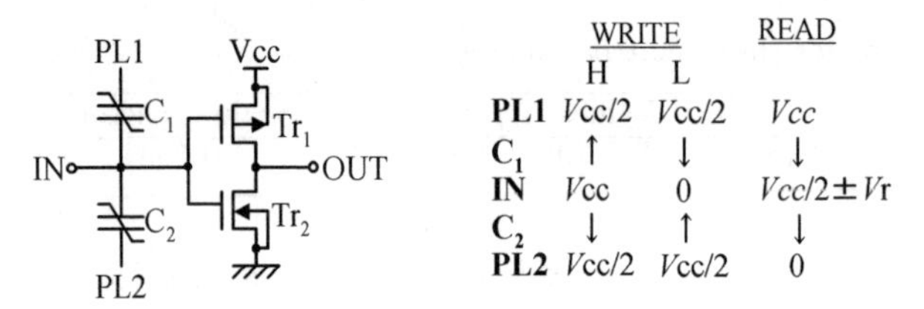

Fig. 19. The operational principle of a nonvolatile logic element

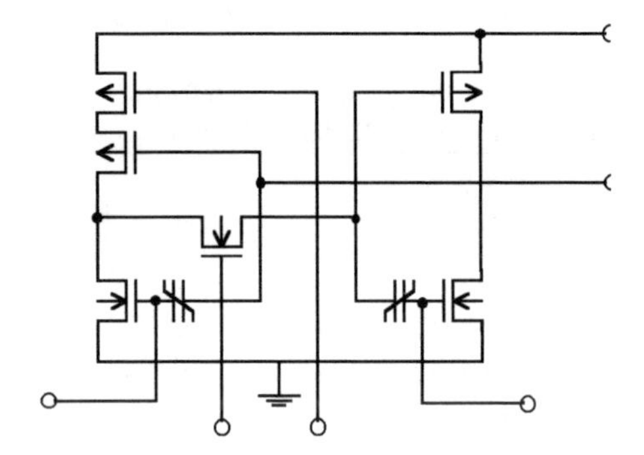

Fig. 20. A ferroelectric latch unit using a ferroelectric FET

3.5 Applications Using the Nonvolatile Latch

Ferroelectric memory technology enables the use of a smaller switching memory cell compared to conventional SRAM-based FPGA (field programmable gate array), and no backup memory is required with a ferroelectric memory based FPGA, as shown in Fig. 21. The small cell of the FPGA with its ferroelectric memory will be applied to more than 500 k gate count of FPGA with fine cell architecture similar to that of a gate array. This ferroelectric memory based FPGA can have a good security function, to protect the circuit from being copied. Furthermore, a more important feature of this FPGA is that it will take self-learning results into the logic circuit configuration during operation in a system, because of its high-speed programming and the fact that there is no additional high-voltage power supply requirement on board.

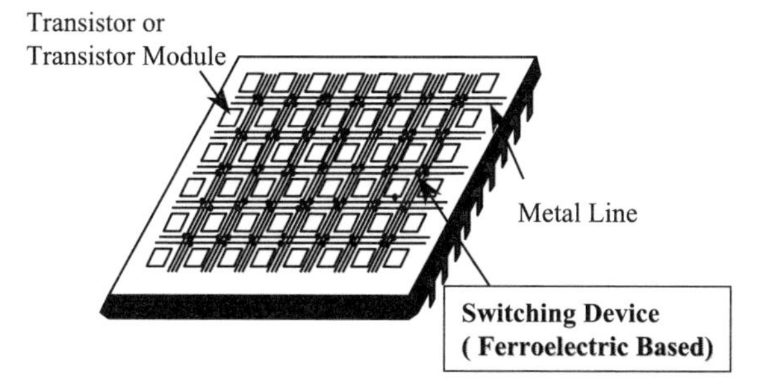

Fig. 21. The ferroelectric-based FPGA

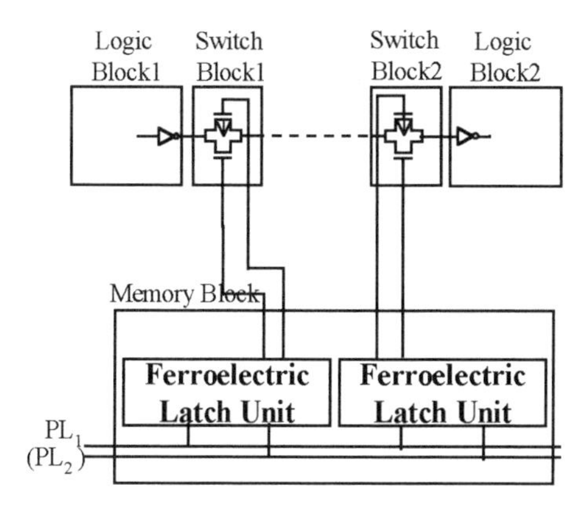

Fig. 22. The DPGA application of ferroelectric nonvolatile latches

By applying this method in a large logic circuit, the electrical power supply is not required for low-power applications, but just for maintaining the logic status, and a local power-off (no electrical power supply on some area of a large logic circuit) approach will take place. This ferroelectric memory FPGA can be used as a DPGA (dynamic PGA), and in the system programming of hardware via telecommunication line it will become available through the use of DPGA technology. This DPGA will open up a new world for the system hardware, which will be reconfigured instantly upon request. Instead of using dual threshold transistors, ferroelectric capacitors can be used to minimize the electric power consumption and the number of transistors used in logic circuits, as shown in Fig. 22. This DPGA needs many logic blocks and selects them. By using a pass logic circuit, the number of transistors can be dramatically reduced, as shown in Fig. 23.

3.6 A Novel Functional Device

Ferroelectric capacitors have been used as simple memory elements for a long time. At present, their arithmetical functions attract a great deal of attention. We consider the write operation of a ferroelectric capacitor as a switching operation under the control of two variables, so that both the storage and the

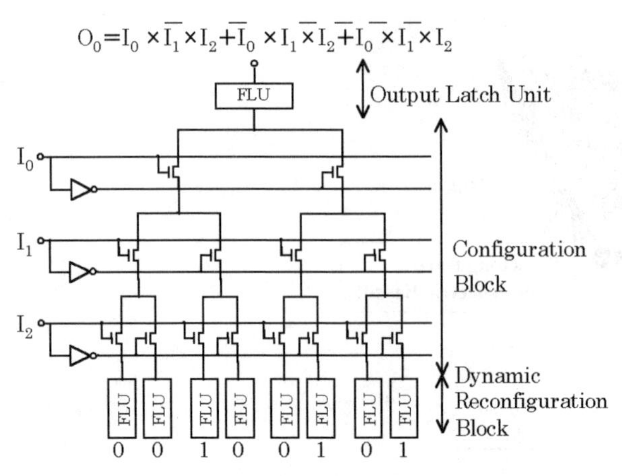

Fig. 23. A reconfigurable logic cell

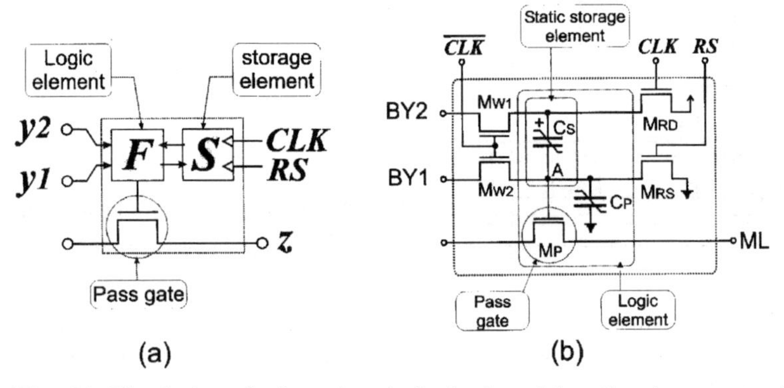

Fig. 24. The design of a ferroelectric device-based functional pass gate. (**a**) A block diagram. (**b**) The circuit diagram

switching functions can be performed simultaneously in a single ferroelectric capacitor.

Figure 24 shows a novel functional pass gate using ferroelectric capacitors. The symbols M_{w1}, M_{w2}, MRD, and MRS represent conventional NMOS transistors that control the voltages at two terminals of the ferroelectric capacitor C_s. C_p is a dummy ferroelectric capacitor that converts the remnant polarization state of C_s into a corresponding voltage level by a capacitive coupling effect. A 54×54 bit pipelined multiplier which has 16 pipeline stages has been designed. The pipeline pitch is determined by the delay of a single functional pass gate. Table 2 provides a comparison of two pipelined multipliers: the conventional static CMOS realization and the proposed one. The power dissipation and effective chip area using the proposed functional pass-gate network are reduced to 46% and 71%, respectively, of a corresponding CMOS implementation.

Table 2. Scaling projections for the circuit parameters

Items	Static CMOS	Proposed
No. of pipeline stages	16	16
Delay per stage (ns)	0.7	1
Power dissipation (W)	1.3	0.6
Area (mm^2)	3.94	2.83

HSPICE simulation based on a 0.6 μm ferroelectric/CMOS technology

4 Summary

Nonvolatile latches were studied for low standby power LSIs of portable systems. Quick-on and quick-off responses of the supply power were realized using ferroelectric nonvolatility. Several reconfigurable circuits have been proposed for protocol-flexible mobile systems. Ferroelectric technology will have a big impact on the existing semiconductor memories, and open up new application fields, which will change our lifestyle, or enable new device architectures for the 21st century.

References

1. R. Scheuerlein, W. Gallagher, S. Parkin, A. Lee, S. Ray, R. Robertazzi, W. Reohr: Tech. Dig. Int. Solid-State Circuits Conf. (2000) p. 128
2. S. Lai, T. Lowrey: Tech. Dig. Int. Electron Devices Meeting (2001) p. 803
3. S. S. Eaton, D. B. Butler, M. Parris, D. Wilson, H. McNeillie: Tech. Dig. Int. Solid-State Circuits Conf. (1988) p. 130
4. T. Nakamura, Y. Nakao, A. Kamisawa, H. Takasu: Jpn. J. Appl. Phys. **33**, 5207 (1994)
5. Y. Fujimori, T. Nakamura, H. Takasu: Jpn. J. Appl. Phys. **38**, 5346 (1999)
6. K. Kijima, H. Ishiwara: Jpn. J. Appl. Phys. **41**, L716 (2002)
7. S. Mutoh, T. Douseki, Y. Matsuya, T. Aoki, S. Shigematsu, J. Yamada: IEEE J. Solid-State Circuits **30**, 847 (1995)
8. K. Kumagai, H. Iwaki, H. Yoshida, H. Suzuki, T. Yamada, S. Kurosawa: Tech. Dig. Symp. VLSI Circuits (1998) p. 44
9. T. Miwa, J. Yamada, H. Koike: Tech. Dig. Symp. VLSI Circuits (2001) p. 129

The Application of FeRAM to Future Information Technology World

Shoichi Masui[1], Shunsuke Fueki[2], Koichi Masutani[2], Amane Inoue[2], Toshiyuki Teramoto[2], Tetsuo Suzuki[2], and Shoichiro Kawashima[2]

[1] Fujitsu Laboratories Limited, 4-1-1 Kamikodanaka, Nakahara-ku, Kawasaki, 211-8588, Japan
masui@fram.ed.fujitsu.co.jp
[2] Fujitsu Limited, 4-1-1 Kamikodanaka, Nakahara-ku, Kawasaki, 211-8588, Japan

Abstract. The future information technology world needs a simple identification and secure information storage medium. The advanced smart card is a promising technology to extend information-providing services from terminal (PC and mobile phone) dominated schemes to more flexible schemes in which the users' intention dominates. In this smart card, unique identification, individual information and values are stored in nonvolatile memories; among the nonvolatile memories, FeRAM presents overwhelming characteristics compared to conventional floating-gate type memories in terms of programming time, energy consumption, and its scalability/affinity to below 0.18 μm standard CMOS technology. FeRAM can improve the transaction time in transportation applications through its fast programming time; moreover, a data protection scheme from power failure can be implemented without the addition of off-chip components, through its low energy consumption. Scalability to advanced CMOS technology enables enhancement services in multi-application cards, which include a mobile phone identity module. A smart card LSI embedded with 32 bit RISC CPU and 64 kB FeRAM is introduced with an associated OS. Further applications of the FeRAM circuit are discussed, along with its technology trend.

1 Introduction – A Prospect for the Future Information Technology World

The upcoming establishment and enhancement of the information network assure fast and secure internet access through the personal computer (PC), personal digital assistance (PDA), and the mobile phone. The realized open information society creates opportunities to improve the effectiveness of government, corporate, and individual activities. Moreover, information technology breakthroughs should provide additional conveniences and diversified services for those who are not familiar with electronic devices, such as young children and older people. The smart card is the simplest device that does not have any input terminals; however, it is a robust medium to identify the person who holds it, as well as to store the values and information of the holder.

H. Ishiwara, M. Okuyama, Y. Arimoto (Eds.): Ferroelectric Random Access Memories, Topics Appl. Phys. **93**, 271–283 (2004)
© Springer-Verlag Berlin Heidelberg 2004

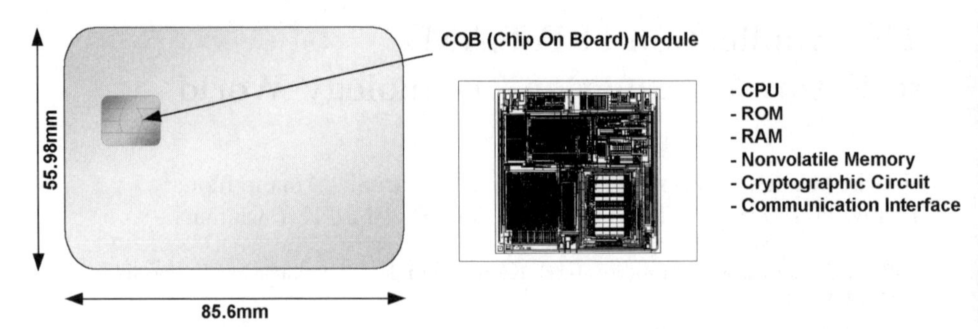

Fig. 1. The structure of a typical smart card

Let us imagine that, in the near future, we are going to visit a foreign country, in the company of young children. We will have to carry identification media, such as a passport, an international driver's license, a credit card, a bank card, and a retail loyalty card for various activities throughout the tour. Possibly, a globally roaming mobile phone card, a hotel key card, a health (medical) card, and a transportation access card will be required. Our children will carry their own cards, in which family information, their medical history, their parents' internationally addressed phone numbers, and a small amount of virtual money might be stored. So, even when they are lost in a large shopping mall or an amusement park, a security person can read their card and notify their location to a mobile phone. In addition, identification technology will extend its market to airline baggage claims, amusement park access control, and so on.

A smart card is defined as a high-end IC card with a cryptographic function. The structure of a smart card is shown in Fig. 1. The smart card LSI is packaged in a COB (chip on board) module, and is implemented in an ID-1 size plastic card (the same physical structure as our credit card) [1]. A socket module is used for a mobile phone identity module. A typical smart card LSI consists of a CPU, ROM, RAM, a nonvolatile memory, a cryptographic coprocessor, and a communications interface. The identification, individual information, and values are stored in nonvolatile memory, and FeRAM is expected to become an enabling nonvolatile memory technology for future smart card LSIs.

As is evident from the above example, it is not practical for us to hold as many cards as are required in the various applications. Our target is to integrate all of the applications into one ID-1 size smart card, except for the mobile phone identity module, as shown in Fig. 2. In this chapter, we first describe smart card market/system requirements, to clarify the technological priorities. Then, we compare the available nonvolatile memory technologies to demonstrate the superiority of FeRAM in smart card applications. Various application examples, including contactless transportation and multi-appli-

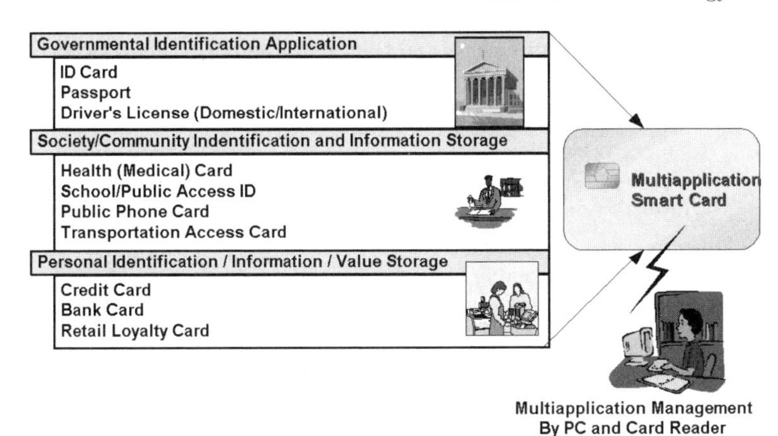

Fig. 2. The multi-application card integrates various services into one smart card

cation smart cards, are illustrated in subsequent sections. Finally, we discuss the future trend of smart card technology and associated FeRAM technology.

2 Smart Card Market/System Requirements

Service providers of smart cards are strategically requesting new technologies to provide high-level systems in terms of security, convenience, reliability, and flexibility. Table 1 summarizes the market/system requirements associated with these aspects. Although the security aspect is not the subject of this publication, due to its independence from the nonvolatile memory characteristics, we have listed typical crypto-circuit specifications for smart cards, as well as listing references providing basic theory and implementation examples in Table 2. No significant differences are discovered with respect to tamper-resistance between FeRAM and conventional EEPROMs.

As can be seen from Table 1, significant service improvements can be obtained through the use of contactless interface and multi-application schemes. Both schemes are associated with architectural design; however, the performance and characteristics obtained are strongly dependent on the nonvolatile memory process. FeRAM has attracted attention for smart card applications [5]; on the other hand, the impacts of FeRAM on smart card characteristics have not yet been investigated in detail. In the following sections, we discuss the characteristics of FeRAM and demonstrate its suitability and superiority to conventional smart card LSIs.

Table 1. Smart card system requirements and their implementations on LSI/OS

Aspects	Market/system requirements
Security	Cryptosystem (private/public key)
	Memory firewalling
	Tamper-resistant design
	Third-party evaluation
Convenience	Contactless application
	Code-reprogrammable multi-application function
Reliability	Maintenance reduction by contactless interface
	Technical maturity
	Infrastructure compatibility/interoperability
Flexibility	Dual interface (contact/contactless)
	Multiple OS support

Table 2. Crypto-circuits used in the smart card

Cryptosystem	Function	Algorithm	Key length (bits)	Ref.
Private key	Encryption/decryption	DES(TDES)	56(112,168)	[2]
Public key	Key distribution	RSA	1024–2048	[3]
	Digital signature	ECC	163–239	[4]
Hash function	Digital signature	SHA-1	160	[2]
	Digital signature	MD5	128	[2]

3 Nonvolatile Memory Characteristics

Nonvolatile memory is used to store unique identification, and individual information and values. Table 3 compares the characteristics of nonvolatile memories that are commonly used for smart cards. EEPROM and NOR Flash memories store information on a floating electrode fabricated on an active device area [6]. In EEPROM, Fowler–Nordheim tunneling mechanism is used to control electron injection to or from the floating gate. On the other hand, the NOR Flash memory uses the power-consuming channel hot electron injection mechanism; therefore, its application is restricted in the contact smart card. Contact and contactless interfaces are illustrated in the next section.

Since these tunneling and injection mechanisms accompany the gradual degradation of the tunneling oxide film, the programming endurance is far inferior to that of FeRAM. This low programming endurance, however, does not become a serious bottleneck in smart card applications as long as the access frequency is once in 10 min (total access time less than 6×10^5 in 10-year operation).

As shown in Table 3, other benefits of FeRAM in smart card applications can be summarized as follows:

- fast programming time;
- low energy consumption;
- scalability and affinity to below 0.18 μm standard CMOS technology.

Table 3. Nonvolatile memory characteristics

Characteristics	FeRAM	EEPROM	NOR Flash
Read cycle time	180 ns	70 ns	70 ns
Programming cycle time	180 ns	2 ms	> 1 s
Read power	< 1 mW/MHz	< 1 mW/MHz	1 mW/MHz
Programming power	< 1 mW/MHz	1 mW/MHz	> 10 mW/MHz
Programming energy	0.18 nJ	2 μJ	8 μJ
Cell area	8–15 F^2	$> 20\,F^2$	6 F^2
Endurance	10^{10}	10^6	10^6
Scalability	Good	Restricted	Restricted
Affinity with CMOS	Good	Restricted	Restricted

The fast programming time and low energy consumption can bring about superior features for contactless applications. Scalability is becoming a critical factor in high-end multi-application cards, since the market requires a large capacity for nonvolatile memory. The power consumption is less significant compared to the RFID (radio frequency identification) application [7], since the CPU and cryptographic coprocessors consume more power than the nonvolatile memory in smart card applications.

MRAM might be considered as a nonvolatile memory candidate for the smart card [8]; however, as is described in the contactless smart card standard [9], the smart card must function after exposure to a static 640 kA/m magnetic field. This magnitude is far higher than the coercive force of the magnetoresistive material.

4 Contactless Cards

A contactless interface can enhance reliability due to the lack of electrical contacts over a conventional contact interface; moreover, its data rate exceeds 106 kbps. Table 4 compares the characteristics of contact and contactless interfaces. The data rate of a contactless interface is more than 10 times higher than that of a contact interface. A fast contactless interface is especially expected for transportation applications, since the total transaction time is restricted to below 200 ms, so as to accomodate a human's walking speed through an interrogator gate.

Table 4. A comparison of the characteristics of contact and contactless interfaces

	Contact	Contactless
ISO/IEC standards	7816	14443
Power/data communication	Wired	Inductive coupling
Clock/carrier frequency	3.57 MHz	13.56 MHz
Communication data rate	9.6(optionally, 19.2/38.4) kbps	106 kbps
Anticollision function	Not implemented	Implemented

A contactless interface needs to have an additional ability to distinguish one card from other cards that exist in the same communications area of an interrogation antenna [10]; otherwise, multiple smart card responses might interfere and result in an unreadable signal on the interrogator's receiving antenna. The technique to avoid multiple-response interference is called anticollision. A drawback of the contactless interface is the relatively inefficient power transfer due to inductive coupling. As the result, power-consuming circuits, such as NOR Flash memory, cannot be used in contactless applications.

Table 5 lists the detailed characteristics of the contactless interface as specified in ISO/IEC 14443 standard [10]. This standard has two schemes, called type A and type B. The most significant difference between type A and type B is the modulation index in the data transfer from an interrogator to cards. Since the modulation index of the type A scheme is 100%, the supplied power is shut down for a specific period of 2.95 μs. Therefore, the type A interface cannot be used for smart card applications, in which an LSI with a CPU and a memory needs to be supplied with continuous power.

Table 5. The detailed characteristics of the international standard contactless interface

	Standards	Type A	Type B
Power/data transfer to card	Carrier frequency	13.56 MHz ± 7 kHz	13.56 MHz ± 7 kHz
	Modulation scheme	ASK	ASK
	Modulation index	100%	10%
	Data rate	106 kbps (fc/128)	> 106 kbps (fc/128)
	Bit cording	Modified miller	NRZ
Data transfer from card	Signal transfer	Load switching	Load switching
	Subcarrier frequency	847.5 kHz (fc/16)	847.5 kHz (fc/16)
	Modulation scheme	ASK	BPSK
	Bit cording	Manchester	NRZ
	Data rate	106 kbps (fc/128)	> 106 kbps (fc/128)

Subsequently, we are going to illustrate the transaction time difference between FeRAM- and EEPROM-based smart card systems, as evaluated with the ISO/IEC 14443 type B interface. Figure 3 shows a presumed transaction sequence of an analyzed transportation application. This sequence consists of anticollision (based on the slotted-ALOHA algorithm), authentication, and write and read with a communication data rate of 212 kbps [11]. For EEPROM, one block of data (128 bits) are programmed at once to overcome a long erase and programming cycle of 2 ms; on the other hand, for FeRAM, 8 bits are simultaneously programmed. This programming circuit and the high-voltage generator circuit required for EEPROM indicates another advantage of FeRAM in terms of circuit area. As is shown in Fig. 4, FeRAM can reduce the transaction time by 32% for eight-block read/write, and by 39% for 16-block read/write. FeRAM can provide an additional timing margin for the trans-

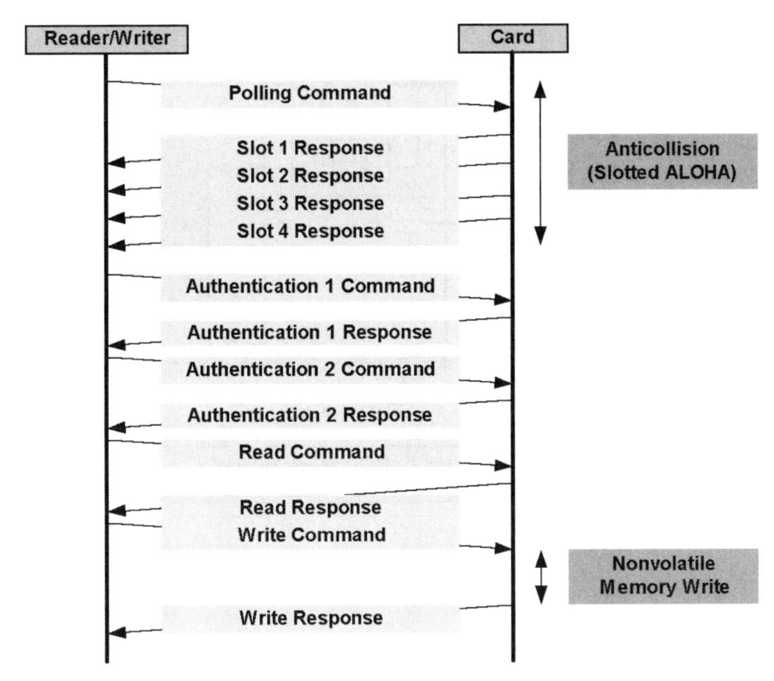

Fig. 3. A typical sequence for a transportation ticketing application

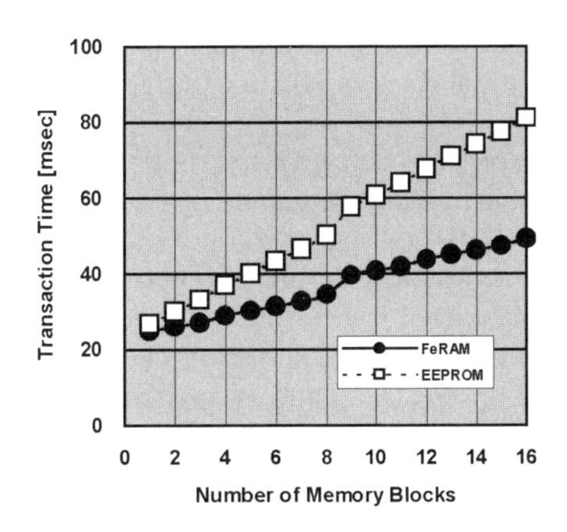

Fig. 4. The dependence of the transaction time on the nonvolatile memory block sizes in the transportation ticketing application

action retries required by communication errors or the programming due to verification failures.

One of the most significant problems associated with a contactless interface is nonvolatile memory data protection from a power failure. There is always the possibility of an unintentional power down just at a moment at which the nonvolatile memory is programming. FeRAM again presents superior characteristics, because of its low energy consumption and the availability of a large-permittivity on-chip capacitor. Figure 5 shows a schematic

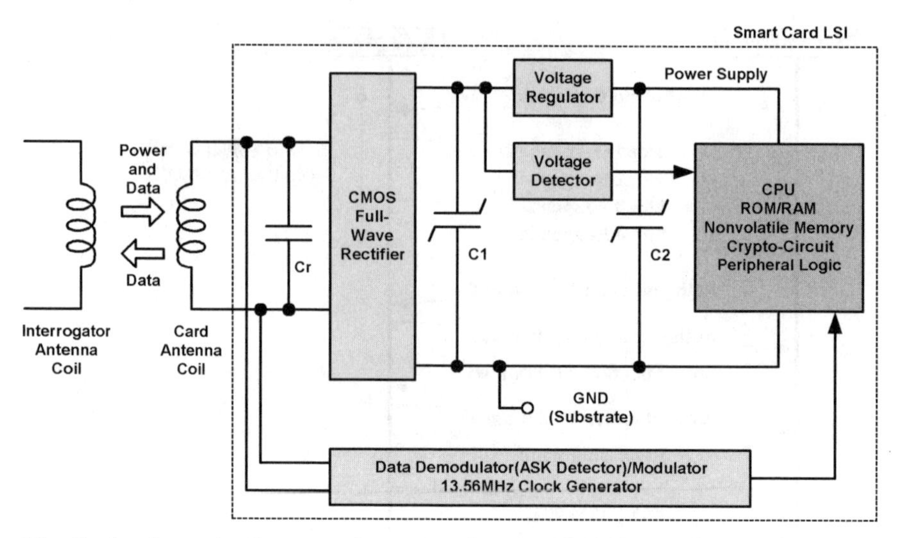

Fig. 5. A schematic diagram of a contactless interface circuit for FeRAM-based smart card LSIs

diagram of the contactless interface circuit used in FeRAM-based smart card LSIs. The transmitted RF power from the interrogator is received by the LC tank, and the DC power is extracted by the CMOS full-wave rectifier. By the use of FeRAM technology, the rectified DC voltage is stored in ferroelectric capacitors C_1 and C_2, whose capacitance per area is 30 times larger than that of the usual oxide capacitor.

C_1 is designed to be smaller than C_2 in order to detect the RF power modulation and decrease. If the voltage detector detects an anomalous decrease in the rectified voltage, it interrupts the CPU and other resources to stop its operation. With this scheme, FeRAM is designed to finish its entire read or write sequence, once this sequence has started. Since the read operation is destructive, it should be followed by writeback. The necessary energy to complete the sequence is supplied by C_2. On the other hand, since EEPROM requires large programming energy, and does not have a large-permittivity capacitor, it causes data destruction or requires an off-chip capacitor. As a result, the cost of a card is increased due to the additional capacitor and its packaging.

5 Code-Reprogrammable Multi-Application Cards

In high-end multi-application smart cards, application programs and associated data stored in nonvolatile memory can be loaded, deleted, and upgraded upon the user's request after cards have been issued, as shown in Fig. 6. In this example, credit and banking applications are initially installed based upon the user's request. Then, this user can add an airline ticket mileage

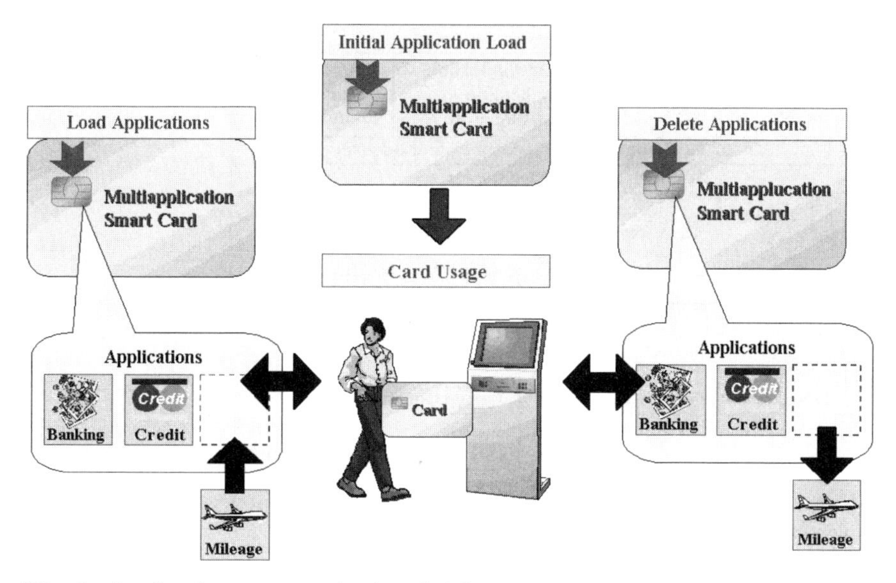

Fig. 6. Application program load and delete operations in the multi-application smart card

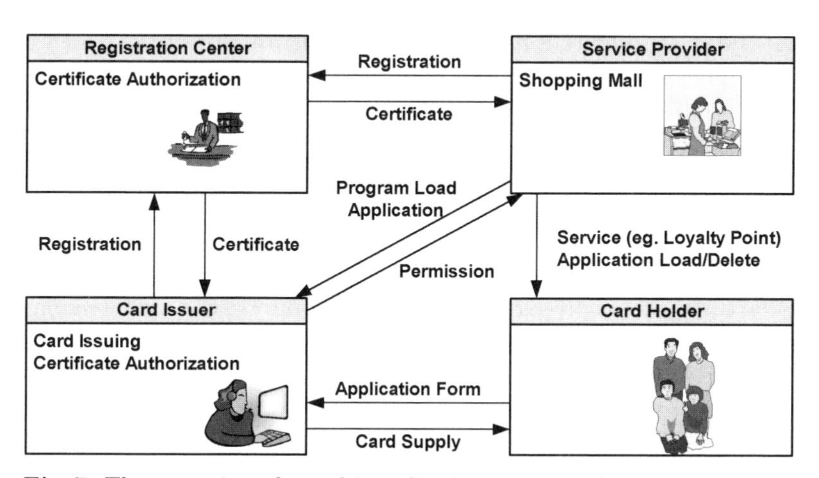

Fig. 7. The operation of a multi-application smart card

application afterwards. As a consequence of integrating multiple services, the user can obtain mileage points immediately after a rental car contract with the use of the credit card application installed on the same card. This kind of service is currently provided with multiple cards; however, it is obvious that users' convenience and advantages can be enormously enhanced by the use of a multi-application scheme.

The multi-application card system is operated by registered service providers and card issuers as shown in Fig. 7. To accommodate multi-appli-

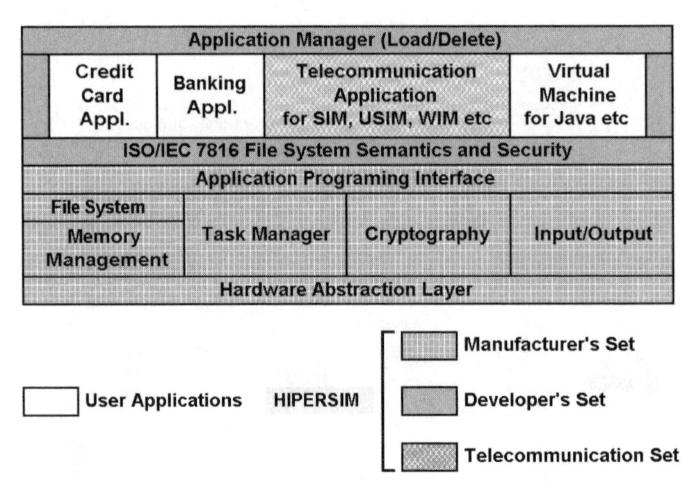

Fig. 8. The modular architecture of OS structure and application programs as implemented with HIPERSIM

cation services, smart cards should have a sophisticated OS to support the load/delete functions and a high-level language virtual machine, such as Java, MULTOS, and so on. Figure 8 represents the modular architecture of smart card OS and application programs as is implemented with HIPERSIM[1] (high-performance smart IC manager). HIPERSIM consists of the following three parts:

- the manufacturer's set;
- the developer's set;
- the telecommunications set.

The manufacturer's set is prepared for system integrators who are intending to develop dedicated multi-application smart card systems. This is a basic OS system including a hardware abstraction layer (a layer to provide an interface between the hardware and the upper-layer operating system), a file system, a memory manager, a task manager, and an encryption/decryption engine, and the API. HIPERSIM features a multi-tasking function that enables secure communication between applications along with application firewalling.

A standard IC card system compliant with the ISO/IEC 7816 file system interface can be developed using the developer's set. The application manager can provide the facility to load and delete applications as well as to implement new protocols.

The telecommunications set provides internet access ability through a mobile phone. This set includes a SIM (subscriber identity module, used to store identification and individual information, such as a phone book, and to establish secure communications between GSM-based 2nd-generation mobile

[1] HIPERSIM and HIFERRON are trademarks of Fujitsu Limited

Fig. 9. A multi-application smart card LSI

phones), a USIM (universal subscriber identity module, evolved from SIM to 3rd-generation mobile phones), and a microbrowser.

The microbrowser is an internet browser that supports WML (Wireless Markup Language, a content markup language specifically designed for wireless devices), and enables efficient internet access with the less expensive GSM mobile phone currently used in Europe. When a user is browsing an internet site with the GSM phone, an application program written with WML is transferred into byte-codes in a wireless internet gateway, and is then transmitted to the phone. As a result, high-speed and low-power browsing is realized with a data rate of 9600 bps.

Table 6. Smart card LSI specifications

CPU	32-bit RISC
Memory	ROM: 96 kB
	Data RAM: 4 kB
	FeRAM: 64 kB
Cryptographic coprocessor	Elliptic curve cryptography – sequential multiplier
	DES (single and triple/ECB and CBC mode)
IC card interface	Dual interface
	ISO/IEC 7816,9.6 k/19.2 k/38.4 kbps
	ISO/IEC 14443 Type B, 106 k/212 kbps
OS support	Java, MULTOS, etc.
Technology	0.35 μm CMOS + FeRAM

HIPERSIM is targeting a smart card LSI called the HIFERRON[1] with 32-bit RISC CPU LSIs and FeRAM. Figure 9 represents a die micrograph of the HIFERRON LSI, and Table 6 summarizes the specification of this LSI.

0.35 µm three-layer metal CMOS technology enables 64 kB FeRAM along with secure and efficient implementation of CPU and crypto-circuits. FeRAM is used not only for the storage of identification, individual information, and values, but also for application programs and cache memory. The fast program time of the FeRAM can mitigate the program load time during the initial application load and the actual operational schemes. As for the security aspect, a triple DES hardware macro and an elliptic curve cryptosystem coprocessor are embedded. To assure tamper-resistance, scrambled wiring and metal covering technologies are carefully utilized. FeRAM has its own hardware firewalling structure to protect it from illegal access.

6 Future Trends

As mentioned in the previous section, FeRAM is an enabling technology for the realization of multi-application smart card technology. This multi-application card would require an increasing capacity of nonvolatile memory storage, from 64 kb to over 256 kb. With the continuous scaling of device technology, FeRAM would be able to establish an unwavering position for nonvolatile memory.

Not only for the smart card application but also for various security systems, secure processing and identification information data storage would be required. In the field of consumer electronics, the set-top box and the games machine are using their own identification module for the protection of intellectual properties. The market for this type of secure LSI is considered to extend toward communications networks. FeRAM would accommodate a secure computing architecture.

Acknowledgements

The authors would like to thank Dr. Jedidi Kamouaa for marketing information, and Dr. Hidetoshi Nishi and Dr. Yoshihito Arimoto for their continuous encouragement.

References

1. ISO/IEC 7816-2 Identification cards – Integrated circuit(s) cards with contacts – Part 2: Dimensions and location of the contacts
2. B. Schneier: *Applied Cryptography*, 2nd edn. (Wiley, New York 1996)
3. C. K. Koc: *RSA Hardware Implementation* (RSA Laboratories, Redwood City 1995)
4. M. Rosing: *Implementing Elliptic Curve Cryptography* (Manning, Greenwich 1999)
5. J. Yamada, T. Miwa, H. Koike, H. Toyoshima, K. Amanuma, S. Kobayashi, T. Tatsumi, Y. Maejima, H. Hada, H. Mori, S. Takahashi, H. Takeuchi, T. Kunio: Tech. Dig. Int. Solid-State Circuits Conf. (2000) p. 270

6. A. Sharma: *Semiconductor Memories* (IEEE Press, New York 1997)
7. U. Kaiser, W. Steinhagan: IEEE J. Solid-State Circuits **30**, 301 (1995)
8. M. Durlam, P. Naji, M. DeHerrera, S. Tehrani, G. Kerszykowski, K. Kyler: Tech. Dig. Int. Solid-State Circuits Conf. (2000) p. 130
9. ISO/IEC 14443-1 Identification cards – Contactless integrated circuit(s) cards – Proximity cards, Part 1: Physical characteristics
10. ISO/IEC 14443-2 Identification cards – Contactless integrated circuit(s) cards – Proximity cards, Part 2: Air interface
11. SONY FeliCa Card User's Manual Rev. 1.52e, available from `http://www.sony.co.jp/en/Products/felica/contents02b.html` (2000)
12. S. Masui, E. Ishii, T. Iwawaki, Y. Sugawara, K. Sawada: Tech. Dig. Int. Solid-State Circuits Conf. (1999) p. 162

Index

Topics in Applied Physics

77 **Photomechanics**
By P. K. Rastogi (Ed.) 2000, 314 Figs. XVI, 472 pages

78 **High-Power Diode Lasers**
By R. Diehl (Ed.) 2000, 260 Figs. XIV, 416 pages

79 **Frequency Measurement and Control**
Advanced Techniques and Future Trends
By A. N. Luiten (Ed.) 2001, 169 Figs. XIV, 394 pages

80 **Carbon Nanotubes**
Synthesis, Structure, Properties, and Applications
By M. S. Dresselhaus, G. Dresselhaus, Ph. Avouris (Eds.) 2001, 235 Figs. XVI, 448 pages

81 **Near-Field Optics and Surface Plasmon Polaritons**
By S. Kawata (Ed.) 2001, 136 Figs. X, 210 pages

82 **Optical Properties of Nanostructured Random Media**
By Vladimir M. Shalaev (Ed.) 2002, 185 Figs. XIV, 450 pages

83 **Spin Dynamics in Confined Magnetic Structures I**
By B. Hillebrands and K. Ounadjela (Eds.) 2002, 166 Figs. XVI, 336 pages

84 **Imaing of Complex Media with Acoustic and Seismic Waves**
By M. Fink, W. A. Kuperman, J.-P. Montagner, A. Tourin (Eds.) 2002, 162 Figs. XII, 336 pages

85 **Solid–Liquid Interfaces**
Macroscopic Phenomena – Microscopic Understanding
By K. Wandelt and S. Thurgate (Eds.) 2003, 228 Figs. XVIII, 444 pages

86 **Infrared Holography for Optical Communications**
Techniques, Materials, and Devices
By P. Boffi, D. Piccinin, M. C. Ubaldi (Eds.) 2003, 90 Figs. XII, 182 pages

87 **Spin Dynamics in Confined Magnetic Structures II**
By B. Hillebrands and K. Ounadjela (Eds.) 2003, 179 Figs. XVI, 321 pages

88 **Optical Nanotechnologies**
The Manipulation of Surface and Local Plasmons
By J. Tominaga and D. P. Tsai (Eds.) 2003, 168 Figs. XII, 212 pages

89 **Solid-State Mid-Infrared Laser Sources**
By I. T. Sorokina and K. L. Vodopyanov (Eds.) 2003, 263 Figs. XVI, 557 pages

90 **Single Quantum Dots**
Fundamentals, Applications, and New Concepts
By P. Michler (Ed.) 2003, 181 Figs. XII, 352 pages

91 **Vortex Electronis and SQUIDs**
By T. Kobayashi, H. Hayakawa, M. Tonouchi (Eds.) 2003, 259 Figs. XII, 302 pages

92 **Ultrafast Dynamical Processes in Semiconductors**
By K.-T. Tsen (Ed.) 2004, 190 Figs. XI, 400 pages

93 **Ferroelectric Random Access Memories**
Fundamentals and Applications
By H. Ishiwara, M. Okuyama, Y. Arimoto (Eds.) 2004, Approx. 200 Figs. XIV, 288 pages

94 **Silicon Photonics**
By L. Pavesi, D.J. Lockwood (Eds.) 2004, 262 Figs. XVI, 397 pages